AP Precalculus Made Easy

Ultimate Study Guide and Test Prep with Key Points, Examples, and Practices. The Best Tutor for Beginners and Pros + Two Practice Tests

Dr. Abolfazl Nazari

PUBLISHED BY EFFORTLESS MATH EDUCATION

EFFORTLESSMATH.COM

Welcome to AP Precalculus Made Easy

2024

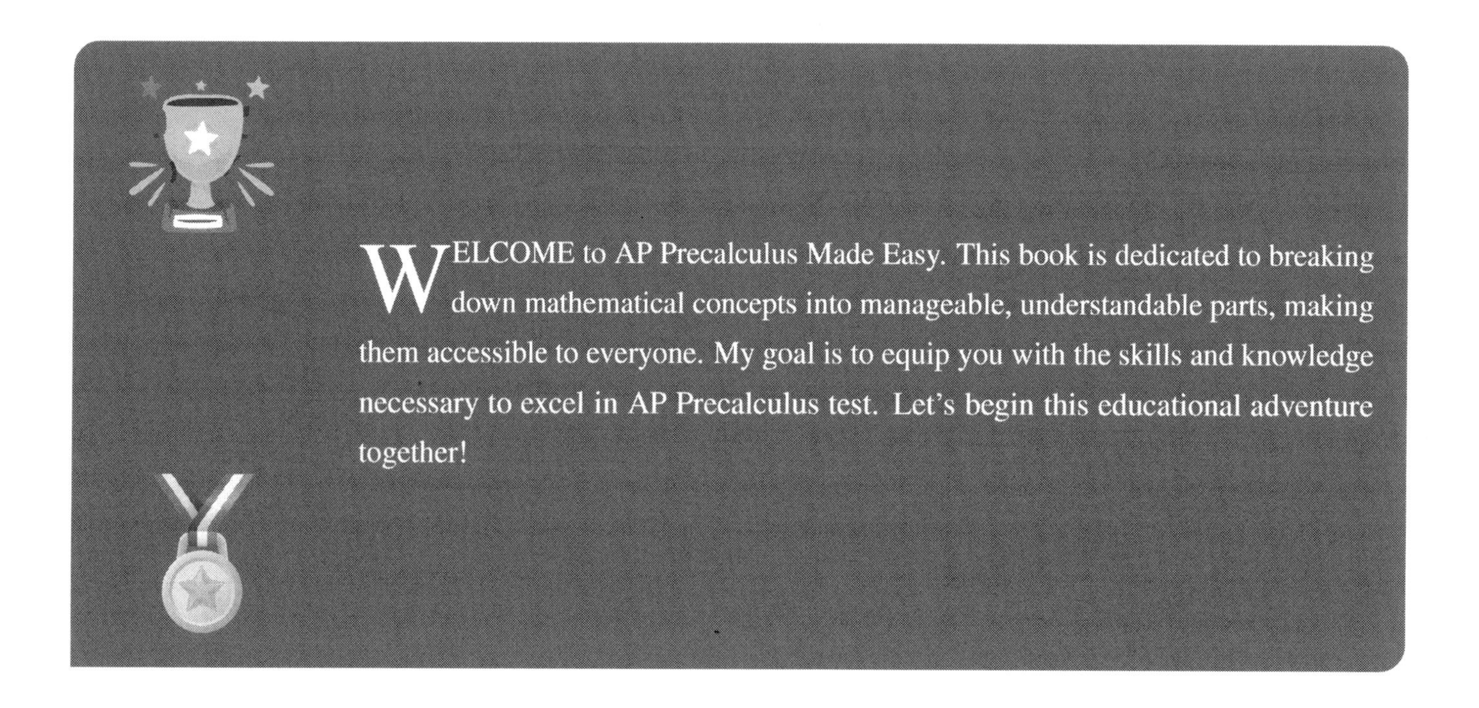

AP Precalculus Made Easy provides comprehensive coverage of the key topics needed for the AP Precalculus test. The book is structured into detailed chapters covering all topics of AP Precalculus. Each chapter starts with basic concepts and gradually moves to more complex ones, ensuring you gain a complete understanding of each topic. The content is tailored to not only prepare you for the AP Precalculus test but also to apply these skills in real-life situations.

In keeping with the *Math Made Easy* series' philosophy, this book adopts an interactive and practice-oriented approach to learning. Each mathematical concept is introduced in a clear and straightforward manner, accompanied by examples to help illustrate its application. A variety of practice problems are provided to mirror the style and challenges of the AP Precalculus test, enabling you to test your knowledge and strengthen your understanding. I am excited to show you what the book contains.

What is included in this book

- ☑ Online resources for additional practice and support.
- ☑ A guide on how to use this book effectively.
- ☑ All AP Precalculus concepts and topics you will be tested on.
- ☑ End of chapter exercises to help you develop the basic math skills.
- ☑ AP Precalculus test tips and strategies.
- ☑ 2 realistic and full-length practice tests with detailed answers.

Effortless Math's Precalculus Online Center

Effortless Math Online Precalculus Center offers a complete study program, including the following:

- ☑ *Step-by-step instructions on how to prepare for the AP Precalculus test*
- ☑ *Numerous AP Precalculus worksheets to help you measure your math skills*
- ☑ *Complete list of AP Precalculus formulas*
- ☑ *Video lessons for all AP Precalculus topics*
- ☑ *Full-length AP Precalculus practice tests*

Visit `EffortlessMath.com/PreCalculus` to find your online AP Precalculus resources.

Scan this QR code

(No Registration Required)

Tips for Making the Most of This Book

This book is all about making mathematics easy and approachable for you. Our aim is to cover everything you need to know, keeping it as straightforward as possible. Here is a guide on how to use this book effectively: First, each math topic has a core idea or concept. It's important to understand and remember this. That's why we have highlighted key points in every topic. These are like mini-summaries of the most important stuff.

Examples are super helpful in showing how these concepts work in real problems. In every topic, we've included a couple of examples. If you feel very smart, you can try to solve them on your own first. But they are meant to be part of the teaching; They show how key concepts are applied to the problems. The main thing is to learn from these examples.

And, of course, practice is key. At the end of each chapter, you will find problems to solve. This is where you can really sharpen your skills.

To wrap it up:

- ***Key Points***: Don't miss the key points. They boil down the big ideas.
- ***Examples***: Try out the examples. They show you how to apply what you're learning.
- ***Practices***: Dive into the practice problems. They're your chance to really get it.

In addition to the material covered in this book, it is crucial to have a solid plan for your test preparation. Effective test preparation goes beyond understanding concepts; it involves strategic study planning and practice under exam conditions.

- **Begin Early.** Start studying well before the exam to avoid rushing, allowing for a thorough review.
- **Daily Study Sessions.** Study regularly for 30 to 45 minutes each day to enhance retention and reduce stress.
- **Active Note-Taking.** Write down key points to internalize concepts and improve focus. Review notes regularly.
- **Review Challenges.** Spend extra time on difficult topics for better understanding and performance.
- **Practice.** Engage in extensive practice using end-of-chapter problems and additional workbooks.

Explore other guides, workbooks, and tests in the series to complement your study, offering extra practice and enhancing understanding, problem-solving skills, and academic preparation.

Contents

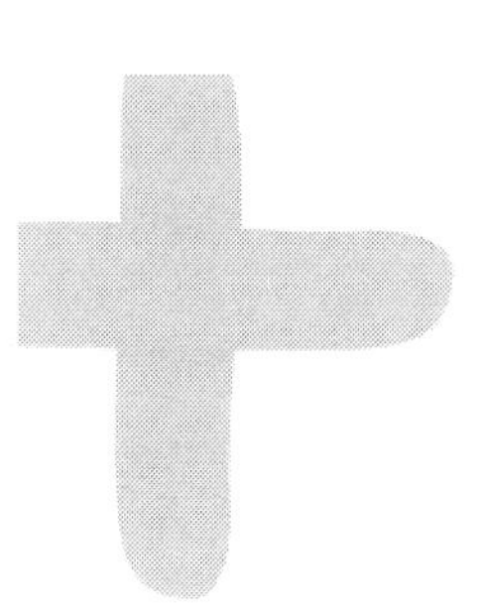

1. Equations and Inequalities

1.1 Solving Multi-Step Equations

Solving multi-step equations involves a few systematic steps, crucial for finding the solution:

1. **Combine Like Terms**: Simplify the equation by adding or subtracting terms with the same variable.
2. **Isolate the Variable**: Move all instances of the variable to one side of the equation using addition or subtraction.
3. **Use Inverse Operations**: Apply the opposite of addition, subtraction, multiplication, or division to further simplify the equation.
4. **Verify the Solution**: Check your answer by substituting the solution back into the original equation to ensure it balances.

Key Point

The goal is to systematically simplify the equation by combining like terms, isolating the variable, using inverse operations for further simplification, and verifying the solution.

Solve this equation for x: $4x + 8 = 20 - 2x$.

Solution: First, bring variables to one side by adding $2x$ to both sides. We can express this step as follows:

$$4x + 8 + 2x = 20 - 2x + 2x \Rightarrow 6x + 8 = 20.$$

Then, simplify the equation by subtracting 8 from both sides:

$$6x + 8 - 8 = 20 - 8 \Rightarrow 6x = 12.$$

Following this, divide both sides by 6 to isolate x:

$$\frac{6x}{6} = \frac{12}{6} \Rightarrow x = 2.$$

Lastly, we check our solution by substituting $x = 2$ into the original equation:

$$4(2) + 8 = 20 - 2(2) \Rightarrow 16 = 16.$$

Thus, $x = 2$ is indeed the correct solution.

Solve this equation for x: $-5x + 4 = 24$.

Solution: Start by subtracting 4 from both sides of the equation:

$$-5x + 4 - 4 = 24 - 4 \Rightarrow -5x = 20.$$

Next, isolate x by dividing both sides by -5:

$$\frac{-5x}{-5} = \frac{20}{-5} \Rightarrow x = -4.$$

Finally, check the solution by substituting $x = -4$ into the original equation:

$$-5(-4) + 4 = 24 \Rightarrow 24 = 24.$$

Hence, $x = -4$ is the correct solution.

1.2 Slope and Intercepts

The slope of a line, denoted by m, quantifies its steepness, representing the ratio of vertical change (rise) to horizontal change (run). Intercepts are points where the line crosses the axes. In the linear equation $y = mx + b$, m stands for the slope, and b denotes the y-intercept, the y value when $x = 0$.

Key Point

The slope m of a line passing through points (x_1, y_1) and (x_2, y_2) can be calculated using the formula: $m = \frac{y_2 - y_1}{x_2 - x_1} = \frac{rise}{run}$.

 Find the slope of the line through points $A=(1,-6)$ and $B=(3,2)$.

Solution: Using the formula for the slope, we get: $m=\frac{2-(-6)}{3-1}=\frac{8}{2}=4$. So, the slope of the line passing through points A and B is 4.

 Find the slope of the line with equation $y=-3x+4$.

Solution: In the line $y=-3x+4$, the slope is -3.

1.3 Using Intercepts

In Pre-calculus, understanding intercepts is essential after learning about slopes. Intercepts are points where a graph crosses the axes on the Cartesian plane.

There are two types of intercepts:

- x-intercept: Where the graph crosses the x-axis, found by setting $y=0$ in the equation and solving for x.
- y-intercept: Where the graph crosses the y-axis, found by setting $x=0$ in the equation and solving for y.

An x-intercept is a point on the graph where the y-value is zero, while a y-intercept is a point on the graph where the x-value is zero.

When the equation of a line is arranged in the slope-intercept form $y=mx+b$, the y-intercept can be directly observed as the value of b. This is because if you replace the value of x with zero, the term mx will also be zero, leaving only b or the y-intercept.

 **Example** Find the x-intercept and y-intercept of the line equation $5x+4y=40$.

Solution: The x-intercept is found by setting $y=0$ in the equation:

$$5x+4(0)=40 \Rightarrow 5x=40 \Rightarrow x=\frac{40}{5} \Rightarrow x=8.$$

So, the x-intercept is 8. The y-intercept is found by setting $x=0$ in the equation:

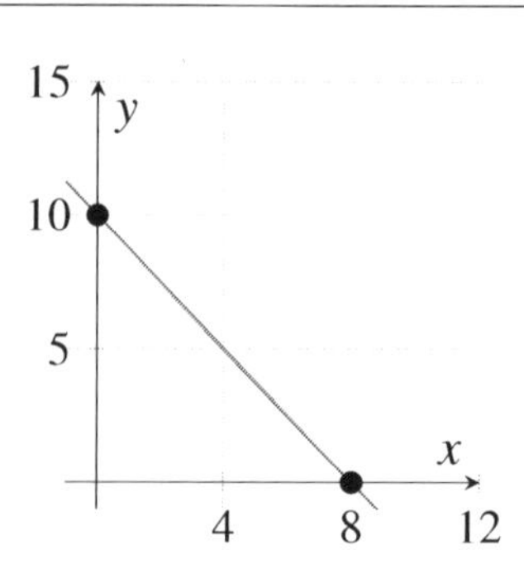

$$5(0)+4y=40 \Rightarrow 4y=40 \Rightarrow y=\frac{40}{4} \Rightarrow y=10.$$

So, the y-intercept is 10.

1.4 Transforming Linear Functions

Linear functions, represented as $y=mx+b$, form the foundation of understanding how equations can depict straight lines on a graph. Here, y represents the dependent variable, influenced by x (the independent variable), with m determining the line's slope and b setting its starting point on the y-axis. Our focus will shift to how variations in m and b transform these linear functions.

Transformation refers to modifications in a linear function that adjust the line's orientation or dimension through alterations in slope (m) or y-intercept (b), while maintaining its linear characteristic.

The notion of a *family of functions* groups equations sharing specific traits, like linear functions that graph as straight lines. The *parent function* within this family is $f(x)=x$, serving as the baseline for comparing transformations.

Types of Transformations:

- **Translation:** Modifying b shifts the graph vertically. Increasing b moves the line up; decreasing it moves the line down.
- **Rotation:** Altering m rotates the line around $(0,b)$, changing its angle to become steeper or flatter.
- **Reflection:** Multiplying m by -1 flips the graph across the y-axis, mirroring its slope on the opposite side.

Through these transformations, linear functions demonstrate versatility in graphing, allowing for a deeper understanding of their applications and characteristics.

Key Point

A transformation of a linear function means we are changing the position or size of the line, either by changing the y-intercept b or the slope m, without altering the line's fundamental shape.

Consider the graph $f(x)=x$. Find the graph $g(x)=x+2$.

Solution: Here, we are translating the parent function $f(x)=x$ by changing the y-intercept b in the function $f(x)=mx+b$.

We can see that the graph of the function $y = x$ has been translated upward by 2 units to form the graph of the function $y = x + 2$.

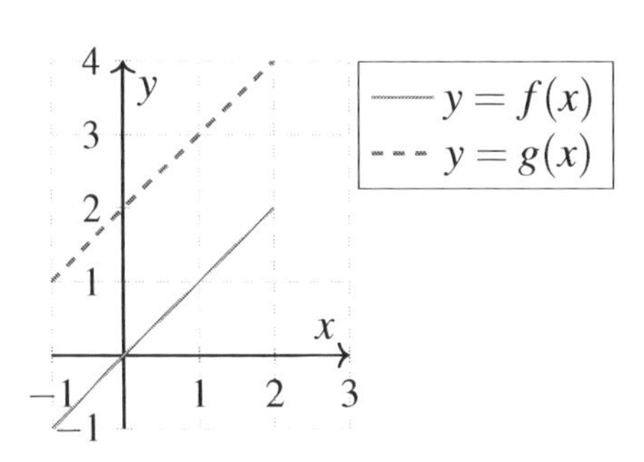

1.5 Solving Inequalities

Solving inequalities is a fundamental skill following the study of linear functions. Unlike equations that denote equality, inequalities express relationships such as less than ($<$), greater than ($>$), less than or equal to ($\leq$), and greater than or equal to ($\geq$).

The process for solving inequalities mirrors that of equations, with a focus on isolating the variable through inverse operations.

Applying the same operation on both sides of an inequality does not alter its truth. Multiplying or dividing both sides by a negative number reverses the inequality sign.

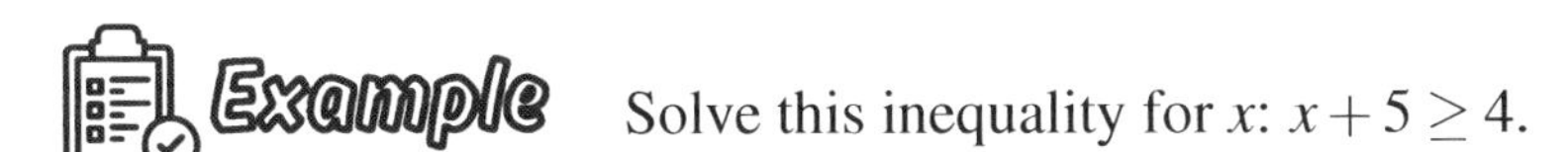

Solve this inequality for x: $x + 5 \geq 4$.

Solution: Here, the inequality is $x + 5 \geq 4$. We notice that 5 is being added to x. The inverse operation of addition is subtraction. Hence, subtract 5 from both sides of the inequality: $x + 5 - 5 \geq 4 - 5$, simplifies to $x \geq -1$. We found our solution: $x \geq -1$.

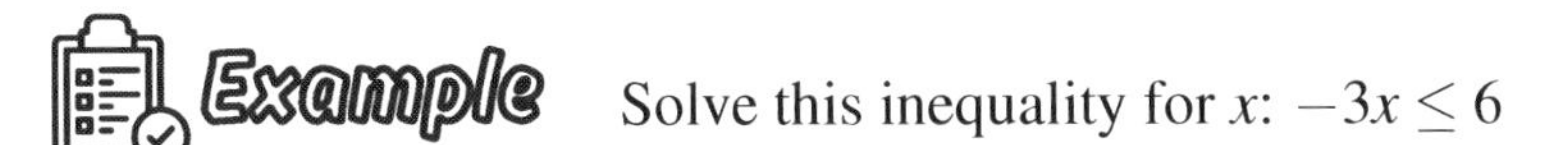

Solve this inequality for x: $-3x \leq 6$

Solution: -3 is multiplied by x. Divide both sides by -3. Remember, when dividing or multiplying both sides of an inequality by negative numbers, flip the direction of the inequality sign. Then:

$$-3x \leq 6 \Rightarrow \frac{-3x}{-3} \geq \frac{6}{-3} \Rightarrow x \geq -2.$$

1.6 Graphing Linear Inequalities

To graph linear inequalities, follow these simplified steps:

1. Plot the line as if the inequality were an equation (e.g., for $y < 2x - 1$, plot $y = 2x - 1$).

2. Use a dashed line for $>$ or $<$, and a solid line for $\geq$ or $\leq$.
3. Pick a test point not on the line, like (0,0), unless it's on the line itself.
4. Check if the test point satisfies the inequality:
    - If yes, shade the side of the line where the test point lies.
    - If no, shade the opposite side.

This method helps visually determine which side of the line contains solutions to the inequality.

Key Point

Graphing linear inequalities involves plotting the corresponding equation's line, deciding on its solidity based on the inequality symbol, selecting a test point to determine which side of the line satisfies the inequality, and shading that side as the solution area.

Illustrate the inequality $y < 2x - 1$.

Solution: To graph the inequality $y < 2x - 1$, follow these steps:

1. Graph the line $y = 2x - 1$ as a dashed line because the inequality symbol is $<$.
2. Choose the origin $(0,0)$ as the test point.
3. Substitute $(0,0)$ into the inequality: $0 < 2(0) - 1$ or $0 < -1$, which is false. So, $(0,0)$ does not satisfy the inequality.
4. Shade the area opposite to $(0,0)$, indicating the solution set.

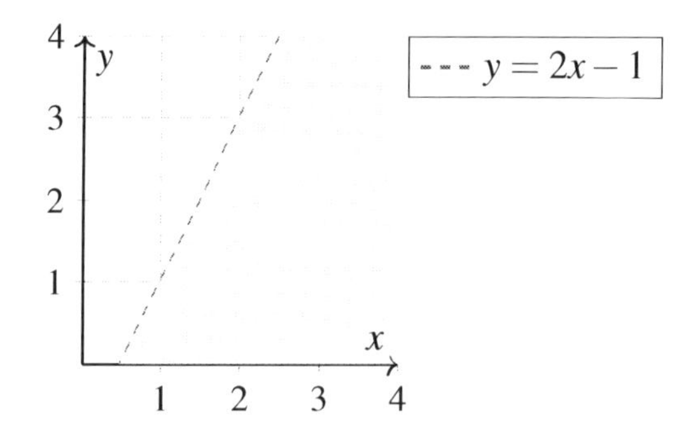

1.7 Solving Compound Inequalities

Compound inequalities, encompassing two or more inequalities linked by "and" or "or," extend the principles of graphing linear inequalities. The equation of a line marks all points on that line, with areas on either side indicating values greater or less than the line. Similarly, compound inequalities are solved by considering both inequalities simultaneously but require an additional step in their solution process.

Methodology Simplified:

1. Combine Like Terms: Initiate by consolidating like terms to one side, ensuring the variable's coefficient remains positive. This step simplifies the inequality for easier analysis.
2. Variable Isolation: Shift all variables to one side using addition or subtraction, aiming for a simplified expression. Further isolate the variable by applying the inverse of these operations, thereby simplifying the inequality to where the variable stands alone.
3. Further Simplification: Utilize the inverse of multiplication or division for additional simplification, paving the way to solve for the variable's value range.

Key Point

Always invert the inequality sign when multiplying or dividing by a negative number.

Example Let us solve the compound inequality $3x-1<x+5<2x+7$.

Solution: First, it is sensible to look at this problem as two separate inequalities:

Inequality 1: $3x-1<x+5$

Inequality 2: $x+5<2x+7$

Next, simplify each inequality. For Inequality 1, subtract x from each side:

$$3x-x-1<x+5-x \Rightarrow 2x-1<5.$$

Then, add 1 to each side:

$$2x-1+1<5+1 \Rightarrow 2x<6.$$

Finally, divide each side by 2:

$$\frac{2x}{2}<\frac{6}{2} \Rightarrow x<3.$$

Hence, the solution to Inequality 1 is $x<3$. Let's solve Inequality 2. Here, subtract x from each side:

$$x+5-x<2x+7-x \Rightarrow 5<x+7.$$

Subtract 7 from each side:

$$5-7<x+7-7 \Rightarrow -2<x.$$

Thus, the solution to Inequality 2 is $x>-2$. The compound inequality $3x-1<x+5<2x+7$ is true when $x>-2$ and $x<3$, or written as $-2<x<3$.

1.8 Solving Absolute Value Equations

Solving equations with absolute values involves creating two separate but related equations, leveraging the non-negative outcome property of absolute values. The process can be broken down into the following steps:

1. Isolate the Absolute Value: Ensure the absolute value expression stands alone on one side of the equation using basic algebraic operations.
2. Create and Solve Two Equations: Form two equations by removing the absolute value notation. For $|x| = a$, write $x = a$ and $x = -a$. Solve each equation separately.
3. Verify Solutions: Confirm solutions by substituting them back into the original equation to ensure accuracy.

Key Point

Isolate the absolute value expression, create and solve two equations considering both positive and negative scenarios, and verify the solutions by substitution into the original equation.

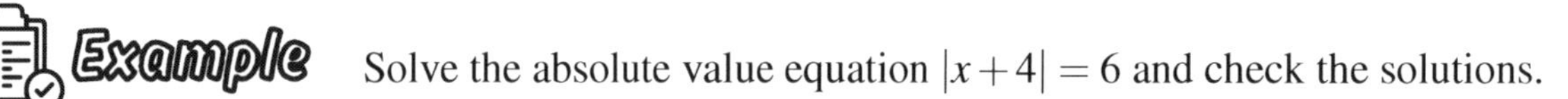

Example Solve the absolute value equation $|x+4| = 6$ and check the solutions.

Solution: Remove the absolute value sign and set up the two equations:

$$x+4 = 6, \text{ or } x+4 = -6.$$

Solving these equations gives us:

$$x = 6-4 = 2, \text{ or } x = -6-4 = -10.$$

Substitute $x = 2$ and $x = -10$ in the original equation, respectively:

If $x = 2$, then $|2+4| = 6$, which simplifies to $|6| = 6$. This is true.

If $x = -10$, then $|-10+4| = 6$, which simplifies to $|-6| = 6$. This is also true.

Hence, both $x = 2$ and $x = -10$ are solutions to the equation.

1.9 Solving Absolute Value Inequalities

Solving absolute value inequalities is similar to solving equations. The key is to isolate the absolute value. Remember, the absolute value is always non-negative, so it cannot be less than a negative number.

Key Point

The concept of an absolute value is that it produces positive outcomes. Hence, an absolute value can never be less than zero.

To effectively solve an absolute value inequality, adhere to the following procedure:

1. Begin by isolating the absolute value on one side of the inequality. This separates it from the other components within the equation.
2. For inequalities of the type $|x| \geq c$, decompose the equation into two separate cases: $x \geq c$ or $x \leq -c$, and then find the solution for each case.
3. Conversely, for inequalities where $|x| < c$, the solution is found by setting up a compound inequality: $-c < x < c$, which captures the range of values that x can take.

Solve the equation $|2x| \geq 24$.

Solution: Here, the inequality sign is greater than or equal. We can write this as:

$$2x \geq 24, \text{ or } 2x \leq -24.$$

Solving both inequalities we get: $x \geq 12$, or $x \leq -12$.

Solve the equation $|1-x| < 4$.

Solution: Here, the inequality sign is less than. We can write this as:

$$1-x < 4, \text{ and } 1-x > -4.$$

Solving both inequalities, we get: $-3 < x < 5$.

1.10 Graphing Absolute Value Inequalities

Absolute value inequalities are solved similarly to absolute value equations. To solve an absolute value inequality, we first isolate the absolute value on one side of the inequality and then use the properties of the inequality to determine the solution.

The Absolute Value Inequality Property simplifies to:

- For $|u| < a$ with $a > 0$, rewrite as $-a < u < a$.
- For $|u| > a$, rewrite as $u > a$ or $u < -a$.

This principle is crucial for identifying the solution boundaries for absolute value inequalities.

Key Point

The solution of the inequality $|u| < a$ is the interval $(-a, a)$, where the numbers $-a$ and a are the key values.

Key Point

The solution of the inequality $|u| > a$ is the union of the intervals $(-\infty, -a) \cup (a, \infty)$. These solutions can easily be represented on a number line.

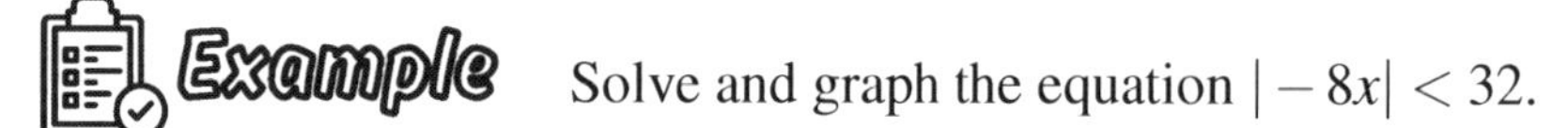

Example Solve and graph the equation $|-8x| < 32$.

Solution: Applying the Property of Inequality, we express the inequality $|-8x| < 32$ as $-32 < -8x < 32$, or equivalently, $-8x > -32$ and $-8x < 32$. Solving these two inequalities we obtain $x < 4$ and $x > -4$. So the solution is $-4 < x < 4$, which can be represented on a number line.

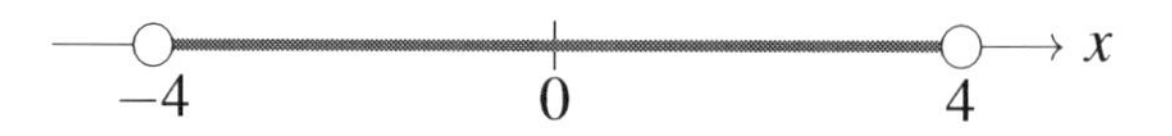

1.11 Solving Systems of Equations

Recall that a system of two equations contains two equations with two variables. Take the following system of equations as an example: $x - y = 1$ and $x + y = 5$. These systems of equations can be solved using various methods, and one of the simplest ways is the elimination method, which uses the addition property of equality. This property allows us to add the same value to each side of an equation without affecting the solution and assists us in solving complex equations.

Consider the above set of equations as the first example. If we add $x + y$ on the left-hand side and 5 on the right-hand side of the first equation, we obtain: $x - y + (x + y) = 1 + 5$. To simplify this equation:

$$2x = 6 \Rightarrow x = 3.$$

The next step involves substituting the calculated value of x (i.e., 3) into the first equation: $3 - y = 1$. This procedure leads us to the solution of the system of equations, which is $x = 3$ and $y = 2$.

Key Point

The elimination method is a technique used to solve systems of equations by manipulating the equations to eliminate one of the variables, allowing for the solution of the remaining variable.

 What is the value of $x+y$ in this system of equations?

$$\begin{cases} 2x+4y=12 \\ 4x-2y=-16 \end{cases}$$

Solution: To solve this system of equations using the elimination method, first, we multiply the first equation by (-2), then add the multiplied first equation to the second equation:

$$\begin{cases} -2(2x+4y)=-24 \\ 4x-2y=-16 \end{cases} \Rightarrow \begin{cases} -4x-8y=-24 \\ 4x-2y=-16 \end{cases} \Rightarrow -10y=-40 \Rightarrow y=4.$$

Next, substitute the value of y into one of the original equations and solve for x:

$$2x+4(4)=12 \Rightarrow 2x+16=12 \Rightarrow 2x=-4 \Rightarrow x=-2.$$

Finally, to find the sum of x and y, just plug the values into the equation $x+y$ which gives: $x+y=-2+4=2$.

1.12 Solving Special Systems

Moving beyond basic systems of equations and methods like elimination, we now explore special systems. Unlike typical systems, these may have no solutions or infinitely many.

In a system of equations with no solution, the graphs of the two equations are parallel and never intersect. This occurs because they have the same slope but different y-intercepts. In the case of infinite solutions, the two equations represent the same line; therefore, all points on the line are solutions to the system. After identifying a special system, you can use the elimination or substitution method to solve it.

 Analyze the system of equations and identify whether it is a special system:

$$\begin{cases} 2x+3y=5 \\ 4x+6y=10 \end{cases}$$

Solution: We may immediately see that the two equations are multiples of each other, and will therefore represent the same line. Hence, this system is special and it has an infinite number of solutions.

Solve the following special system using the substitution method.

$$\begin{cases} 3x+5y=10 \\ 6x+10y=20 \end{cases}$$

Solution: Rewrite the first equation in terms of x to get: $x=\frac{10-5y}{3}$. Then substitute this into the second equation to get:

$$\frac{60-30y}{3}+10y=20.$$

You will find that the equation will hold true for all real numbers of y, so the system has an infinite number of solutions.

1.13 Systems of Equations Word Problems

To solve word problems involving systems of equations effectively, adhere to the following steps:

1. **Define Variables:** Begin by determining the unknown quantities you're solving for. These are your variables.
2. **Build Equations:** Read the problem thoroughly to convert the given information into mathematical equations. Typically, a system of equations problem requires two equations.
3. **Check Equations:** Make sure your equations accurately represent the problem. Validate them against the problem's description to ensure precision.
4. **Solve the System:** Apply an appropriate method, such as the elimination method, to solve the system. This method is especially useful when the coefficients of one variable in both equations are the same or opposites, allowing for straightforward addition or subtraction to eliminate the variable.

Practice and attention to accuracy are crucial for mastering the solving of such problems.

Example Tickets to a movie cost $8 for adults and $5 for students. A group of friends purchased 20 tickets for $115.00. How many adult tickets did they buy?

Solution: Let x represent adult tickets and y represent student tickets. Given that there are 20 tickets purchased, we have: $x+y=20$. Additionally, knowing that adult tickets cost $8 and student tickets $5, and that the total cost for all tickets is $115, we can form another equation: $8x+5y=115$. We now have a system of equations:

$$\begin{cases} x+y=20 \\ 8x+5y=115 \end{cases}$$

Multiplying the first equation by -5 gives us: $-5x-5y=-100$. Adding this result to the second equation clears out the y:

$$8x+5y+(-5x-5y)=115-100 \Rightarrow 3x=15.$$

Solving for x, we find $x=5$. Substituting x into the first equation, we get:

$$5+y=20 \Rightarrow y=15.$$

The group bought 5 adult tickets and 15 student tickets.

1.14 Practices

1) Solve for x: $3x-7=2x+5$.

☐ A. $x=-12$

☐ B. $x=12$

☐ C. $x=-5$

☐ D. $x=5$

2) What value of x satisfies the equation $5x+(4x-7)=4(2x+1)$?

☐ A. $x=11$

☐ B. $x=13$

☐ C. $x=-11$

☐ D. $x=-8$

3) If the slope of a line passing through the points $(0,b)$ and $(1,-3)$ is -5, what is the value of b?

☐ A. $b=2$

☐ B. $b=-2$

☐ C. $b=-8$

☐ D. $b=3$

4) What is the y-intercept of the line with the equation $3x-4y=12$?

☐ A. $y=3$

☐ B. $y=4$

☐ C. $y=-3$

☐ D. $y=-4$

5) Which of the following is the y-intercept of the line represented by the equation $-2x+3y=6$?

☐ A. $(-2,0)$ ☐ C. $(-3,0)$
☐ B. $(0,-2)$ ☐ D. $(0,2)$

6) For the line $y = \frac{1}{2}x - 3$, what are the x-intercept and y-intercept, respectively?

☐ A. $(-3,0)$ and $(0,-3)$ ☐ C. $(6,0)$ and $(0,-3)$
☐ B. $(0,-3)$ and $(-3,0)$ ☐ D. $(0,-6)$ and $(3,0)$

7) Given the parent function $f(x) = x$, which of the following represents a vertical shift downward by 3 units?

☐ A. $f(x) = x - 3$ ☐ C. $f(x) = -x$
☐ B. $f(x) = x + 3$ ☐ D. $f(x) = \frac{x}{3}$

8) If the slope of a linear function is tripled, which of the following new functions reflects this change starting from the parent function $f(x) = x$?

☐ A. $f(x) = 3x + 1$ ☐ C. $f(x) = 3x$
☐ B. $f(x) = \frac{x}{3}$ ☐ D. $f(x) = x - 3$

9) If $2 < x < 5$, which of the following inequalities also must be true?

☐ A. $2x < 10$ ☐ C. $-1 < x - 3 < 2$
☐ B. $2x > 4$ ☐ D. All of the above

10) Which interval represents the solution to the inequality $-3(x-1) > 6$?

☐ A. $x > 3$ ☐ C. $x < -1$
☐ B. $x < 3$ ☐ D. $x > -1$

11) Which of the following regions should be shaded for the inequality $y > -\frac{1}{2}x + 3$?

☐ A. Above the line $y = -\frac{1}{2}x + 3$ ☐ C. The entire plane
☐ B. Below the line $y = -\frac{1}{2}x + 3$ ☐ D. None of the above

12) Which inequality corresponds to the graph with a solid line on $y = 3x + 1$ and shading on the side below the line?

☐ A. $y > 3x+1$
☐ B. $y < 3x+1$
☐ C. $y \leq 3x+1$
☐ D. $y \geq 3x+1$

13) Solve the compound inequality $-2 \leq 5-3x < 8$.

☐ A. $-1 \leq x \leq \frac{7}{3}$
☐ B. $x \leq -1$ or $x > \frac{7}{3}$
☐ C. $-1 < x \leq \frac{7}{3}$
☐ D. $x > -1$ and $x < \frac{7}{3}$

14) Which interval represents the solution set of the compound inequality $1 < \frac{x}{4} - 2 \leq 3$?

☐ A. $(12,20]$
☐ B. $(12,28]$
☐ C. $(-8,20]$
☐ D. $(4,24]$

15) Solve the equation $|3x-9| = 0$.

☐ A. $x = 0$
☐ B. $x = 3$
☐ C. No solution
☐ D. $x = -3$

16) For which value of x is the equation $|2x+1| = 7$ true?

☐ A. $x = 3$
☐ B. $x = -4$
☐ C. $x = 3$ or $x = -4$
☐ D. $x = 6$

17) Solve the inequality $|3-5x| > 12$.

☐ A. $x < -\frac{9}{5}$ or $x > 3$
☐ B. $x > -\frac{9}{5}$ or $x < 3$
☐ C. $x < -\frac{9}{5}$ and $x > 3$
☐ D. $x > -\frac{9}{5}$ and $x < 3$

18) What interval represents the solution for the inequality $|x+2| \leq 5$?

☐ A. $(-7,3)$
☐ B. $[-7,3]$
☐ C. $(-\infty,-7] \cup [3,\infty)$
☐ D. $(-\infty,-7) \cup (3,\infty)$

19) Solve the inequality $3|2x+1| > 15$. Which of the following represents the solution on a number line?

☐ A. $(-\infty, -8) \cup (2, \infty)$

☐ B. $(-\infty, -3) \cup (2, \infty)$

☐ C. $(-5, 5)$

☐ D. $(-8, -2) \cup (2, 8)$

20) Which graph represents the solution set of the inequality $|x-4| \leq 3$?

☐ A.  1 7

☐ B.  1 7

☐ C.  1 7

☐ D.  1 7

21) Solve the following system of equations and select the correct solution:

$$\begin{cases} x + 3y = 9 \\ x - y = 1 \end{cases}$$

☐ A. $x = 3,\ y = 2$

☐ B. $x = 2,\ y = 3$

☐ C. $x = -3,\ y = -2$

☐ D. $x = -2,\ y = -3$

22) Determine the value of x from the following system of equations:

$$\begin{cases} 6x - 2y = 6 \\ 5x + y = 13 \end{cases}$$

☐ A. $x = 3$

☐ B. $x = 2$

☐ C. $x = 1$

☐ D. $x = 0$

23) Given the following system of equations, determine whether it has a single solution, no solution, or infinitely many solutions:

$$\begin{cases} x + 2y = 3 \\ 2x + 4y = 6 \end{cases}$$

☐ A. A single solution

☐ B. No solution

☐ C. Infinitely many solutions

☐ D. Cannot be determined

24) Which of the following systems of equations represents parallel lines and therefore has no solution?

☐ A. $\begin{cases} y = 2x + 1 \\ y = 2x - 1 \end{cases}$

☐ B. $\begin{cases} y = -x + 4 \\ 2y = -2x + 8 \end{cases}$

☐ C. $\begin{cases} 2y = 6x - 2 \\ 4y = 12x - 4 \end{cases}$

☐ D. $\begin{cases} y = \frac{1}{3}x + 2 \\ 3y = x + 6 \end{cases}$

25) A catering service charges a 15 fee to set-up for an event and 12 per person for food. If the total bill for an event was 99, how many people attended the event?

☐ A. 5

☐ B. 7

☐ C. 10

☐ D. 12

26) Two numbers add up to 26. Three times the larger number plus twice the smaller number equals 70. What is the larger number?

☐ A. 10

☐ B. 14

☐ C. 16

☐ D. 18

Answer Keys

1) B. $x = 12$
2) A. $x = 11$
3) A. $b = 2$
4) C. $y = -3$
5) D. $(0,2)$
6) C. $(6,0)$ and $(0,-3)$
7) A. $f(x) = x - 3$
8) C. $f(x) = 3x$
9) D. All of the above
10) C. $x < -1$
11) A. Above the line $y = -\frac{1}{2}x + 3$
12) C. $y \leq 3x + 1$
13) C. $-1 < x \leq \frac{7}{3}$
14) A. $(12,20]$
15) B. $x = 3$
16) C. $x = 3$ or $x = -4$
17) A. $x < -\frac{9}{5}$ or $x > 3$
18) B. $[-7,3]$
19) B. $(-\infty,-3) \cup (2,\infty)$
20) B.
21) A. $x = 3,\ y = 2$
22) B. $x = 2$
23) C. Infinitely many solutions
24) A.
25) B. 7
26) D. 18

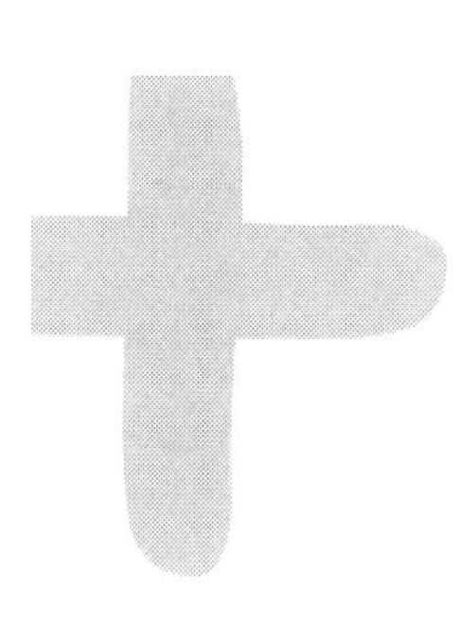

2. Functions

2.1 Evaluating Functions: Function Notation

Functions transform each unique input into a unique output, defining their fundamental nature. We use function notation to concisely express these operations, offering clarity and brevity in mathematical communication.

Key Point

Functions assign a unique output to every given input. A function notation is a concise way to describe a mathematical operation.

The most common function notation, $f(x)$, read as "f of x," uses any letter, like $g(x)$ or $h(x)$, for the function's name. To evaluate a function, replace the variable x with the given input value or expression.

For the function $f(x) = x + 6$, evaluate $f(2)$.

Solution: Substitute x with 2 in the function. Then:

$$f(x) = x + 6 \Rightarrow f(2) = 2 + 6 \Rightarrow f(2) = 8.$$

For the function $h(x) = 4x^2 - 1$, evaluate $h(2a)$.

Solution: Substitute x with $2a$ in the function. Then:

$$h(x) = 4x^2 - 1 \Rightarrow h(2a) = 4(2a)^2 - 1 \Rightarrow h(2a) = 16a^2 - 1.$$

2.2 Modeling Real-world Situations with Functions

A *function* is a relation between variables, defined by *inputs* (independent variables or the *domain*) and *outputs* (dependent variables or the *range*), that maps each input to exactly one output. This mapping is governed by the *Function Rule*, an algebraic expression that delineates the relationship between inputs and outputs.

Functions are instrumental in modeling real-world scenarios because they provide a systematic way to predict outputs for given inputs. The function rule can transform inputs into outputs, encapsulating the essence of the function's behavior.

Functions can be represented through:

- Graphs
- Tables
- Algebraic Expressions

In tables, columns are dedicated to inputs and outputs, illustrating how the function operates on inputs to produce outputs. This visualization aids in understanding the function's application and its rule's effect on transforming inputs into outputs.

Key Point

Functions, defined by a rule, map inputs to outputs, providing a fundamental tool for interpreting and predicting real-world phenomena.

Let us consider the relationship with the given values of x and y:

$$\{(1,4),(2,8),(3,12),(4,16)\}.$$

What is the corresponding equation for this relationship based on the inputs and outputs?

Solution: To find the function that describes this relationship, we will examine the first pair of values: $(1,4)$. In this pair, the y-value is 4 times the x-value ($1 \times 4 = 4$). So, we can guess that the function rule here is $y = 4x$.

To verify our guess, we can use the remaining pairs to confirm that y is indeed four times x for every pair. We see that $2 \times 4 = 8$, $3 \times 4 = 12$, and $4 \times 4 = 16$, which matches up with the y-values for the other pairs. Hence, the correct function for this relationship is $y = 4x$.

According to the values of x and y in the following relationship, find the right equation:

$$\{(1,3),(2,4),(3,5),(4,6)\}.$$

Solution: Find the relationship between the first x-value and first y-value, $(1,3)$. The value of y is 2 more than the x-value: $1+2=3$. So, the equation is:

$$y=x+2.$$

Now check the other values: $2+2=4$, $3+2=5$, and $4+2=6$. Therefore, each x-value and y-value satisfies the equation $y=x+2$.

2.3 Adding and Subtracting Functions

As with numbers and expressions, we can add and subtract two functions, simplify or evaluate them to form a new function. The process of adding or subtracting functions involves operating on the functions' formulas individually just like ordinary algebraic expressions and then combining the results.

Key Point

When two functions, $f(x)$ and $g(x)$, are given, we can create two new functions which are expressed as: $(f+g)(x)=f(x)+g(x)$, and $(f-g)(x)=f(x)-g(x)$.

If $g(x)=2x-2$ and $f(x)=x+1$, find $(g+f)(x)$.

Solution: Firstly, we recall the general formula:

$$(g+f)(x)=g(x)+f(x).$$

Substituting our functions into this formula, we get:

$$(g+f)(x)=(2x-2)+(x+1)=3x-1.$$

If $f(x)=4x-3$ and $g(x)=2x-4$, find $(f-g)(x)$.

Solution: Here, we apply the formula:

$$(f-g)(x) = f(x) - g(x).$$

On substituting our given functions in the formula, we get:

$$(f-g)(x) = (4x-3) - (2x-4) = 2x+1.$$

Given $g(x) = x^2 + 2$ and $f(x) = x + 5$, find $(g+f)(0)$.

Solution: We use the general formula:

$$(g+f)(x) = g(x) + f(x).$$

Substitute our given functions into the formula:

$$(g+f)(0) = (0^2+2) + (0+5) = 7.$$

2.4 Multiplying and Dividing Functions

Similar to how we can add and subtract functions, we can also multiply and divide functions. The multiplication or division of functions involves respectively multiplying or dividing the functions' expressions to create a new function.

Key Point

Given functions $f(x)$ and $g(x)$, we can create two new functions through multiplication and division as follows:

The product function $(f \cdot g)(x)$ is found by multiplying: $(f \cdot g)(x) = f(x) \cdot g(x)$.

The quotient function $\left(\frac{f}{g}\right)(x)$ is found by dividing: $\left(\frac{f}{g}\right)(x) = \frac{f(x)}{g(x)}$, where $g(x) \neq 0$.

For the functions $g(x) = x+3$ and $f(x) = x+4$, find the product $(g \cdot f)(x)$.

Solution: To find $(g \cdot f)(x)$, we multiply the expressions of $f(x)$ and $g(x)$ by applying the formula

$(g.f)(x) = g(x) \cdot f(x)$, then we get:

$$(x+3)(x+4) = x^2 + 4x + 3x + 12 = x^2 + 7x + 12.$$

Therefore, $(g \cdot f)(x) = x^2 + 7x + 12$.

For the functions $f(x) = x+6$ and $g(x) = x-9$, find the quotient $\left(\frac{f}{g}\right)(x)$.

Solution: To get $\left(\frac{f}{g}\right)(x)$, we divide the expression of $f(x)$ by the expression of $g(x)$, according to $(\frac{f}{g})(x) = \frac{f(x)}{g(x)}$. So, for $g(x) \neq 0$ we have:

$$\left(\frac{f}{g}\right)(x) = \frac{x+6}{x-9}.$$

If $f(x) = x+3$ and $g(x) = 2x-4$, find $\left(\frac{f}{g}\right)(3)$.

Solution: To get $(\frac{f}{g})(3)$, we use the formula $(\frac{f}{g})(x) = \frac{f(x)}{g(x)}$. Then,

$$\left(\frac{f}{g}\right)(3) = \frac{3+3}{(2)(3)-4} = 3.$$

2.5 Composite Functions

The composition of functions involves combining two or more functions in a way where the output of one function is used as the input for another. We denote this as $(f \circ g)(x) = f(g(x))$, which means applying function g to x first, then applying function f to the result.

The notation we use for composition is $(f \circ g)(x) = f(g(x))$. This can be read as "f composed with g of x" or "f of g of x".

Key Point

Composition of functions is not commutative. That is, $f(g(x))$ is generally not equal to $g(f(x))$.

Given $f(x) = 2x+3$ and $g(x) = 5x$, find $(f \circ g)(x)$ and $(g \circ f)(x)$.

Solution: We use the definition of function composition, $(f \circ g)(x) = f(g(x))$. Substituting $g(x) = 5x$

into $f(x)$, we get:

$$f(g(x)) = f(5x) = 2(5x) + 3 = 10x + 3.$$

Thus:

$$(f \circ g)(x) = 10x + 3.$$

Now, substituting $f(x) = 2x + 3$ into $g(x)$, we get:

$$g(f(x)) = g(2x + 3) = 5(2x + 3) = 10x + 15.$$

Thus:

$$(g \circ f)(x) = 10x + 15.$$

Suppose $f(x) = x^2 + 5x - 1$ and $g(x) = 3x - 2$. Find $f(g(x))$.

Solution: First, we substitute x with $g(x)$ in $f(x)$:

$$f(g(x)) = g(x)^2 + 5g(x) - 1.$$

Replacing $g(x)$ with $3x - 2$ we get:

$$f(g(x)) = (3x - 2)^2 + 5(3x - 2) - 1.$$

Simplify to get:

$$f(g(x)) = 9x^2 - 12x + 4 + 15x - 10 - 1.$$

Combine like terms to obtain:

$$f(g(x)) = 9x^2 + 3x - 7.$$

2.6 Zeros of a Function

The zeros, or roots, of a function $f(x)$ are the real numbers x for which $f(x) = 0$. These zeros can be visualized on a graph as the x-intercepts, where the graph intersects or touches the x-axis. A polynomial of degree n can have up to n real zeros, but it is possible to have fewer. Understanding the real zeros of a function is crucial for graphing, as they indicate where the graph will touch or cross the x-axis.

Key Point

The zeros of a function $f(x)$, where $f(x) = 0$, correspond to its x-intercepts

To find these zeros, we solve the equation $f(x) = 0$ for x.

Find the zeros of $f(x) = x^2 - 4$.

Solution: To find the zeros of a function (i.e., the values x for which the function is equal to zero), we set the function equal to zero:

$$x^2 - 4 = 0.$$

This equation is a difference of squares and can be solved by factoring:

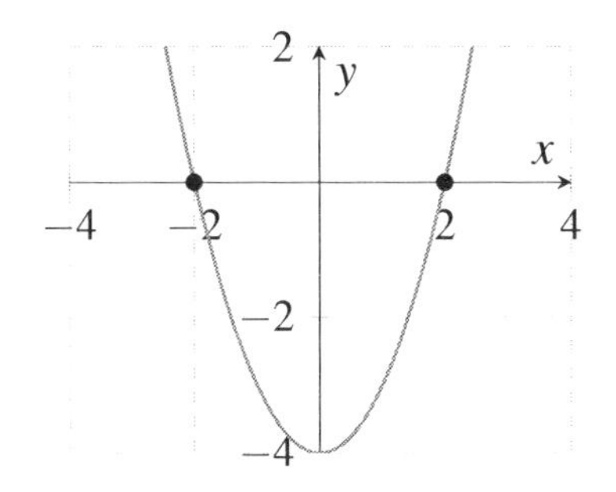

$$f(x) = (x-2)(x+2).$$

The function is zero when both $x-2=0$ and $x+2=0$, giving us $x=2$ and $x=-2$. These are the zeros of the function.

Determine the zeros of the function shown in the graph below.

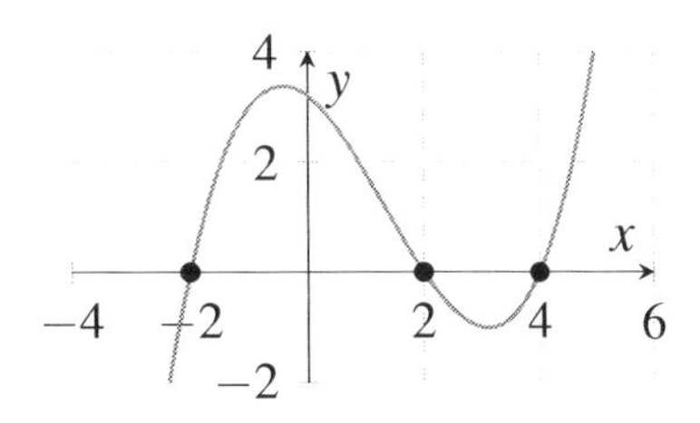

Solution: The zeros of the displayed function are the intersection of the graph and x-axis. That is, points $x = -2$, $x = 2$, and $x = 4$.

2.7 Average Rate of Changes

Understanding how things change at an average pace is key as you move towards learning calculus. This concept, known as the 'average rate of change,' shows how much one thing changes in relation to another. In terms of a function, the 'average rate of change' is calculated as the ratio of the change in the output (or the y values) to the change in the input (or the x values). In simple terms, it describes how much y changes for each unit increase in x. The average rate of change of a function between two points is the slope of the secant line passing through those points.

For a function $f(x)$, the average rate of change between $x = a$ and $x = b$ is given by:

$$\frac{f(b) - f(a)}{b - a}.$$

Example Given the function $f(x) = 2x^2 + 3x + 1$, find the average rate of change of the function on the interval $[1,3]$.

Solution: Recall the formula for the average rate of change is $\frac{f(b)-f(a)}{b-a}$. Substituting $a = 1$ and $b = 3$ in the formula, we get:

$$\frac{f(3) - f(1)}{3 - 1} = \frac{(2(3)^2 + 3(3) + 1) - (2(1)^2 + 3(1) + 1)}{3 - 1} = \frac{28 - 6}{2} = 11.$$

So, the average rate of change of the function $f(x) = 2x^2 + 3x + 1$ on the interval $[1,3]$ is 11.

2.8 Codomain

In studying functions, two critical sets are discussed: the domain and the codomain. The domain comprises all possible inputs for a function, while the codomain, represented by B in the function $f : A \to B$, includes all potential outputs, even those not produced by inputs from the domain.

Key Point

The codomain is a comprehensive set of possible outcomes from a function, regardless of actual occurrence, contrasting with the range, which consists of outputs genuinely produced from the domain's inputs.

The range is a subset of the codomain, containing only the actual outputs generated by the function. Understanding the difference between codomain and range is essential for grasping function concepts, and this distinction becomes clearer with practice and application to various examples.

Example Let $f : \mathbb{R} \to \mathbb{R}$ be defined by $f(x) = x^2$. Here $\mathbb{R}$ is the set of real numbers. What are the domain, codomain and the range of this function?

Solution: The domain and codomain of this function are both the set of all real numbers, $\mathbb{R}$. However, since the function's output (which is x^2) for any real number x is never negative, the range of this function is the set of all nonnegative real numbers.

Consider the function $f: \mathbb{Z} \rightarrow \mathbb{R}$ defined by $f(n) = n^3 - n$, where $\mathbb{Z}$ is the set of integers. What are the codomain and the range of this function?

Solution: The codomain of this function is $\mathbb{R}$ because the function is defined to map to all real numbers. The function can output any integer since for every $n \in \mathbb{Z}$, $f(n) \in \mathbb{Z}$. Hence the range of the function is $\mathbb{Z}$.

2.9 Practices

1) For the function $g(t) = 3t^3 - 2t + 5$, evaluate $g(-3)$.

☐ A. -70
☐ B. -74
☐ C. 74
☐ D. 70

2) Given the function $p(x) = \frac{2}{x} - 7$, what is $p(4)$?

☐ A. -5.5
☐ B. -6
☐ C. -6.5
☐ D. -7.5

3) Given a set of values representing the weekly sales x (in hundreds) and the income y (in thousands) of a store as $\{(1,2),(2,3.5),(3,5),(4,6.5)\}$, which of the following is the correct function rule for this relationship?

☐ A. $y = 1.5x - 0.5$
☐ B. $y = 1.5x + 0.5$
☐ C. $y = x + 1.5$
☐ D. $y = x + 0.5$

4) The height y (in meters) of a ball thrown upwards with time x (in seconds) can be modeled by $y = -4.9x^2 + 10x$. Which of the following represents the height of the ball after 1 second?

☐ A. 5.1 meters
☐ B. 10 meters
☐ C. 15.1 meters
☐ D. 5.9 meters

5) Let $h(x) = x^2 - 3x + 1$ and $k(x) = 2x - 5$. Find $(h-k)(x)$.

☐ A. $x^2 - 5x + 6$
☐ B. $x^2 - x - 4$
☐ C. $x^2 + x + 4$
☐ D. $x^2 - x - 6$

6) Given two functions $p(x) = 3x + 2$ and $q(x) = -2x^2 + x - 3$, what is $(p+q)(1)$?

☐ A. 4
☐ B. 0
☐ C. 1
☐ D. -4

7) If $h(x) = 3x^2 - x$ and $k(x) = x - 5$, then what is $(h \cdot k)(x)$?

☐ A. $3x^3 - 5x^2 - x - 5$
☐ B. $3x^3 - 16x + 5$
☐ C. $3x^2 - 16x + 5$
☐ D. $3x^3 - 16x^2 + 5x$

8) Given the functions $f(x) = \sqrt{x}$ and $g(x) = x^2 - 1$ where $x \neq \pm 1$, the division $\left(\frac{f}{g}\right)(x)$ is

☐ A. $\frac{1}{x^2-1}$
☐ B. $\frac{\sqrt{x}}{x^2-1}$
☐ C. $\sqrt{x}(x^2 - 1)$
☐ D. $x^{\frac{3}{2}} - 1$

9) Given $h(x) = x^2 - 4x + 4$ and $k(x) = 2 - x$, calculate $(h \circ k)(x)$.

☐ A. $x^2 - 8x + 8$
☐ B. x^2
☐ C. $-x^2 + 8x - 12$
☐ D. $4 - 4x$

10) If $p(x) = \sqrt{x+1}$ and $q(x) = 3x - 7$, what is $(p \circ q)(2)$?

☐ A. $\sqrt{2}$
☐ B. $\sqrt{-1}$
☐ C. $\sqrt{5}$
☐ D. 0

11) Given the function $f(x) = x^3 - 3x^2 - 4x + 12$, which of the following is a real zero of $f(x)$?

☐ A. 0
☐ B. 2
☐ C. -1
☐ D. 1

12) What is a possible number of real zeros of the polynomial $g(x) = x^4 - 5x^2 + 4$?

☐ A. 0
☐ B. 1
☐ C. 3
☐ D. 4

13) If $f(x) = x^3 - 3x + 2$, what is the average rate of change of the function on the interval $[0, 2]$?

☐ A. 2
☐ B. 1
☐ C. 4
☐ D. 6

14) Consider the function $f(x) = -x^2 + 2x + 3$. What is the average rate of change of this function on the interval $[-1, 1]$?

☐ A. -2
☐ B. 1
☐ C. 2
☐ D. -1

15) Let $g : \mathbb{R} \to \mathbb{R}$ be the function defined by $g(x) = 3x - 2$. Which of the following represents the codomain of g?

☐ A. $\{3x - 2 \mid x \in \mathbb{Z}\}$
☐ B. All real numbers less than 2
☐ C. $\mathbb{R}$
☐ D. All real numbers greater than -2

16) Consider the function f defined by $f(x) = \sqrt{x} - 1$. What is the range of the function?

☐ A. $[-1, \infty)$
☐ B. $[0, \infty)$
☐ C. $[-1, 0)$
☐ D. $(0, \infty)$

Answer Keys

1) A. -70
2) C. -6.5
3) B. $y=1.5x+0.5$
4) A. 5.1 meters
5) A. x^2-5x+6
6) C. 1
7) D. $3x^3-16x^2+5x$
8) B. $\frac{\sqrt{x}}{x^2-1}$
9) B. x^2
10) D. 0
11) B. 2
12) D. 4
13) B. 1
14) C. 2
15) C. $\mathbb{R}$
16) A. $[-1,\infty)$

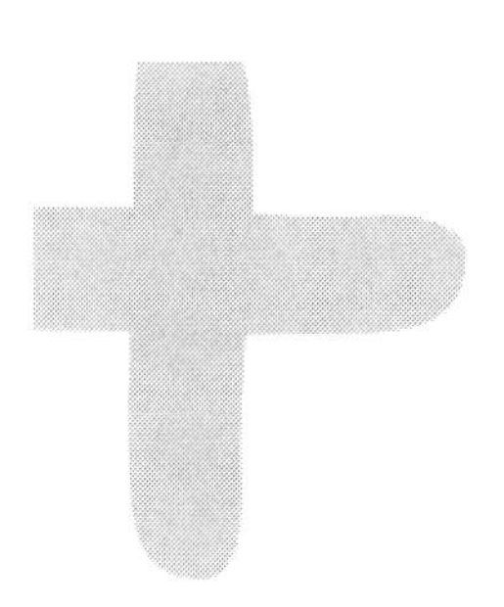

3. Polynomial and Rational Functions

3.1 The Simplest Functions: Constant and Identity

Two of the basic functions in mathematics are the constant function and the identity function. These functions play a key role in creating a foundation for comprehending complex functions and their transformations.

Key Point

A constant function is a function with a constant output, irrespective of the input. It is symbolically represented as $f(x) = c$, where c is a constant.

The graph of a constant function $f(x) = c$, is seen as a horizontal line at the height corresponding to the value c.

Key Point

An identity function is a function where the output is identical to the input. Symbolically, it is written as $f(x) = x$.

On a graph, the identity function is depicted as a straight line inclined at an angle of 45°, with the origin as its intercept.

Let us consider $f(x) = 2$ as an example of a constant function.

Solution: Irrespective of the input, the output for this function will always be 2.

As seen below, the graph of this function appears as a horizontal line along $y = 2$.

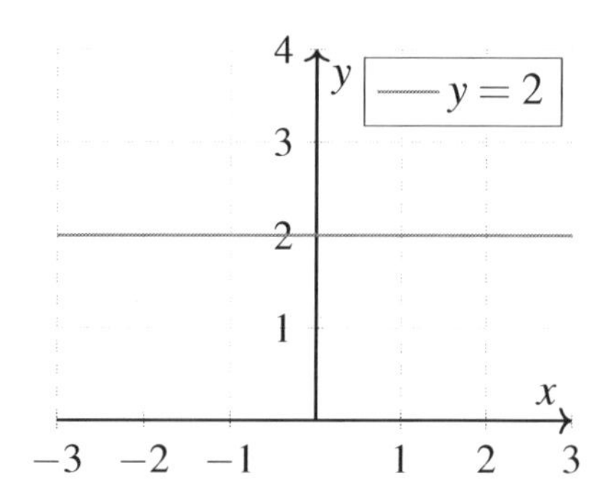

Example Draw the graph of the identity function.

Solution: If the input is 2, the output is 2; if the input is -3, the output is -3.

The graph of this function is a straight line with a slope of 1 that passes through the origin.

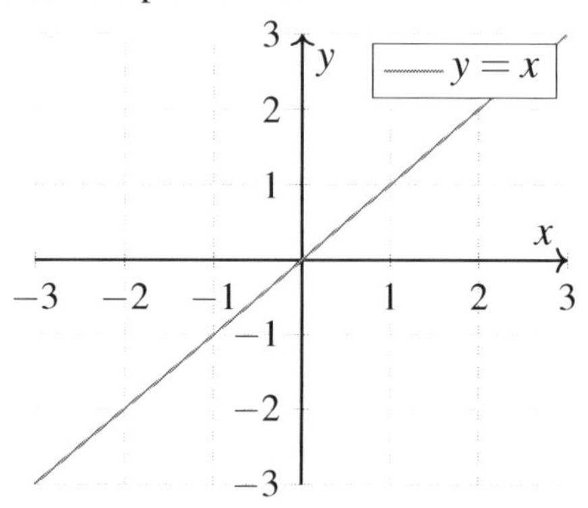

3.2 Polynomial Functions

A *polynomial function* is a type of mathematical function where the terms are all composed of variables raised to whole number exponents and multiplied by coefficients. It can be expressed in the following general form:

$$P(x) = a_n x^n + a_{n-1} x^{n-1} + \cdots + a_2 x^2 + a_1 x + a_0,$$

where a_n, a_{n-1}, ..., a_1, a_0 are constants and n is a non-negative integer. The highest power of x (i.e., n) is called the degree of the polynomial. One significant aspect of the polynomial function is its graph. The *graph of a polynomial* is smooth and continuous, without any breaks, jumps, or holes. This is one of the crucial points that distinguishes them from many other types of functions.

Polynomial functions are further divided into types based on their degree:

1. Constant (Degree 0): $P(x) = a$
2. Linear (Degree 1): $P(x) = ax + b$
3. Quadratic (Degree 2): $P(x) = ax^2 + bx + c$
4. Cubic (Degree 3): $P(x) = ax^3 + bx^2 + cx + d$ and so on.

The **domain** of a polynomial function is $\mathbb{R}$ (all real numbers), represented as $(-\infty, \infty)$. There are no restrictions on the input values for x, emphasizing the universality of polynomial functions.

Understanding the **range** of polynomial functions can be somewhat tricky. The range represents all possible output values. Unlike other functions, determining the range for polynomial functions can be more nuanced.

1. Polynomial functions of *odd degree* have a range of $(-\infty, \infty)$ because they exhibit opposite end behavior: as x approaches positive or negative infinity, y follows suit.
2. Polynomial functions of *even degree* may have a maximum or minimum value based on their leading coefficient, potentially limiting the range to values greater or less than a certain number.

Consider a polynomial $P(x) = -2x^3 + 5x^2 - 8x + 10$. What is its degree and leading coefficient?

Solution: The degree is the highest power of x, which is 3 in this case. The leading coefficient is the coefficient of the term with the highest power, which is -2. Therefore, the degree is 3 and the leading coefficient is -2.

Recognize the type of the polynomial $P(x) = -2x + 8$ and sketch its graph.

Solution: This is a linear polynomial (Degree 1) with a leading coefficient of -2.

The linear polynomial graph is a straight line with a slope equal to the leading coefficient.

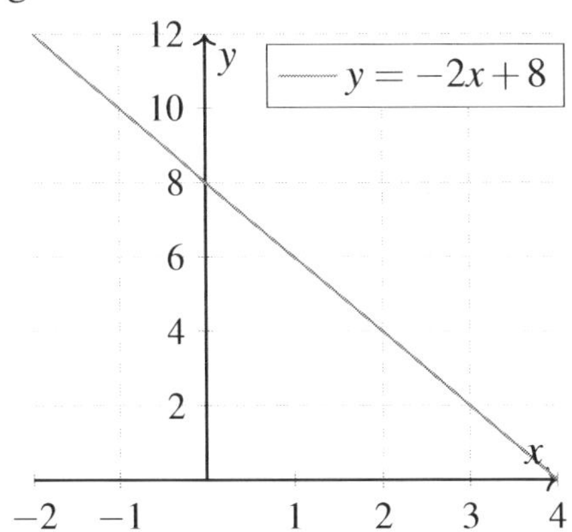

3.3 Graphing Polynomial Functions

A polynomial function is given by $f(x) = a_n x^n + a_{n-1} x^{n-1} + \cdots + a_1 x + a_0$, where n is the degree and a_n is the leading coefficient. The degree and leading coefficient play significant roles in determining the graph's characteristics.

Key Points to Remember:

- **Degree and Shape**: The degree of the polynomial, n, influences its maximum number of turning points and general shape. Odd-degree polynomials have graphs with ends pointing in opposite directions, whereas even-degree polynomials have ends that point in the same direction (either up or down).

- **Leading Coefficient and Direction**:
  - For even-degree polynomials, a positive a_n means the graph opens upward; a negative a_n means it opens downward.
  - For odd-degree polynomials, a positive a_n results in the right end of the graph pointing up and the left down; a negative a_n reverses this direction.
- **Roots or Zeros**: The x-values where the polynomial equals zero are its roots or zeros. These points are where the graph touches or crosses the x-axis.
- **Turning Points**: The graph changes direction at turning points. A polynomial of degree n can have up to $n-1$ turning points.
- **End Behavior**: This describes how the graph behaves as x approaches infinity or negative infinity. Polynomials of even degrees have ends pointing in the same direction, while those of odd degrees have ends pointing in opposite directions.

Key Point

Polynomials with even degrees have graphs with ends pointing in the same direction, up or down based on the leading coefficient. Odd-degree polynomials have ends that point in opposite directions, also determined by the leading coefficient's sign.

Example Let us graph the polynomial function $f(x) = x^3 - 3x^2 - 4x + 12$.

Solution: First, we find the roots of the polynomial by setting $f(x) = 0$ and solving for x. In this case, the roots are $x = -2, 2, 3$. Next, observe that 3 is an odd degree, and the leading coefficient is positive; thus, as x approaches $-\infty$, $f(x)$ approaches $-\infty$, and as x approaches $+\infty$, $f(x)$ approaches $+\infty$. Lastly, as it is a third-degree polynomial, it will have at most 2 turning points. Taking these aspects into account, we can now sketch the graph.

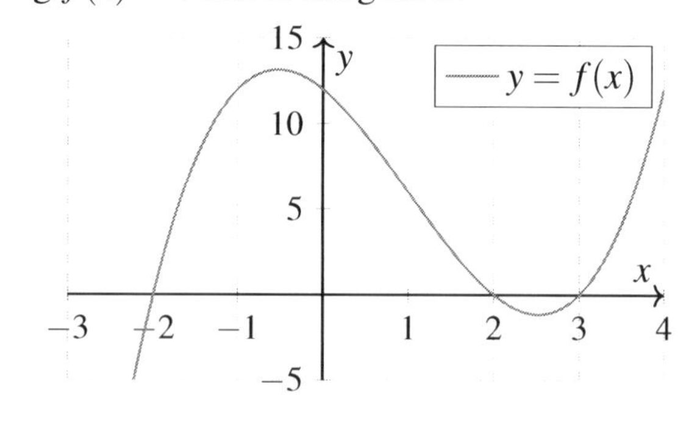

Graph the polynomial $g(x) = x^2 - 6x + 9$.

Solution: This is a perfect square trinomial, $g(x) = (x-3)^2$.

It has a vertex at $(3,0)$ and opens upwards, touching but not crossing the x-axis at $x=3$.

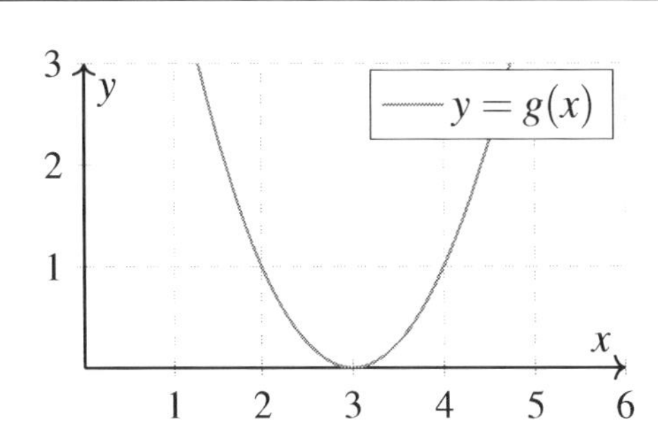

3.4 Writing Polynomials in Standard Form

A polynomial is written in standard form when the terms are ordered by degree in descending order from left to right. For any polynomial of the form:

$$f(x) = a_n x^n + a_{n-1} x^{n-1} + \ldots + a_1 x + a_0,$$

$a_n x^n$ is the term with the highest degree.

Key Point

The standard form of a polynomial starts with the term with the highest degree and ends with the constant term.

Example Arrange the terms in the polynomial $8+5x^2-3x^3$ in standard form.

Solution: The largest exponent is 3, which means the term $-3x^3$ is written first. The next highest exponent is 2, so $5x^2$ comes next. The constant term, 8, comes last. Therefore, the polynomial in standard form is $-3x^3+5x^2+8$.

3.5 Simplifying Polynomials

Simplifying polynomials involves combining *like terms* and using the FOIL method for multiplying binomials.

Key Point

The "*FOIL*" acronym stands for First, Outer, Inner, Last. It is used to remember how to multiply two binomials.

Remember: *"Like terms"* are terms in an expression with the same variable raised to the same exponent. For example, $3x^2$ and $2x^2$ are like terms.

The FOIL process is as follows:

$$\begin{aligned}(x+a)(x+b) = &\text{First Terms: } x\cdot x = x^2\\ &\text{Outer Terms: } x\cdot b = bx\\ &\text{In Terms: } a\cdot x = ax\\ &\text{Last Terms: } a\cdot b = ab\end{aligned}$$

Putting it all together, we get the simplified equation:

$$(x+a)(x+b) = x^2 + (a+b)x + ab.$$

To simplify the polynomial expressions, always add or subtract the like terms. The order of the operation plays a crucial role here as it follows the rules of basic arithmetic.

Simplify the expression $x(4x+7) - 2x$.

Solution: First, use the distributive property:

$$x(4x+7) = 4x^2 + 7x.$$

Now, combine like terms:

$$4x^2 + 7x - 2x = 4x^2 + 5x.$$

So, we get:

$$x(4x+7) - 2x = 4x^2 + 5x.$$

Simplify the expression $(x+3)(x+5)$.

Solution: First, apply the FOIL method:

$$(x+3)(x+5) = x^2 + 5x + 3x + 15.$$

Now, combine like terms:

$$x^2 + 5x + 3x + 15 = x^2 + 8x + 15.$$

3.6 Factoring Trinomials

Factoring trinomials simplifies expressions by applying the distributive property and FOIL method in reverse. It often uses the Difference of Squares, which is particularly useful for perfect square trinomials.

The following identities are often helpful:

- $a^2 - b^2 = (a+b)(a-b)$
- $a^2 + 2ab + b^2 = (a+b)(a+b)$
- $a^2 - 2ab + b^2 = (a-b)(a-b)$

The Reverse FOIL is often used in factoring trinomials. It uses the product-sum concept to find two numbers with a sum and product equal to the coefficients in the trinomial.

Key Point

In using these identities, always remember the *Reverse FOIL* procedure: given a trinomial in the form $x^2 + (b+a)x + ab$, you can always factor it to $(x+a)(x+b)$.

Let us consider the trinomial $x^2 - 2x - 8$. Factorize this trinomial.

Solution: To find the factors of the trinomial, we need to think of two numbers whose product equals -8 (the last term of the trinomial), and their sum equals -2 (the coefficient of the middle term). Remember "Reverse FOIL", these two numbers are 2 and -4. Then:

$$x^2 - 2x - 8 = (x-4)(x+2).$$

3.7 Solving a Quadratic Equation

In this section, we will discuss how to solve a quadratic equation by factorization and by quadratic formula. A quadratic equation is described by the general expression:

$$ax^2 + bx + c = 0,$$

where a, b, and c are constant numbers, and x is the variable we want to find.

Key Point

For a quadratic equation $ax^2 + bx + c = 0$, factorization allows setting each factor to zero to find x. This is the zero-product property.

However, not all quadratic equations can be factorized easily. If we cannot factorize a quadratic equation, we can resort to the quadratic formula: $x = \frac{-b \pm \sqrt{b^2-4ac}}{2a}$.

Key Point

The quadratic formula, $x = \frac{-b \pm \sqrt{b^2-4ac}}{2a}$, provides the solution for a quadratic equation of the form $ax^2 + bx + c = 0$.

Example Find the solutions of the quadratic equation $x^2 + 7x + 12 = 0$.

Solution: We try to factor the quadratic by grouping. We need to find two numbers such that they sum up to 7 (which is the coefficient of x) and whose product is 12 (which is the constant term). The numbers that satisfy these conditions are 3 and 4. So we write:

$$x^2 + 7x + 12 = (x+3)(x+4).$$

Setting each factor equal to zero and solving for x yields two solutions: $(x+3) = 0$, implies $x = -3$. Similarly $(x+4) = 0$, implies $x = -4$.

Example Find the solutions of the quadratic equation $x^2 + 5x + 6 = 0$ with quadratic formula.

Solution: Here, we have $a = 1$, $b = 5$, and $c = 6$. So by quadratic formula, we get:

$$x = \frac{-5 \pm \sqrt{5^2 - 4(1)(6)}}{2(1)}.$$

Finally, we obtain two solutions for this equation, which are $x = -2$ and $x = -3$.

3.8 Graphing Quadratic Functions

Quadratic functions, represented by $y = ax^2 + bx + c$ where a, b, and c are constants, form parabolas that may open upwards or downwards depending on the value of a. These functions can also be expressed in the vertex form, $y = a(x-h)^2 + k$, pinpointing the parabola's vertex at (h,k), which denotes its peak or trough. The vertex form simplifies locating the turning point and the line of symmetry, $x = h$. In the standard form, the vertex's x-coordinate is determined by $x = -\frac{b}{2a}$, assisting in graphing the function's precise peak or trough.

Key Point

The standard form of a quadratic function, $y = ax^2 + bx + c$, is essential for finding the vertex $(x = -\frac{b}{2a})$, which aids in graphing.

Key Point

The vertex form of a quadratic function, $y = a(x-h)^2 + k$, directly reveals the vertex (h,k), simplifying the process of graphing by identifying the parabola's peak or trough and its axis of symmetry at $x = h$.

The vertex form of a quadratic function simplifies determining the parabola's orientation and its maximum or minimum point.

Sketch the graph of function $y = (x+2)^2 - 3$.

Solution: This function is in the vertex form. The vertex here is at $(-2,-3)$ and the axis of symmetry is $x = -2$. We need to find more points to plot the function's graph. Setting $x = 0$ in the function, we find $y = 1$. So, the y-intercept is $(0,1)$. Using the vertex form of the quadratic function, the axis of symmetry, and the y-intercept, we can plot these points on a graph. The graph's symmetry allows us to reflect our points about the axis of symmetry to obtain additional points, for example $(-4,1)$. The final touch will be connecting the points to create the U-shaped curve.

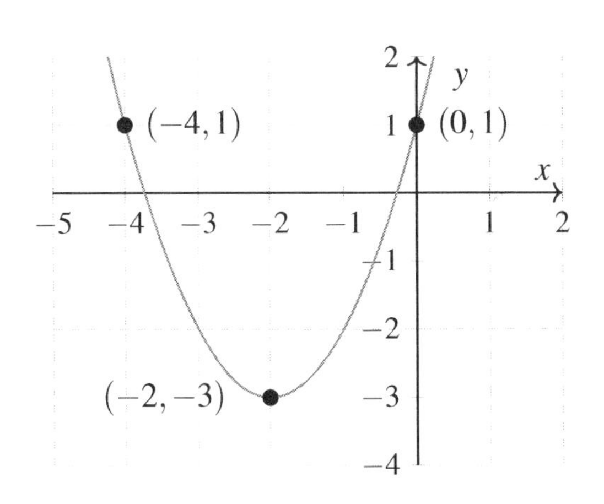

3.9 Solving Quadratic Inequalities

Quadratic inequalities, such as $ax^2 + bx + c > 0$, where the symbol $>$ can be replaced with $<$, $\leq$, or $\geq$, are solved similarly to quadratic equations. The key steps involve identifying the roots or zeros and then testing values between these roots to determine the intervals where the inequality holds true.

Key Point

To address quadratic inequalities, first solve the quadratic equation for its roots, termed critical values or zeros. These roots segment the number line into regions where the inequality sign remains consistently either positive or negative.

Solve the following quadratic inequality $x^2 + x - 6 > 0$.

Solution: We start by finding the roots of the equivalent equation $x^2 + x - 6 = 0$. Factoring gives

$(x-2)(x+3)=0$. Setting each factor equal to zero gives the roots $x=2$ and $x=-3$. Next, we choose a test point in each of the regions determined by $(-3,2)$ to see where the inequality holds. For example, we can choose $x=0$ between -3 and 2. After substituting $x=0$ into the inequality, we find that $-6>0$, which is false. Therefore, the values between 2 and -3 do not satisfy the inequality. The solution to the inequality is thus $x>2$ or $x<-3$. In interval notation, the solution set is $(-\infty,-3)\cup(2,\infty)$.

3.10 Graphing Quadratic Inequalities

Graphing quadratic inequalities starts by understanding the structure of quadratic inequalities. They are generally represented in the form $y>ax^2+bx+c$ (or substitute $<$, $\leq$, or $\geq$ in place of $>$). Here, a,b,c are real numbers where $a\neq 0$.

Key Point

Quadratic inequalities can be graphed by sketching the parabola and testing a point to identify the solution region.

Graph quadratic inequalities by first plotting the corresponding parabola of $y=ax^2+bx+c$, where $a\neq 0$. Use a dashed line for $<$ or $>$ inequalities and a solid line for $\leq$ or $\geq$. Then, test a point not on the parabola using the inequality. If true, the area including the point is the solution region; if false, the opposite region is the solution.

Sketch the graph for the inequality $y<2x^2$.

Solution: First, graph the quadratic: $y=2x^2$.

Quadratic functions in vertex form: $y=a(x-h)^2+k$, where (h,k) is the vertex. Then, the vertex of $y=2x^2$ is: $(h,k)=(0,0)$. Since the inequality sign is $<$, we need to use dashed lines. Now, choose a testing point inside the parabola. Let us choose $(0,2)$, which implies $2<0$. This is not true. So, the area outside the parabola is the solution region.

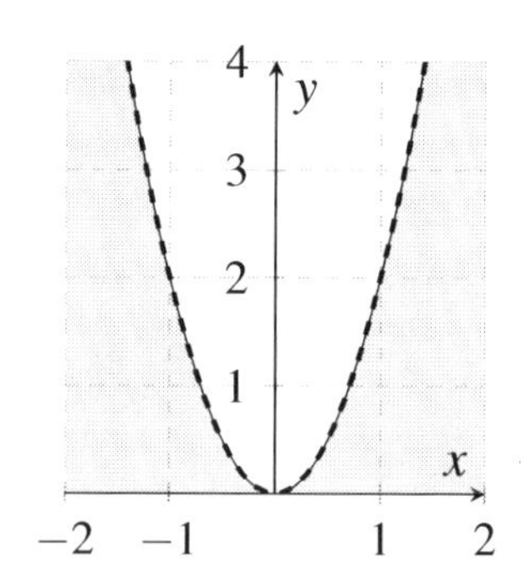

3.11 Rational Equations

Rational equations involve ratios of polynomials and are essential for solving real-world problems in motion, work, and fluid dynamics. The two main strategies for solving these equations are:

- Finding a common denominator.
- Cross-multiplying.

Key Point

Select the solving method based on the equation's structure: use cross-multiplying for equations with a single fraction on each side; otherwise, opt for finding a common denominator.

To solve using a common denominator, identify the least common denominator (LCD) of all fractions, multiply each term by the LCD to eliminate fractions, and solve the resulting equation.

For cross-multiplying, applicable when each side has one fraction, multiply across the equation diagonally and solve the resulting equation, ensuring no solution makes a denominator zero. Here, you multiply the first numerator by the second denominator and equate it to the product of the second numerator and first denominator, following the formula:

$$\frac{a}{b} = \frac{c}{d} \Rightarrow ad = bc.$$

Then, solve for the variable. Be careful that the solutions obtained for the equation do not make the denominators of the fractions zero.

Solve the following equation: $\frac{x-2}{x+1} = \frac{x+4}{x-2}$.

Solution: To solve this, we cross-multiply as we have a fraction on either side. Thus, following $\frac{a}{b} = \frac{c}{d} \Rightarrow ad = bc$, we multiply: $(x-2)(x-2) = (x+4)(x+1)$. Next, we expand to:

$$(x-2)^2 = x^2 - 4x + 4, \text{ and } (x+4)(x+1) = x^2 + 5x + 4.$$

This equates to: $x^2 - 4x + 4 = x^2 + 5x + 4$, and simplifies to: $x^2 - 4x = x^2 + 5x$. Subtracting both sides, we get: $x^2 - 4x - (x^2 + 5x) = 0$, which gives us $x = 0$. Since $x = 0$ does not make the denominator of the fractions zero, it is considered an acceptable answer.

Solve the following equation by converting to a Common Denominator method:

$$\frac{2x}{x-3} = \frac{2x+2}{2x-6}.$$

Solution: Multiply the numerator and denominator of the rational expression on the left by 2 to get a common denominator $(2x-6)$: $\frac{2(2x)}{2(x-3)} = \frac{2x+2}{2x-6}$. Now, the denominators on both sides of the equation are equal. Therefore, their numerators must be equal too: $4x = 2x + 2$, which implies $x = 1$. Since $x = 1$ does not make the denominator of the fractions zero, it is considered an acceptable answer.

3.12 Multiplying and Dividing Rational Expressions

The multiplication and division of rational expressions follow the same rules as those used for regular fractions. Let us now dive into the concepts of multiplying and dividing rational expressions.

Multiplication of rational expressions is similar to multiplication of fractions. We multiply the numerators together and likewise, the denominators.

Key Point

To multiply rational expressions: $\frac{a}{b} \times \frac{c}{d} = \frac{a \times c}{b \times d}$. After multiplying, always try to simplify the result by factoring and canceling out common terms in the numerator and the denominator.

For division, we follow the concept of "Keep, Change, Flip" which is applicable for the division of fractions. That is, we keep the first fraction, change the division to multiplication, and flip the second fraction.

Key Point

To divide rational expressions: $\frac{a}{b} \div \frac{c}{d} = \frac{a}{b} \times \frac{d}{c} = \frac{a \times d}{b \times c}$. Just like multiplication, after division, try to simplify by factoring and canceling out common terms.

Multiply the following rational expressions: $\frac{x+6}{x-1} \times \frac{x-1}{5}$.

Solution: Following the multiplication principle:

$$\frac{(x+6)(x-1)}{5(x-1)} = \frac{x^2 - x + 6x - 6}{5x - 5} = \frac{x^2 + 5x - 6}{5x - 5}.$$

The numerator factors to $(x+6)(x-1)$, and $x-1$ is a common term in the numerator and the denominator, so it can be simplified: $\frac{(x+6)(x-1)}{5(x-1)} = \frac{x+6}{5}$.

Solve the following rational expressions: $\frac{5x}{x+3} \div \frac{x}{2x+6}$.

Solution: Using the fraction division rule, we have:

$$\frac{5x}{x+3} \div \frac{x}{2x+6} = \frac{5x}{x+3} \times \frac{2x+6}{x}.$$

Then, using the rules of multiplication of fractions, we get: $\frac{5x}{x+3} \div \frac{x}{2x+6} = 10$.

3.13 Simplifying Rational Expressions

Simplifying rational expressions involves making them easier to understand, similar to simplifying numerical fractions. The process typically includes:

Key Point

The process of simplification usually involves four steps:

1. Factorize the numerator and denominator if possible,
2. Identify common factors between the numerator and the denominator,
3. Cancel out the common factors in both,
4. Simplify the expression further if needed.

Simplify the rational expression $\frac{9x^2y}{3y^2}$.

Solution: First, we should cancel the common factor 3:

$$\frac{9x^2y}{3y^2} = \frac{3x^2y}{y^2}.$$

Now, we see that y is a common factor in the numerator and denominator. We should cancel this as well:

$$\frac{3x^2y}{y^2} = \frac{3x^2}{y}.$$

Therefore, we have:

$$\frac{9x^2y}{3y^2} = \frac{3x^2}{y}.$$

This is the simplified form of the given expression.

3.14 Rational Functions

A Rational Function $R(x) = \frac{P(x)}{Q(x)}$ is defined as the quotient of two polynomials, where $Q(x) \neq 0$. Its domain includes all real numbers, except where $Q(x) = 0$.

A Vertical Asymptote at $x = a$ is where the function tends to infinity or negative infinity, occurring when $Q(x) = 0$ but $P(x) \neq 0$.

A Horizontal Asymptote at $y = b$ is where the function's graph approaches as x goes to positive or negative infinity.

Key Point

In terms of behavior, the graph will either approach or recede from these asymptotes as we increase or decrease the value of x. However, the function will never actually reach the value of the asymptote.

Example Find the vertical and horizontal asymptotes of $f(x) = \frac{x+1}{x-2}$.

Solution: The vertical asymptote occurs where the denominator equals zero but the numerator does not, which is at $x = 2$. To find the horizontal asymptote, observe the function as x becomes very large or very small. The $+1$ in the numerator and the -2 in the denominator have negligible effects, leading to a horizontal asymptote of $y = 1$.

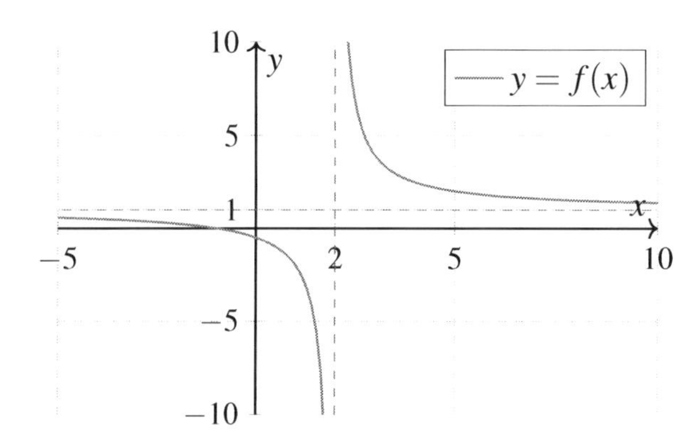

3.15 Graphing Rational Expressions

Rational expressions are fractions with polynomials in both the numerator and denominator, denoted as $\frac{P(x)}{Q(x)}$ where $P(x)$ and $Q(x)$ are polynomials, ensuring $Q(x) \neq 0$. To graph these expressions, identifying asymptotes (vertical, horizontal, or slant) and intercepts (x and y) is crucial.

1. **Vertical Asymptotes:** These occur at zeros of the denominator that do not cancel with the numerator. Simplifying the expression helps locate them faster.
2. **Horizontal or Slant Asymptotes:** Horizontal asymptotes are determined by the degrees of $P(x)$ and $Q(x)$. If the degree of $P(x)$ is less than or equal to that of $Q(x)$, and the degree of $Q(x)$ is higher, the asymptote is $y = 0$. If both degrees are equal, it is the ratio of their leading coefficients. A slant asymptote occurs when the degree of $P(x)$ is greater than that of $Q(x)$, found via long division.
3. **Intercepts:** Set $x = 0$ for the y-intercept and solve $\frac{P(x)}{Q(x)} = 0$ for the x-intercept.

Through these steps, one can accurately graph rational functions.

Key Point

Graphing rational expressions involves identifying vertical asymptotes (zeros of the denominator), horizontal or slant asymptotes (based on polynomial degrees), and intercepts (by setting $x = 0$ for y-intercept and solving $\frac{P(x)}{Q(x)} = 0$ for x-intercept), facilitating the sketching of the function.

Example Graph the function $y = \frac{1}{x-2}$.

Solution: First, let us find the vertical asymptote. The denominator is zero when $x = 2$, but the numerator is not. So, we have a vertical asymptote at $x = 2$. The degree of the numerator is 0, and the degree of the denominator is 1. So, the horizontal asymptote is at $y = 0$. The x-intercept is found by setting $y = 0$ (which yields no solution), and the y-intercept is found by setting $x = 0$ to get $y = -\frac{1}{2}$. Hence, the graph of the function $y = \frac{1}{x-2}$ will have a vertical asymptote at $x = 2$, a horizontal asymptote at $y = 0$, and y-intercept as $(0, -\frac{1}{2})$.

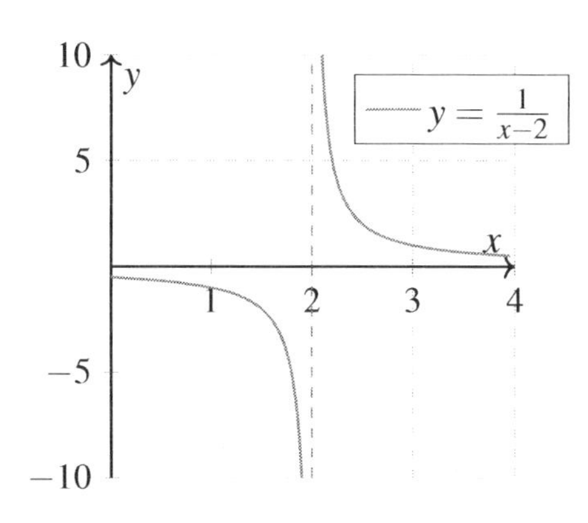

Example Graph the rational function $f(x) = \frac{x^2-x+2}{x-1}$.

Solution: The first step is to find asymptotes. The vertical asymptote is $x = 1$ while the slant asymptote is $y = x$. There is no horizontal asymptote in this graph. Next, we determine the intercepts. For the y-intercept, we have: $y = \frac{x^2-x+2}{x-1} = \frac{0^2-0+2}{0-1} = -2$.

This gives us y-intercept at the point $(0, -2)$. Also, the graph has no x-intercept. Lastly, we test some x values and determine the corresponding y values to plot some additional points.

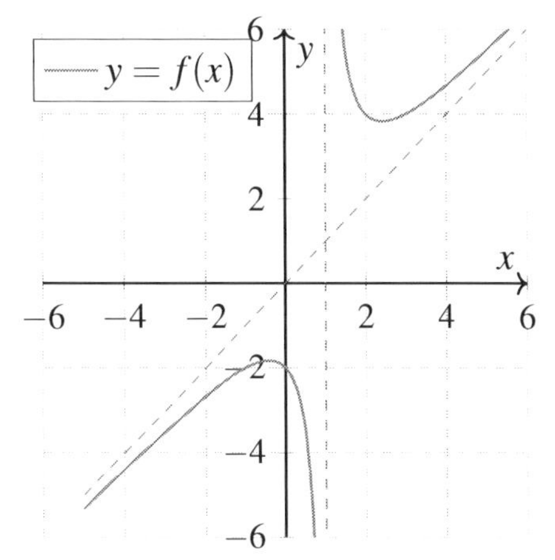

3.16 Practices

1) Which of the following equations represents a constant function?

☐ A. $f(x) = 3x+1$ ☐ C. $f(x) = 0$
☐ B. $f(x) = x^2$ ☐ D. $f(x) = \frac{1}{x}$

2) If $g(x)$ is the identity function, which of the following is true for any real number a?

☐ A. $g(a) = 0$ ☐ C. $g(a) = 1$
☐ B. $g(a) = a^2$ ☐ D. $g(a) = a$

3) What is the degree of the polynomial function $P(x) = 7x^4 - 3x^2 + 9x - 1$?

☐ A. 1 ☐ C. 3
☐ B. 2 ☐ D. 4

4) If a polynomial function $P(x)$ has a degree of 5 and positive leading coefficient, which of the following could be the end behavior of the polynomial's graph?

☐ A. As $x \to \infty$, $P(x) \to \infty$; as $x \to -\infty$, $P(x) \to \infty$
☐ B. As $x \to \infty$, $P(x) \to \infty$; as $x \to -\infty$, $P(x) \to -\infty$
☐ C. As $x \to \infty$, $P(x) \to -\infty$; as $x \to -\infty$, $P(x) \to -\infty$
☐ D. Cannot be determined.

5) What is the end behavior of the polynomial function $P(x) = -2x^4 + 3x^3 - x + 1$?

☐ A. As $x \to \infty$, $P(x) \to \infty$; as $x \to -\infty$, $P(x) \to \infty$.
☐ B. As $x \to \infty$, $P(x) \to -\infty$; as $x \to -\infty$, $P(x) \to \infty$.
☐ C. As $x \to \infty$, $P(x) \to \infty$; as $x \to -\infty$, $P(x) \to -\infty$.
☐ D. As $x \to \infty$, $P(x) \to -\infty$; as $x \to -\infty$, $P(x) \to -\infty$.

6) How many turning points can the polynomial function $Q(x) = x^5 - x^4 + x^3 - x^2 + x - 1$ have at most?

☐ A. 3 ☐ C. 5
☐ B. 4 ☐ D. 6

7) Write the polynomial $2 - 6x^4 + x - 3x^2$ in standard form.

☐ A. $-3x^2 + x - 6x^4 + 2$ ☐ C. $2 - 6x^4 + x - 3x^2$
☐ B. $-6x^4 - 3x^2 + x + 2$ ☐ D. $-6x^4 + x - 3x^2 + 2$

8) Which of the following represents the polynomial $3x^2 - 7 + 4x^3 - 9x$ in standard form?

☐ A. $3x^2+4x^3-9x-7$
☐ B. $-9x+3x^2-7+4x^3$
☐ C. $4x^3+3x^2-9x-7$
☐ D. $4x^3-7-9x+3x^2$

9) Simplify the expression $3x^2+2x-5x^2+7x$.

☐ A. $-2x^2+5x$
☐ B. $-2x^2+9x$
☐ C. x^2+9x
☐ D. x^2+5x

10) Which of the following is the result of multiplying $(2x-4)(x+3)$ using the FOIL method?

☐ A. $2x^2+2x-12$
☐ B. $2x^2+6x-4$
☐ C. $2x^2-2x-12$
☐ D. $2x^2+6x+12$

11) Factor completely: x^2+6x+9.

☐ A. $(x+3)(x+2)$
☐ B. $(x+3)(x+3)$
☐ C. $(x-3)(x+3)$
☐ D. $(x-3)(x-3)$

12) Which of the following represents the factors of the trinomial $x^2-10x+25$?

☐ A. $(x-5)(x+5)$
☐ B. $(x-10)(x-2.5)$
☐ C. $(x-5)(x-5)$
☐ D. $(x+5)(x+5)$

13) Which of the following is the solution to the equation $x^2-5x+6=0$?

☐ A. $x=2$ and $x=-3$
☐ B. $x=2$ and $x=3$
☐ C. $x=-2$ and $x=3$
☐ D. $x=-2$ and $x=-3$

14) What are the solutions to the quadratic equation $3x^2+x-4=0$ using the quadratic formula?

☐ A. $x=-4$ and $x=\frac{1}{3}$
☐ B. $x=-1$ and $x=-\frac{4}{3}$
☐ C. $x=1$ and $x=-\frac{4}{3}$
☐ D. $x=-1$ and $x=\frac{4}{3}$

15) Which of the following is the vertex of the quadratic function $y=3(x-1)^2+2$?

☐ A. $(1,2)$ ☐ C. $(-1,2)$
☐ B. $(3,2)$ ☐ D. $(1,-3)$

16) What is the axis of symmetry for the function $y = -\frac{1}{4}(x+4)^2 + 5$?

☐ A. $x = -4$ ☐ C. $x = -5$
☐ B. $x = 4$ ☐ D. $x = 1$

17) Which of the following is the solution set for the quadratic inequality $x^2 - 4x - 5 < 0$?

☐ A. $(-\infty, -1) \cup (5, \infty)$ ☐ C. $(-1, 5)$
☐ B. $(-\infty, -1] \cup [5, \infty)$ ☐ D. $(-\infty, -1] \cup (5, \infty)$

18) Find the solution set for the inequality $2x^2 - 3x > 0$.

☐ A. $(0, \frac{3}{2})$ ☐ C. $(-\infty, 0) \cup (\frac{3}{2}, \infty)$
☐ B. $(0, \infty)$ ☐ D. $(-\infty, \frac{3}{2})$

19) Which region represents the solution set for the inequality $y \leq -x^2 + 2x + 3$?

☐ A. Inside the parabola. ☐ C. Above the parabola.
☐ B. Outside the parabola. ☐ D. Below the parabola.

20) Which of the following points is in the solution set of the inequality $y > x^2 - 4x + 3$?

☐ A. $(0,0)$ ☐ C. $(1,1)$
☐ B. $(1,0)$ ☐ D. $(0,-1)$

21) Solve the rational equation: $\frac{2}{x-5} + \frac{3}{x+2} = \frac{x+1}{x^2-3x-10}$.

☐ A. $x = 1$ ☐ C. $x = -2$
☐ B. $x = 3$ ☐ D. $x = 5$

22) If the equation $\frac{1}{x-1} - \frac{2}{x+3} = \frac{3}{x^2+2x-3}$ has a solution $x = a$, what is the value of a?

☐ A. $x = -2$ ☐ C. $x = 1$

☐ B. $x = -1$ ☐ D. $x = 2$

23) Simplify the multiplication of the following rational expressions: $\frac{2x+4}{3x^2-12} \times \frac{x-2}{4}$.

☐ A. $\frac{2}{3x+12}$ ☐ C. $\frac{x-2}{6x}$

☐ B. $\frac{1}{6x}$ ☐ D. $\frac{1}{6}$

24) Find the simplified form of the divided rational expressions: $\frac{x^2-9}{x^2-4x+4} \div \frac{x+3}{x-2}$.

☐ A. $\frac{x+1}{x-2}$ ☐ C. $\frac{x-1}{x-2}$

☐ B. $\frac{x-3}{x-2}$ ☐ D. $\frac{x+3}{(x-2)^2}$

25) Simplify the rational expression $\frac{35m^3n^2}{14mn^4}$.

☐ A. $5m^2$ ☐ C. $\frac{5m^2}{2n^2}$

☐ B. $5m^2n^2$ ☐ D. $\frac{5mn}{n^3}$

26) Which of the following is a simplified form of the expression $\frac{x^3-1}{x-1}$?

☐ A. $x^2 + x$ ☐ C. $x^2 - x + 1$

☐ B. $x^2 + x + 1$ ☐ D. $x + 1$

27) Identify the vertical asymptote(s) of the rational function $R(x) = \frac{3x-1}{x^2-4}$.

☐ A. $x = 2$ and $x = -2$ ☐ C. $x = 4$

☐ B. $x = \frac{1}{3}$ ☐ D. No vertical asymptotes

28) What is the horizontal asymptote of the rational function $f(x) = \frac{2x^2+3x+4}{5x^2+x-1}$?

☐ A. $y = 0$ ☐ C. $y = \frac{5}{2}$

☐ B. $y = \frac{2}{5}$ ☐ D. $y = 1$

29) What is the horizontal asymptote of the function $y = \frac{x+1}{x^2+x+1}$?

☐ A. $y=1$ ☐ C. $y=x$

☐ B. $y=0$ ☐ D. $y=\frac{1}{2}$

30) For the rational function $y=\frac{x^3-4x}{x-2}$, what type of asymptote does the graph have?

☐ A. Vertical only ☐ C. Slant

☐ B. Horizontal only ☐ D. No asymptote

Answer Keys

1) C. $f(x) = 0$

2) D. $g(a) = a$

3) D. 4

4) B. As $x \to \infty$, $P(x) \to \infty$; as $x \to -\infty$, $P(x) \to -\infty$

5) D. As $x \to \infty$, $P(x) \to -\infty$; as $x \to -\infty$, $P(x) \to -\infty$.

6) B. 4

7) B. $-6x^4 - 3x^2 + x + 2$

8) C. $4x^3 + 3x^2 - 9x - 7$

9) B. $-2x^2 + 9x$

10) A. $2x^2 + 2x - 12$

11) B. $(x+3)(x+3)$

12) C. $(x-5)(x-5)$

13) B. $x = 2$ and $x = 3$

14) C. $x = 1$ and $x = -\frac{4}{3}$

15) A. $(1,2)$

16) A. $x = -4$

17) C. $(-1,5)$

18) C. $(-\infty,0) \cup (\frac{3}{2},\infty)$

19) A. Inside the parabola.

20) C. $(1,1)$

21) B. $x = 3$

22) D. $x = 2$

23) D. $\frac{1}{6}$

24) B. $\frac{x-3}{x-2}$

25) C. $\frac{5m^2}{2n^2}$

26) B. $x^2 + x + 1$

27) A. $x = 2$ and $x = -2$

28) B. $y = \frac{2}{5}$

29) B. $y = 0$

30) D. No asymptote

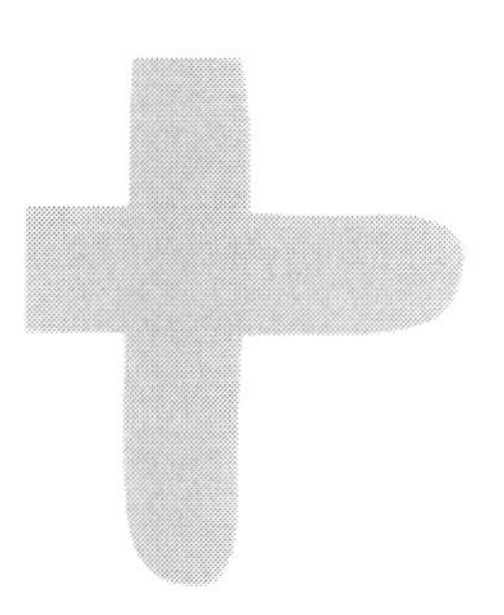

4. Exponential Functions and Logarithms

4.1 Exponential Functions: Definition and Properties

An exponential function is a specific type of function defined by the general form $f(x) = a^x$, where a is a positive real number and $a \neq 1$. This real number a is also known as the base of the exponential function.

Key Point

Exponential functions, denoted by $f(x) = a^x$ with $a > 0$ and $a \neq 1$, yield diverse real outputs for any real input, ensuring the function varies and avoids non-real values.

Exponential functions can represent two types of exponential behavior: growth and decay. If $a > 1$, then the function $f(x) = a^x$ is an exponential growth function. On the other hand, $f(x) = a^x$ represents exponential decay if the base a is between 0 and 1 ($0 < a < 1$).

Refer to Figure: for $a > 1$, the exponential function rises as x increases (growth) and for $0 < a < 1$, the exponential function falls as x increases (decay).

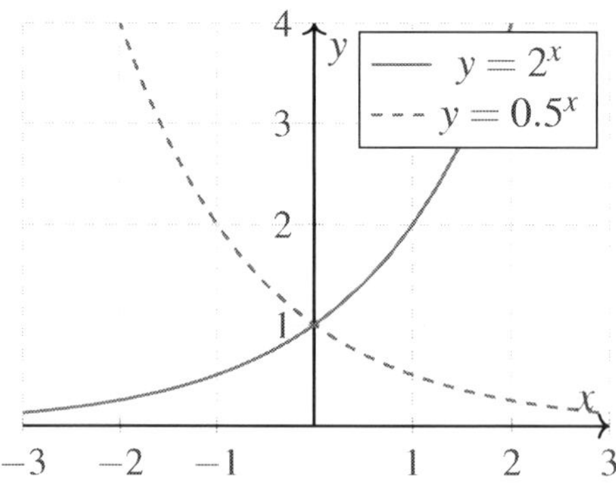

All exponential functions of the form $f(x) = a^x$ share the same domain, range, and y-intercept.

Key Point

For any exponential function $f(x) = a^x$, the domain is all real numbers $\mathbb{R}$, the range is all positive real numbers $(0, +\infty)$, and the y-intercept is 1.

Calculate $f(3)$ and $f(-2)$ for the function $f(x) = 2^x$.

Solution: Substituting into $f(x)$ gives $f(3) = 2^3 = 8$ and $f(-2) = 2^{-2} = 0.25$.

What is the range and y-intercept for the function $f(x) = 10^x$?

Solution: For the function $f(x) = 10^x$, the range is all positive numbers $(0, +\infty)$, and the y-intercept, which occurs at $x = 0$, is at 1 because $f(0) = 10^0 = 1$.

4.2 Logarithmic Functions: Definition and Properties

We first discuss the definition of a logarithmic function. Mathematically, a logarithmic function is defined as $f(x) = \log_b x$, where b is the base of the logarithm and $b > 0$, and $b \neq 1$. This means that a logarithm essentially gives us the exponent, to which we need to raise the base to get a number x. For example, if $b = 2$ and $x = 8$, we have $f(x) = \log_2 8 = 3$, meaning that 2 raised to the power of 3 equals 8.

Key Point

The logarithmic function $y = \log_b(x)$ is the inverse of the exponential function $b^y = x$, relating by $y = \log_b(x) \Leftrightarrow b^y = x$.

Next, we will explore the graph of a logarithmic function. The graph of $y = \log_b x$ is a curve that:

- Passes through the point $(1,0)$ since any base raised to the power of 0 equals 1.

- Has a vertical asymptote at $x = 0$.

- Is increasing for all $x > 0$, if $b > 1$ and decreasing if $0 < b < 1$.

As seen from the Figure, the graph of a logarithm function always passes through the point $(1,0)$, increases for all $x > 0$ if $b > 1$, decreases for all $x > 0$ if $0 < b < 1$, and has a vertical asymptote at $x = 0$.

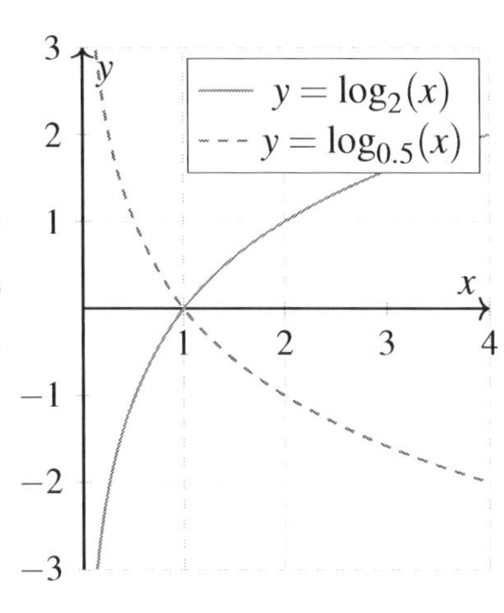

Key Point

The domain of a logarithmic function is for $x > 0$, log is undefined for $x \leq 0$. The output, or the range of a logarithmic function is all real numbers, i.e., $(-\infty, \infty)$.

Example Sketch the graph for the function $f(x) = \log_2(x+1)$.

Solution: A lateral shift of one unit to the left of $f(x) = \log_2 x$ results in $f(x) = \log_2(x+1)$.

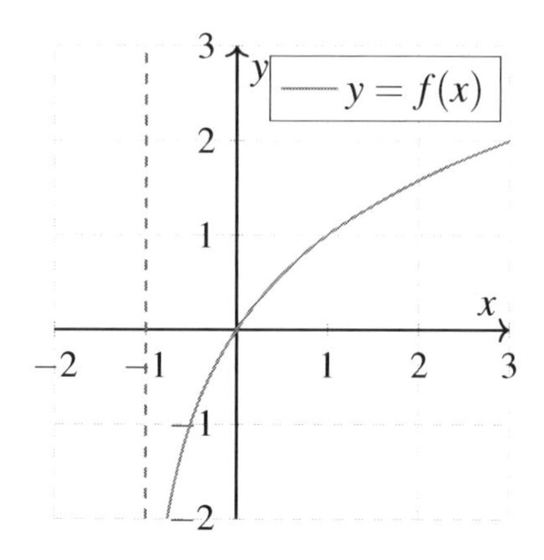

4.3 Evaluating Logarithms

Evaluating logarithms effectively relies on understanding and applying their fundamental rules, which simplifies the process.

Key Point

Assuming all numbers are positive, and a and b do not equal zero or one, the rules are expressed as follows:

- Rule 1: $\log_a(M \cdot N) = \log_a M + \log_a N$.
- Rule 2: $\log_a \frac{M}{N} = \log_a M - \log_a N$.
- Rule 3: $\log_a M^k = k \log_a M$.
- Rule 4: $\log_a a = 1$.
- Rule 5: $\log_a 1 = 0$.
- Rule 6: $a^{\log_a k} = k$.
- Rule 7: $\log_{a^k} M = \frac{1}{k} \log_a M$, for $k \neq 0$.

Evaluate $\log_2 32$.

Solution: First, we write 32 as a power of 2: $32 = 2^5$. Hence: $\log_2 32 = \log_2(2)^5$. Applying rule 3, we get: $\log_2(2)^5 = 5\log_2 2$. Using rule 4: $\log_2 2$ is 1. Therefore, $5\log_2 2 = 5$ and we get: $\log_2 32 = 5$.

Evaluate $3\log_5 125$.

Solution: First, we write 125 as a power of 5: $125 = 5^3$. Hence: $\log_5 125 = \log_5(5)^3$. Applying rule 3: $\log_5(5)^3 = 3\log_5 5$. Using rule 4: $\log_5 5$ is 1. Therefore, $3\log_5 5 = 3$ and we get: $\log_5 125 = 3$.

4.4 Properties of Logarithms

Following our exploration of evaluating logarithms, we now delve into understanding the properties of logarithms. This knowledge will enable us to expand or condense logarithmic expressions, enhancing our proficiency in simplifying and solving logarithmic equations.

Key Point

Here are some important properties of logarithms:

1. $a^{\log_a b} = b$
2. $\log_a 1 = 0$
3. $\log_a a = 1$
4. $\log_a \frac{1}{x} = -\log_a x$
5. $\log_a x = \frac{1}{\log_x a}$
6. $\log_a x^p = p\log_a x$ (power rule)
7. $\log_a (x \cdot y) = \log_a x + \log_a y$ (product rule)
8. $\log_a \frac{x}{y} = \log_a x - \log_a y$ (quotient rule)
9. $\log_{a^k} x = \frac{1}{k}\log_a x$, where $k \neq 0$
10. $\log_a x = \log_{a^c} x^c$

The properties of logarithms, namely the product rule, quotient rule, and power rule, are crucial mathematical tools used to simplify or condense logarithmic expressions. Armed with an understanding of these properties, we can manipulate and simplify logarithmic expressions.

Expand the logarithmic expression $\log_a(15)$.

Solution: In this problem, we can apply the product rule, employing this rule leads to:

$$\log_a(15) = \log_a(3 \times 5) = \log_a 3 + \log_a 5.$$

Calculate the following expression using the properties of logarithms:

$$\log_2 24 - \log_2 6.$$

Solution: Using the quotient rule, we have:

$$\log_2 24 - \log_2 6 = \log_2 \frac{24}{6} = \log_2 4.$$

Now, using the power rule and the rule $\log_a a = 1$, we get:

$$\log_2 4 = \log_2 2^2 = 2\log_2 2 = 2.$$

Evaluate the logarithm expression, $\log_{7^2}(\frac{1}{7}) =?$

Solution: Use log rules, $\log_a \frac{1}{x} = -\log_a x$, and $\log_{a^k} x = \frac{1}{k}\log_a x$. Then:

$$\log_{7^2}(\tfrac{1}{7}) = -\tfrac{1}{2}\log_7 7 = -\tfrac{1}{2}.$$

4.5 Natural Logarithms

The natural logarithm, represented as $\ln x$ or $\log_e x$, is a logarithm with a distinctive base of the mathematical constant e. The value of e is an irrational number approximately equal to 2.71.

The following figure shows the graph of the natural logarithm function $y = \ln x$ or $y = \log_e x$.

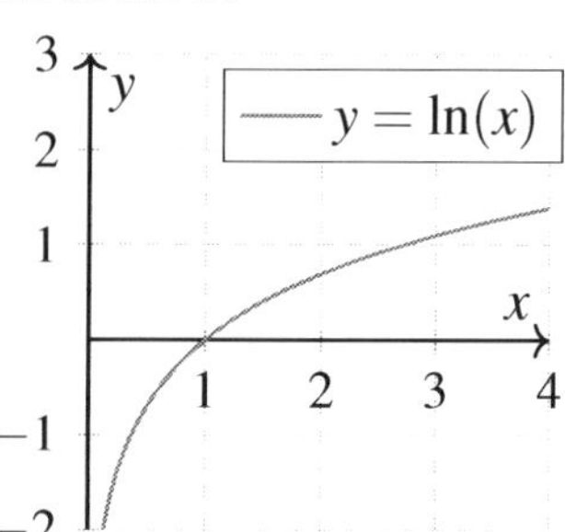

The natural logarithm, denoted $\ln x$ or $\log_e x$, has a base of e (approximately 2.71).

After mastering the properties of logarithms in the previous section, we can readily apply those properties here, explicitly focusing on the base-e scenario.

Expand the natural logarithm for the expression $\ln 4x^2$.

Solution: We use the logarithm property $\log_a(x \cdot y) = \log_a x + \log_a y$. Applying this to our equation, we obtain:

$$\ln 4x^2 = \ln 4 + \ln x^2.$$

We also remember from our logarithm properties that $\log_a M^k = k\log_a M$. Applying this rule, we obtain:

$$\ln 4 + \ln x^2 = \ln 4 + 2\ln x.$$

So, we have successfully expanded the natural logarithm $\ln 4x^2$.

Condense the following expression to a single logarithm: $\ln x - \log_e(2y)$.

Solution: We can use the log property:

$$\log_a x - \log_a y = \log_a(\frac{x}{y}).$$

Hence, we get:

$$\ln x - \log_e(2y) = \ln\left(\frac{x}{2y}\right).$$

4.6 Solving Exponential Equations

Exponential equations, which involve variables in the exponent, can be solved using methods such as equalizing bases and logarithms. It is important to remember that the logarithmic function $y = \log_b(x)$ is equivalent to the exponential equation $x = b^y$.

Key Point

When an exponential equation has the *same base* on both sides, the exponents must also be equal. Mathematically, for an equation $b^m = b^n$, it is obvious that $m = n$.

Logarithms can be used to simplify and solve exponential equations. By taking the log on both sides, the exponent (where the variable often resides) can be brought down, and the equation can then be solved algebraically.

Follow these steps to solve an exponential equation using logarithms:

1. Take the natural logarithm (ln) or logarithm (base 10, log) of both sides of the equation.
2. Utilize logarithmic properties to bring the exponent down.
3. Rearrange the equation to solve for the variable.

Solve the equation $2^x = 8$.

Solution: To solve this equation, we can use the method of equalizing bases. We know that 8 can be expressed as 2^3. So, the equation becomes $2^x = 2^3$. Since the bases are equal, we can equate the exponents.

Therefore, $x = 3$.

Solve the equation $5^x = 20$.

Solution: The method of equalizing bases does not work here, since 20 cannot be expressed as 5^n where n is a natural number. Thus, we should apply logarithms to the equation. Starting with: $5^x = 20$. We take the natural logarithm of both sides to bring down the exponent as a multiplier and use the properties of logarithms: $x \cdot \ln(5) = \ln(20)$. Finally, to solve for x, divide by $\ln(5)$: $x = \frac{\ln(20)}{\ln(5)}$.

Suppose we have the equation $3^x = 81$, and find the value of x using the definition of a logarithm.

Solution: To solve for x, we need to write the equation in a logarithmic form. Hence, the equation can be expressed as: $x = \log_3 81$. We know that $3^4 = 81$, hence, $x = 4$ is the solution to the equation.

4.7 Solving Logarithmic Equations

The process of solving logarithmic equations involves simplifying them using logarithm properties and, if possible, converting them to exponential form.

To solve a logarithmic equation, there are three steps:

1. Conversion to an exponential equation: The first step in solving a logarithmic equation usually involves converting it into an equivalent exponential equation, if applicable. Remember: if no base is indicated, the base of the logarithm is 10.

2. Condensing logarithms: If there are multiple logarithms on one side of the equation, condense them into a single logarithm using the laws of logarithms.

3. Solution check: After obtaining possible solutions for x, always plug them back into the original equation and check if they do not lead to a logarithm of a negative number.

It is crucial to always check the solutions back into the original logarithmic equation since logarithms only accept positive arguments.

Find the value of x in this equation $\log_2(36 - x^2) = 4$.

Solution: Given that the equation is already a simple logarithmic equation, we can convert it to an exponential equation. According to the logarithm rule, when $\log_b x = \log_b y$, then $x = y$. Thus, we can rewrite 4 as a logarithm as follows: $4 = \log_2\left(2^4\right)$. So, the equation changes to:

$$\log_2\left(36 - x^2\right) = \log_2\left(2^4\right) = \log_2 16.$$

Then:

$$36 - x^2 = 16 \Rightarrow 36 - 16 = x^2 \Rightarrow x^2 = 20 \Rightarrow x = \pm\sqrt{20}.$$

Now, check the solutions in the original equation.
For $x = \sqrt{20}$: $\log_2\left(36 - \left(\sqrt{20}\right)^2\right) = 4 \Rightarrow \log_2(36 - 20) = 4 \Rightarrow \log_2 16 = 4.$
For $x = -\sqrt{20}$: $\log_2\left(36 - \left(-\sqrt{20}\right)^2\right) = 4 \Rightarrow \log_2(36 - 20) = 4 \Rightarrow \log_2 16 = 4.$
Both solutions work in the original equation satisfying the log properties.

4.8 Practices

1) If $f(x) = 3^x$, what is $f(4)$?

☐ A. 12
☐ B. 81
☐ C. 64
☐ D. 27

2) Consider the function $f(x) = \left(\frac{1}{2}\right)^x$. Which of the following is true for $f(-1)$?

☐ A. 0.5
☐ B. −0.5
☐ C. 2
☐ D. −2

3) If $\log_3(x) = 5$, what is the value of x?

☐ A. 125
☐ B. 64
☐ C. 243
☐ D. 81

4) For which value of x is $\log_{\frac{1}{2}}(x) = -3$ true?

☐ A. 8
☐ B. $\frac{1}{8}$
☐ C. −8
☐ D. $\frac{1}{-8}$

5) If $\log_4 64 = x$, what is the value of x?

☐ A. 2
☐ B. 3
☐ C. 4
☐ D. 9

6) Simplify $2\log_{10} 100 - \log_{10} 10$.

☐ A. 1
☐ B. 2
☐ C. 3
☐ D. 4

7) Simplify the logarithmic expression $\log_5(125)$.

☐ A. 3
☐ B. $3\log_5 2$
☐ C. $2\log_5 4$
☐ D. $2\log_5 27$

8) If $\log_b(5) = x$ and $\log_b(2) = y$, express $\log_b(10)$ in terms of x and y.

☐ A. $x - y$
☐ B. $2x$
☐ C. $x + y$
☐ D. $\frac{x}{y}$

9) Solve for x: $\ln(x) + \ln(2) = 3$.

☐ A. $e^3 - 2$
☐ B. $\frac{e^3}{2}$
☐ C. e^3
☐ D. $2e^3$

10) If $f(x) = \ln(x^2)$, find $f(e)$.

☐ A. 1
☐ B. 2
☐ C. e
☐ D. e^2

11) If $4^{x+1} = 64$, then what is the value of x?

☐ A. 1
☐ B. 2
☐ C. 3
☐ D. 4

12) Solve the exponential equation $3^{2x+1} = 27$.

☐ A. $x = \frac{1}{2}$
☐ B. $x = 1$
☐ C. $x = \frac{3}{2}$
☐ D. $x = 2$

13) Solve the logarithmic equation $\log(x) + \log(x+3) = 1$.

☐ A. 5
☐ B. 2
☐ C. 10
☐ D. 12

14) What is the solution of the equation $\log_3(x+2) = 4$?

☐ A. 79
☐ B. 81
☐ C. 83
☐ D. 85

Answer Keys

1) B. 81
2) C. 2
3) C. 243
4) A. 8
5) B. 3
6) C. 3
7) A. 3
8) C. $x+y$
9) B. $\frac{e^3}{2}$
10) B. 2
11) B. 2
12) B. $x=1$
13) B. 2
14) A. 79

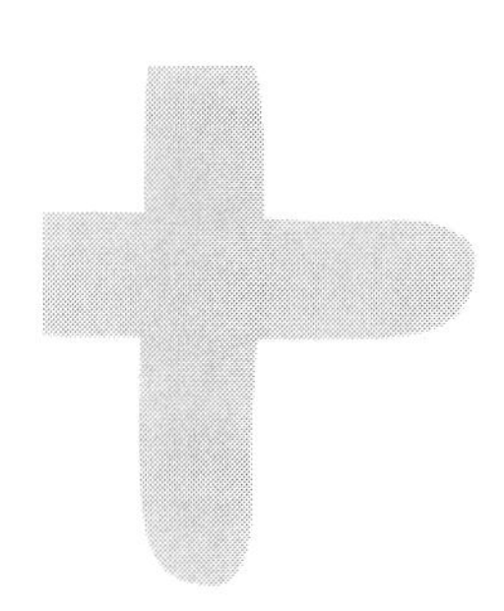

5. Radical and Absolute Value Functions

5.1 Simplifying Radical Expressions

Simplifying radicals makes the manipulation of equations or functions involving radical expressions easier by identifying prime factors within the radical and applying radical properties for expression simplification. Initially, determine the prime factors of the numbers or expressions inside the radical, utilizing the concept of factorization to simplify the contents of the radical.

Key Point

To simplify radical expressions further, apply these key radical properties:

1. $\sqrt[n]{x^a} = x^{\frac{a}{n}}$
2. $\sqrt[n]{xy} = \sqrt[n]{x} \times \sqrt[n]{y}$
3. $\sqrt[n]{\frac{x}{y}} = \frac{\sqrt[n]{x}}{\sqrt[n]{y}}$
4. $\sqrt[n]{x} \times \sqrt[n]{y} = \sqrt[n]{xy}$

Using these properties, we can systematically simplify expressions inside radicals. A simplified radical has no factor that can be extracted from the radical, and its denominator is not a radical.

Write the radical $\sqrt[3]{x^4}$ in exponential form.

Solution: To write a radical in exponential form, we can use this rule $\sqrt[n]{x^a} = x^{\frac{a}{n}}$. Thus, $\sqrt[3]{x^4}$ can be written as $x^{\frac{4}{3}}$.

Simplify the square root of $\sqrt{144x^2}$.

Solution: Firstly, find the factor of the expression $144x^2$: $144 = 12 \times 12$ and $x^2 = x \times x$. We can use the radical rule $\sqrt[n]{a^n} = a$. Then we have: $\sqrt{12^2} = 12$ and $\sqrt{x^2} = |x|$. Finally, we obtain:

$$\sqrt{144x^2} = \sqrt{12^2} \times \sqrt{x^2} = 12 \times |x| = 12|x|.$$

 Simplify $\sqrt{12a^5b^4}$.

Solution: Break down $12a^5b^4$ into its prime factors: $12a^5b^4 = 2^2 \times 3 \times a^5 \times b^4$. Identify perfect squares to obtain: $12a^5b^4 = 2^2 \times 3 \times a^4 \times a \times b^4$. Then rewrite the radical expression:

$$\sqrt{12a^5b^4} = \sqrt{2^2 \times a^4 \times b^4} \times \sqrt{3a}.$$

Using the rule $\sqrt[n]{a^n} = a$, we have:

$$\sqrt{2^2 \times a^4 \times b^4} \times \sqrt{3a} = 2 \times a^2 \times b^2 \times \sqrt{3a} = 2a^2b^2\sqrt{3a}.$$

Therefore, the simplified form of $\sqrt{12a^5b^4}$ is $2a^2b^2\sqrt{3a}$.

5.2 Simplifying Radical Expressions Involving Fractions

Simplifying expressions with radicals in the fraction's denominator involves additional steps beyond the basic rules for radical simplification. Our goal is often to eliminate these radicals to simplify calculation. This is achieved by multiplying the numerator and the denominator by the radical, effectively multiplying the expression by 1, to remove the radical from the denominator. When the denominator includes both a radical and an integer, multiply the numerator and the denominator by the denominator's conjugate ($a+b$ becomes $a-b$ and vice versa) to simplify further.

Key Point

Radical expressions should not be in the denominator of a fraction. To eliminate the radical in the denominator, multiply the numerator and denominator by the radical in the denominator or conjugate of the denominator.

Example Simplify the expression $\frac{2}{\sqrt{3}-2}$.

Solution: We have a fraction with a radical in the denominator. To eliminate this, we multiply the numerator and denominator by the conjugate of the denominator, which in this case is $\sqrt{3}+2$. Therefore, our expression becomes: $\frac{2}{\sqrt{3}-2} \times \frac{\sqrt{3}+2}{\sqrt{3}+2}$. Solving $\sqrt{3}-2$ times $\sqrt{3}+2$, we get -1, because: $(\sqrt{3})^2 - (2)^2 = 3 - 4 = -1$. Our expression becomes $\frac{2(\sqrt{3}+2)}{-1}$, which simplifies to $-2\left(\sqrt{3}+2\right)$.

Example Simplify the expression $\frac{2}{\sqrt{5}}$.

Solution: We can multiply both the numerator and the denominator by the denominator, which in this case is $\sqrt{5}$: $\frac{2}{\sqrt{5}} \times \frac{\sqrt{5}}{\sqrt{5}}$. Simplifying the denominator, we get: $\frac{2\sqrt{5}}{5}$. So, $\frac{2}{\sqrt{5}}$ simplifies to $\frac{2\sqrt{5}}{5}$.

5.3 Adding and Subtracting Radical Expressions

In algebra, terms with identical variables can be combined; similarly, radical expressions can be added or subtracted if they have the same radical part. For example, $5\sqrt{5}$ and $7\sqrt{5}$ share the radical $\sqrt{5}$, allowing them to be combined into a single term. In contrast, terms like $\sqrt{3}$ and $\sqrt{5}$, which have different radicals, cannot be combined due to their distinct radical parts.

Key Point

Only terms with the same radical part can be added or subtracted. This operation involves combining the coefficients of these like radical terms.

When combining like radical terms, the coefficients are the portions combined. If there are no coefficients explicitly stated, one is assumed. Remember that if the radical parts of two terms are different, the terms cannot be combined.

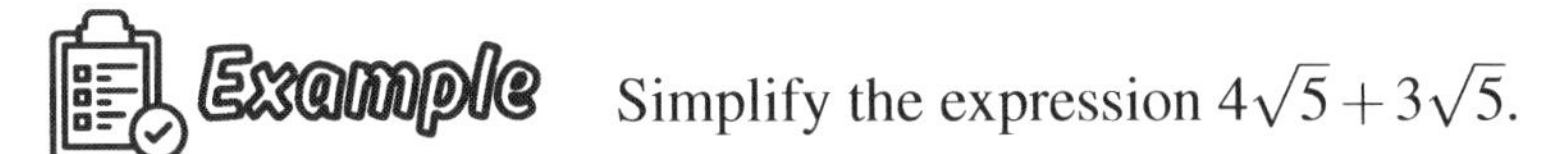

Example Simplify the expression $4\sqrt{5}+3\sqrt{5}$.

Solution: In this case, both terms share the same radical part, which is $\sqrt{5}$. Therefore, we can combine these terms by adding the coefficients. The calculation is as follows:

$$4\sqrt{5}+3\sqrt{5}=7\sqrt{5}.$$

So, $4\sqrt{5}+3\sqrt{5}$ simplifies to $7\sqrt{5}$.

Example Simplify $3\sqrt{2}+2\sqrt{5}+5\sqrt{2}$.

Solution: In this instance, $\sqrt{2}$ is our recurring radical part, so we add their multipliers together:

$$3\sqrt{2}+5\sqrt{2}=8\sqrt{2}.$$

However, the $2\sqrt{5}$ term remains as it is because this term does not have the same radical part as the others. Hence, the result is: $8\sqrt{2}+2\sqrt{5}$.

5.4 Multiplying and Dividing Radical Expressions

Multiplying and dividing radical expressions is an essential skill. Here is a concise guide to effectively handle these operations.

Multiplication of Radicals: Follow three steps:

1. Multiply the coefficients (numbers outside the radicals).
2. Multiply the radicands (numbers inside the radicals).
3. Simplify the resulting expression.

Division of Radicals: To simplify a fraction with radicals:

1. Simplify the radical in the numerator.
2. Simplify the radical in the denominator.
3. Rewrite and reduce the fraction, removing common factors.

Both processes aim at simplifying expressions to their most elementary form, employing prime factorization for fractions and leveraging the radical rule for efficient simplification.

Key Point

Multiply coefficients and inside radicals separately, then simplify. Divide by simplifying numerator and denominator radicals, then reduce.

Key Point

For any real a and positive integer n, the expression $\sqrt[n]{a^n}$ simplifies to: $\sqrt[n]{a^n} = \begin{cases} |a| & \text{if } n \text{ is even,} \\ a & \text{if } n \text{ is odd.} \end{cases}$

Example Multiply the radical expressions $\sqrt{16} \times \sqrt{9}$.

Solution: To solve this, we first factor the numbers: $16 = 4^2$ and $9 = 3^2$. Consequently, we can rewrite the expression:

$$\sqrt{16} \times \sqrt{9} = \sqrt{4^2} \times \sqrt{3^2}.$$

We then use the radical rule, $\sqrt[n]{a^n} = a$, which states that the nth root of a number raised to the power n equals the number itself. Thus, we get:

$$\sqrt{4^2} \times \sqrt{3^2} = 4 \times 3 = 12.$$

So, the multiplication of $\sqrt{16}$ and $\sqrt{9}$ equals to 12.

 Evaluate $\sqrt{25} \times \sqrt{4}$.

Solution: Here, we again factor in the numbers: $25 = 5^2$ and $4 = 2^2$. Now, rewrite the expression:

$$\sqrt{25} \times \sqrt{4} = \sqrt{5^2} \times \sqrt{2^2}.$$

By applying the radical rule ($\sqrt[n]{a^n} = a$), we get:

$$\sqrt{5^2} \times \sqrt{2^2} = 5 \times 2 = 10.$$

So, the multiplication of $\sqrt{25}$ and $\sqrt{4}$ equals to 10.

Example Simplify the expression $\sqrt{\frac{32}{50}}$.

Solution: Start by simplifying both the numerator and the denominator under the square root:

$$\sqrt{\frac{32}{50}} = \sqrt{\frac{2^5}{2 \times 5^2}}.$$

This can then be simplified by taking the square root of the numerator and the denominator separately:

$$\sqrt{\frac{2^5}{2 \times 5^2}} = \frac{\sqrt{2^5}}{\sqrt{2 \times 5^2}}.$$

Further simplification gives:

$$\frac{\sqrt{2^5}}{\sqrt{2 \times 5^2}} = \frac{4\sqrt{2}}{5\sqrt{2}}.$$

So, $\sqrt{\frac{32}{50}} = \frac{4}{5}$.

 Simplify the expression $\sqrt{\frac{45}{125}}$.

Solution: Starting with the given expression:

$$\sqrt{\frac{45}{125}} = \sqrt{\frac{3^2 \times 5}{5^3}}.$$

We simplify by taking the square root of the numerator and the denominator separately:

$$\sqrt{\frac{3^2 \times 5}{5^3}} = \frac{\sqrt{3^2 \times 5}}{\sqrt{5^3}}.$$

Simplifying further yields:

$$\frac{\sqrt{3^2 \times 5}}{\sqrt{5^3}} = \frac{3\sqrt{5}}{5\sqrt{5}}.$$

This can be simplified by removing the common factor of $\sqrt{5}$ in the numerator and the denominator:

$$\frac{3\sqrt{5}}{5\sqrt{5}} = \frac{3}{5}.$$

So, $\sqrt{\frac{45}{125}} = \frac{3}{5}$.

5.5 Radical Equations

Radical equations contain variables under a radical sign. To solve them, follow these steps:

1. Isolate the radical on one side.
2. Square both sides to remove the radical.
3. Solve the resulting equation for the variable.
4. Verify the solution to avoid extraneous results.

Isolating the radical simplifies the equation, making it solvable via algebraic manipulations. For instance, in $a + \sqrt{b} = c$, isolate $\sqrt{b}$ to obtain $\sqrt{b} = c - a$. Squaring both sides yields $b = (c - a)^2$, eliminating the radical.

Key Point

Squaring the equation can introduce extraneous solutions. Always validate your solution by substituting it back into the original equation to confirm its correctness.

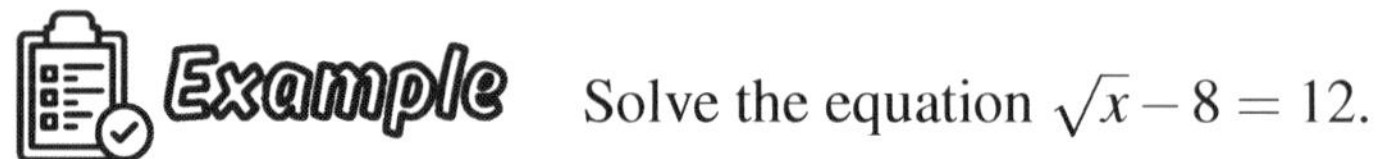

Solve the equation $\sqrt{x} - 8 = 12$.

Solution: Start by isolating the radical on one side. This means moving 8 to the other side by adding 8 to both sides:

$$\sqrt{x} = 20.$$

Once the radical is isolated, you can square both sides in order to remove the radical, resulting in $x = 400$.

You then need to check your answer by substituting it back into the original equation:

$$\sqrt{400}-8=20-8=12.$$

which is the same as the other side of the equation, confirming that $x=400$ is indeed the solution.

 Solve the equation $2\sqrt{x-5}+1=5$.

Solution: Begin by isolating the radical. Subtract 1 from each side to obtain:

$$2\sqrt{x-5}=4.$$

Next, divide by 2 on both sides to find:

$$\sqrt{x-5}=\frac{4}{2}.$$

Square both sides to remove the square root. This yields:

$$x-5=\left(\frac{4}{2}\right)^2,$$

or $x-5=2^2$. Now, solve for x: $x=4+5$ or $x=9$. Finally, fill in $x=9$ into the original equation to double-check our answer. We find that it indeed does make the original equation true.

5.6 Solving Radical Inequalities

Radical inequalities involve expressions under a radical. Unlike solving radical equations, solving inequalities requires careful attention to the inequality's direction. The steps to solve are as follows:

1. **Isolation**: Position the radical expression on one side of the inequality, separate from other terms. This step is vital for a clear path to the solution.
2. **Elimination of the Radical**: Remove the radical by raising both sides of the inequality to a power that matches the radical's index.
3. **Solving**: After eliminating the radical, simplify and solve the resulting inequality for the variable.

Key Point

Graphing the inequality provides a visual method to understand and verify the solution.

Key Point

When dealing with a radical expression with an even index, ensure that the result does not make the expression inside the radical negative.

Solve for x in the given inequality $2\sqrt[3]{1-x}-3 \geq 0$.

Solution: The first step is to isolate the radical expression: $2\sqrt[3]{1-x}-3 \geq 0$ becomes $2\sqrt[3]{1-x} \geq 3$, which simplifies to: $\sqrt[3]{1-x} \geq \frac{3}{2}$. Next, we raise both sides of the inequality to the power equal to the index of the radical: $\left(\sqrt[3]{1-x}\right)^3 \geq \left(\frac{3}{2}\right)^3$, which simplifies to: $1-x \geq \frac{27}{8}$. Finally, we solve for x: $1-x \geq \frac{27}{8}$ simplifies to $1-\frac{27}{8} \geq x$, and on further reducing we obtain: $x \leq -\frac{19}{8}$. Therefore, the solution to the inequality is $x \in (-\infty, -\frac{19}{8}]$.

5.7 Radical Functions

Radical functions are defined as $f(x) = \sqrt[n]{g(x)}$, where $g(x)$ is the radicand, and n is the index of the root. The domain and range of these functions are influenced by the index n and the nature of $g(x)$.
To determine the domain of $f(x)$, the following considerations are made:

1. **Even Index:** The domain is determined by the condition $g(x) \geq 0$, implying that $g(x)$ must be non-negative. Therefore, the domain is $\{x : g(x) \geq 0\}$.
2. **Odd Index:** Since there are no restrictions on $g(x)$, the domain of $f(x)$ is the same as the domain of $g(x)$.

The range of radical functions depends on the index:

- Functions with an even index yield non-negative outputs, influencing their range, except in cases of vertical shifts.
- The range of functions with an odd index is determined by the properties of $g(x)$.

Key Point

Understanding the index and radicand is essential for defining a function's domain and range. For the domain, all valid values within the radicand must be considered, while the radicand's minimum and maximum values are crucial for establishing the function's range.

Determine the domain of the function $f(x) = \sqrt{x^2-4}$.

Solution: For this function, the radicand is x^2-4. Since the index of the root is even, the radicand must be non-negative. Therefore, $x^2-4 \geq 0$ which leads to $x^2 \geq 4$. Which gives $-2 \geq x$ or $x \geq 2$. Hence, the domain of this function is $(-\infty,-2] \cup [2,\infty)$.

Example Find the domain and range of the radical function $y = \sqrt{x-2}$.

Solution: To find the domain, ensure non-negative values under the square root. Solve $x-2 \geq 0$ to get $x \geq 2$. Thus, the domain is $x \geq 2$. Now, we need to find the range. Clearly, the radical function provides non-negative outputs, so considering 2 as minimum of domain, the range of function is $y \geq 0$.

Example Find the domain and range of the radical function $y = 3\sqrt{2x-3}+1$.

Solution: To secure non-negative values under the square root, we solve $2x-3 \geq 0$ to get $x \geq \frac{3}{2}$. The domain is $x \geq \frac{3}{2}$. For the range, we have $y \geq 1$.

5.8 Graphing Radical Functions

A typical radical function can be expressed as $f(x) = \sqrt[n]{g(x)}$. The behavior and properties of these functions are determined by the root's degree and the radicand. To effectively graph radical functions, follow these steps:

1. Determine the function's domain.
2. Locate key points, particularly where $g(x) = 0$ to find x-intercepts.
3. Plot these points and draft the graph, noting that even roots yield curves starting at a point and ascending, while odd roots may cross the x-axis.

Key Point

Graph shapes vary with the radical's degree and undergo transformations like shifts, stretches, or reflections depending on $g(x)$. The function's domain and outcome's sign heavily depend on the radical's degree (even/odd) and its radicand.

Example What does the graph look like for the square root function $f(x) = \sqrt{x}$?

Solution: Following the steps mentioned above:

1. The domain is $x \geq 0$, as we are dealing with an even root.
2. The key point to identify here is $x = 0$.
3. We plot these on a graph.

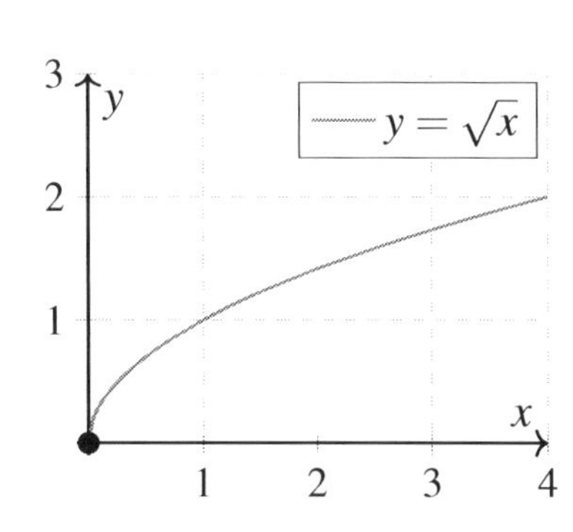

Example Consider $f(x) = \sqrt{x-3}$. What does this function's graph look like in comparison to $f(x) = \sqrt{x}$?

Solution: In this function, we notice that the shift factor in -3, causes a horizontal shift in the function compared to graph $f(x) = \sqrt{x}$. The graph of $f(x) = \sqrt{x-3}$ will be a shift of 3 units to the right of the graph of $f(x) = \sqrt{x}$. The steps involved in graphing remain the same:

1. The domain for this function is $x \geq 3$.
2. Identify $x = 3$ as the key point in this case.
3. Plot the graph as shown here.

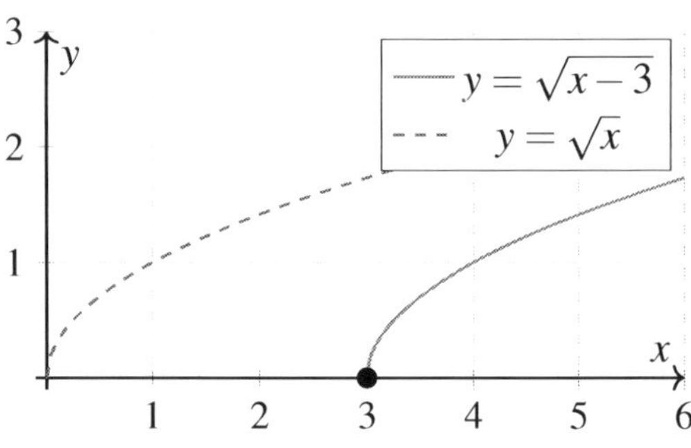

Graph $g(x) = \sqrt[3]{x}$.

Solution: There is no domain restriction.

For key points $g(-1) = -1$, $g(0) = 0$, and $g(1) = 1$, plot and sketch.

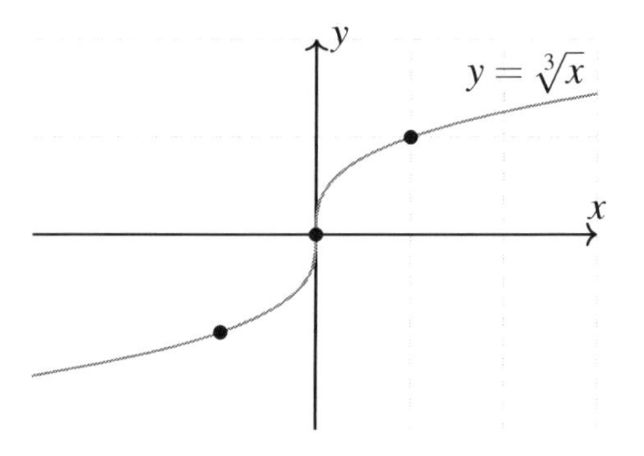

5.9 Absolute Value Functions

Absolute value functions, represented as $f(x) = |g(x)|$, measure the "distance" of $g(x)$ from zero regardless of its sign. These functions maintain the domain of $g(x)$, indicating no restriction on input values and ensuring that outputs are non-negative, with the range starting from 0 to positive infinity, barring any vertical shifts.

Key Point

The domain of $f(x) = |g(x)|$ mirrors that of $g(x)$. The range is non-negative, commencing at 0 and reaching towards infinity, unless altered by vertical shifts.

Graphically, parts of $g(x)$ below the x-axis are mirrored above, creating a distinctive "V" or "W" shape. Transformations such as shifts, stretches, or compressions of $g(x)$ similarly impact $f(x)$. The x-intercepts remain identical to those of $g(x)$, while y-intercepts may vary. Further details on graphing these functions will follow.

Consider we need to identify the range of $f(x) = |x^2 - 4|$.

Solution: Firstly, let us identify the minimum value of the function $g(x) = x^2 - 4$. Given its domain as all real numbers and recognizing that a square is always non-negative ($x^2 \geq 0$), we conclude that the minimum value $g(x)$ is -4. Hence, the range of $g(x)$ is $[-4, \infty)$. However, we are interested in $f(x) = |g(x)|$. The absolute value function will make all values of $g(x)$ non-negative, transforming the range to $[0, \infty)$.

Find the x-intercepts of $f(x) = |x^3 - x^2 - 2x|$.

Solution: The x-intercepts occur where $g(x) = x^3 - x^2 - 2x = 0$. So, from $x^3 - x^2 - 2x = 0$, we have: $x(x^2 - x - 2) = x(x+1)(x-2) = 0$. Thus, $x = 0, -1, 2$, are the x-intercepts of $f(x)$.

5.10 Graphing Absolute Value Functions

Graphing $|g(x)|$ involves reflecting $g(x)$'s negative outputs to positive ones, essentially mirroring parts of $g(x)$ that lie below the x-axis to above it.

The process for graphing the absolute value function $|g(x)|$ can be compactly enumerated as follows:

1. Plot or identify the graph of $g(x)$. This step involves understanding the basic shape and key features of $g(x)$.
2. Locate the points where $g(x) = 0$. These points act as the "hinges" or "corners" of the graph, indicating where the direction of the graph changes due to the absolute value operation.
3. Reflect segments of the $g(x)$ graph that are below the x-axis to above the x-axis.

Key Point

The absolute value function reflects negative $g(x)$ values to positive, with parts of $g(x)$ below the x-axis mirrored above, acting as a visual "mirror".

Graph $f(x) = |x|$.

Solution: The function $g(x) = x$ is a straight line passing through the origin.

Since it is already non-negative for all $x \geq 0$, the graph of $f(x)$ is identical to $g(x)$ for $x \geq 0$. For $x < 0$, the graph is reflected upwards, resulting in a V-shaped graph as depicted in the figure.

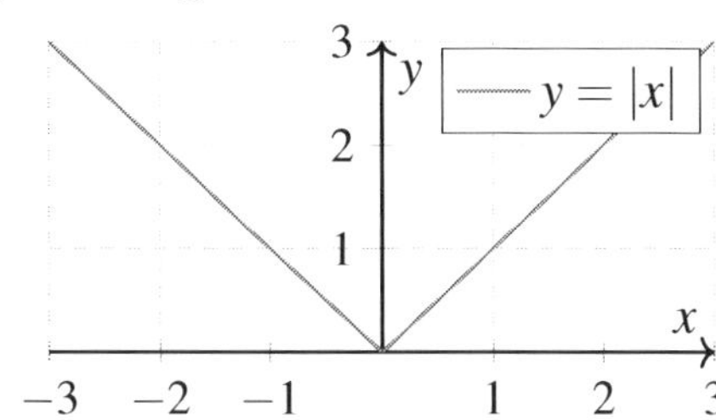

Graph $g(x) = |x^2 - 1|$.

Solution: The parabola $g(x) = |x^2 - 1|$ intersects the x-axis at $x = -1$ and $x = 1$.

Since the parabola is above the x-axis everywhere except between these points, only the segment between -1 and 1 gets reflected upwards.

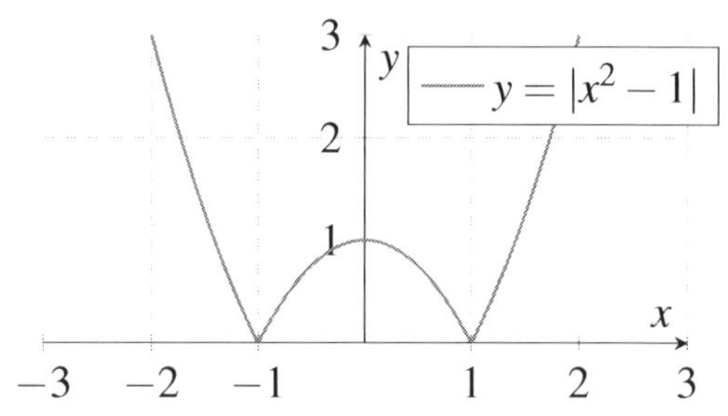

5.11 Absolute Value Properties

Building upon our understanding of graphing absolute value functions, we delve into the fundamental properties of absolute values, crucial for simplifying expressions, solving problems, and graphing functions involving absolute values.

The absolute value $|x|$, for any real number x, represents its distance from zero on the number line, disregarding its sign, effectively its magnitude without direction.

Key Point

Absolute value denotes magnitude, ignoring the sign of a real number.

Key properties of absolute value include:

1. $|x| \geq 0$ for any real number x, indicating that absolute values are always non-negative.
2. $|x| = x$ if $x \geq 0$ and $|x| = -x$ if $x < 0$, showing absolute value equals x itself when non-negative, or the negation of x when negative.
3. $|-x| = |x|$ for all real x, demonstrating the sign-insensitivity of absolute value.
4. $|xy| = |x||y|$ for any real numbers x and y, signifying that the absolute value of a product is the product of the absolute values.
5. $\left|\frac{x}{y}\right| = \frac{|x|}{|y|}$ for all real x and y (with $y \neq 0$), meaning absolute value distributes over division.
6. $|x+y| \leq |x| + |y|$ for all real x and y, highlighting the triangle inequality, which states that the absolute value adheres to the geometric principle regarding the lengths of sides in a triangle.

Key Point

The triangle inequality is a critical absolute value property with wide-ranging applications in mathematics.

 Evaluate $|7|$ and $|-7|$.

Solution: From the property $|x| = x$ if $x \geq 0$ and $|x| = -x$ if $x < 0$, we find that $|7| = 7$, and $|-7| = -(-7) = 7$.

 If $x = -4$ and $y = 6$, find $|x+y|$.

Solution: First, put the given value in the expression $|x+y|$. So, we have:

$$|x+y| = |(-4)+6| = |-4+6| = |2| = 2.$$

 For $x = -3$ and $y = 4$, prove the Triangle Inequality.

Solution: Use the Triangle Inequality, $|x+y| \leq |x|+|y|$. Plug the given value in the inequality. Next: $|x+y| = |(-3)+4| = 1$, on the other hand: $|x|+|y| = |-3|+|4| = 3+4 = 7$. Clearly, $|x+y| \leq |x|+|y|$, because $(1 \leq 7)$.

5.12 Floor Value

The floor function, denoted as $[x]$, rounds a real number x down to its nearest integer below or equal to it. This function finds widespread use across various fields, from theoretical mathematics to practical applications in computing and numerical analysis, facilitating a consistent method for downward rounding.

Key Point

Several key properties of the floor function allow us to manipulate and work with it easily:

1. **Idempotence:** This property ensures that for any integer n, $[n] = n$.
2. **Additivity with integers:** If you have a real number x and an integer n, the floor function satisfies $[x+n] = [x]+n$.
3. **Floor of negative values:** This property is especially significant and states that $[-x] = -[x+1]$, where $[x]$ represents the smallest integer greater than or equal to x.

 **Example** Find the floor value of $[-1.5]$.

Solution: This question asks us to find the largest integer which is less than or equal to -1.5. As per the definition of the floor function, we are rounding down to the nearest integer. Therefore, $[-1.5] = -2$, because -2 is the largest integer less than -1.5.

What is the floor value of $[-5]$?

Solution: Since -5 is an integer, as per the idempotence property, we have $[-5] = -5$.

5.13 Floor Function

The standard form of the floor function, $f(x) = a[g(x-b)] + k$, also indicates linear transformations of the floor function itself with the elements a, b, and k:

- a stands for a vertical stretch or compression.
- b stands for a horizontal shift.
- k stands for a vertical shift.

Keep in mind that $[x]$ symbolizes the floor function. The $[g(x-b)]$ term is a placeholder for inputting a function - the inner workings of the floor function. Mathematically, $f(x)$ is the largest integer that is less than or equal to x. In simpler terms, this means that $f(x)$ takes a real number x and rounds it *down* to the nearest integer.

Key Point

The domain of the floor function, $f(x) = a[g(x-b)] + k$, is the domain of $g(x)$ as the floor function can handle any real number inputs.

Key Point

The range of the floor function will depend on the nature of $g(x)$. Typically, for a continuous function spanning all the real numbers like a line or a parabola, the range of $f(x)$ will be all integers (due to the floor function), multiplied by the factor of a, then shifted by k.

Suppose we are given the function $f(x) = 2[x-1] + 3$, and we want to determine the value at $x = 3.5$.

Solution: First, we input $x = 3.5$ into our function equation. With $x = 3.5$, we have $f(3.5) = 2[3.5-1] + 3$. Simplifying inside the brackets, we get $f(3.5) = 2[2.5] + 3$. Now, we need to remember that the floor function rounds down. So, $2[2.5]$ becomes $2[2]$, which equals to 4. Adding 3, we arrive at $f(3.5) = 7$. So, when $x = 3.5$, $f(x) = 7$.

Determine the domain and range for $h(x) = [x^2] + 1$.

Solution: The domain is $(-\infty, \infty)$. For the range, as $k(x) = x^2$ is quadratic and covers all non-negative real numbers, the range of the floor function is all non-negative integers. When adding 1, the range is all integers greater than 1.

5.14 Graphing Floor Function

In the generic form $f(x) = a[g(x-b)] + k$:

- a represents vertical stretch (if $a > 1$) or compression (if $0 < a < 1$), with negativity implying a reflection across the x-axis.
- b shifts the graph horizontally to the right for positive b and to the left for negative b.
- k shifts the graph vertically up for positive k and down for negative k.

Let us break down the process of graphing $f(x) = a[g(x-b)] + k$ into steps:

1. Start by plotting the function $y = g(x)$ without applying the floor operation.
2. Once you have this, apply the floor operation to $g(x)$. This results in horizontal steps whose height corresponds to the integer values.
3. Apply the vertical transformation a to stretch or compress the steps.
4. After that, proceed to shift the graph horizontally by b.
5. Finally, apply the vertical shift k to move the entire graph either up or down.

Key Point

The floor function generates a step-like graph with jumps at each integer value. When graphing, note that the left endpoint of each "step" is included, while the right endpoint is not.

Example Let us graph the function $f(x) = 2[x-1] + 3$.

Solution: This function has three main components. First, the base function $y = [x]$ which is a simple step function with steps of height 1 at every integer value. Second, the expression $2[x-1]$ which vertically stretches the base function by a factor of 2 and shifts it 1 unit to the right. Finally, the "+3" term vertically shifts the entire graph by 3 units upwards.

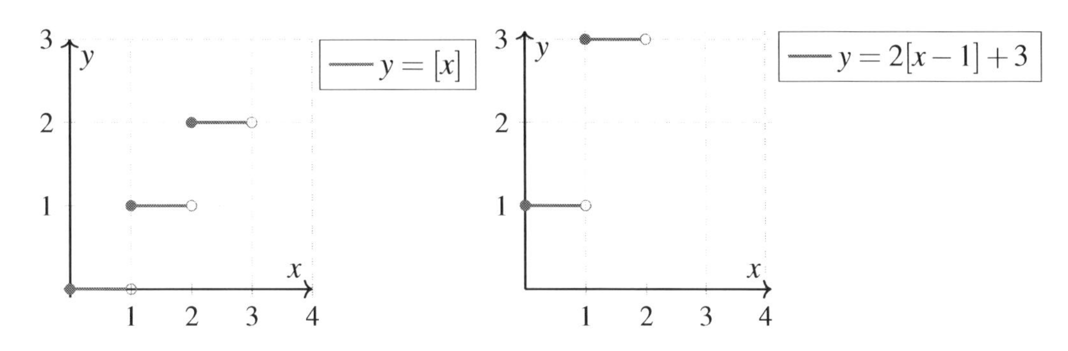

You can see from figures that the steps for the function $f(x) = 2[x-1] + 3$ are twice the height of the steps for the function $y = [x]$ because of the multiplication by 2. In addition, the function $f(x) = 2[x-1]+3$ is shifted one unit to the right and three units up compared to $y = [x]$.

5.15 Practices

1) Simplify the expression $\sqrt{50}$.

☐ A. $5\sqrt{2}$
☐ B. 10
☐ C. $25\sqrt{2}$
☐ D. $7\sqrt{2}$

2) Which of the following is the simplified form of $\sqrt[3]{40x^3y^6z^9}$?

☐ A. $2\sqrt[3]{5xy^3z^3}$
☐ B. $2xy^2z^3\sqrt[3]{5}$
☐ C. $2xy^2z^3\sqrt[3]{5x}$
☐ D. $2x^2y^2z^3$

3) Simplify the expression $\frac{5}{3+\sqrt{2}}$.

☐ A. $\frac{5\sqrt{2}}{7}$
☐ B. $\frac{15-5\sqrt{2}}{7}$
☐ C. $\frac{15+5\sqrt{2}}{7}$
☐ D. $-5\left(3-\sqrt{2}\right)$

4) Simplify the expression $\frac{\sqrt{3}}{2\sqrt{3}+4}$.

☐ A. $\frac{3\sqrt{2}-2}{3}$
☐ B. $\frac{3\sqrt{2}+2}{3}$
☐ C. $\frac{2\sqrt{3}-3}{2}$
☐ D. $\frac{2\sqrt{3}+3}{2}$

5) Simplify the expression $6\sqrt{7}-2\sqrt{7}+\sqrt{7}$.

☐ A. $5\sqrt{7}$
☐ B. $4\sqrt{7}$
☐ C. $7\sqrt{14}$
☐ D. 9

6) Simplify the expression $3\sqrt[3]{5}-\sqrt[3]{5}+4\sqrt[3]{5}$.

☐ A. $8\sqrt[3]{5}$
☐ B. $6\sqrt[3]{10}$
☐ C. $2\sqrt[3]{5}$
☐ D. $6\sqrt[3]{5}$

7) Simplify the product of radical expressions $\sqrt{49}\times\sqrt{14}$.

☐ A. $\sqrt{685}$

☐ B. 14

☐ C. $7\sqrt{14}$

☐ D. 21

8) What is the simplified form of $\sqrt[3]{\frac{125}{27}}$?

☐ A. $\sqrt[3]{\frac{5}{9}}$

☐ B. $\frac{5}{3}$

☐ C. $\frac{25}{9}$

☐ D. $\frac{125}{27}$

9) Solve the equation $\sqrt{3x+9}-3=0$.

☐ A. $x=0$

☐ B. $x=1$

☐ C. $x=4$

☐ D. $x=9$

10) Solve the equation $\sqrt{x+20}+\sqrt{x}=10$.

☐ A. $x=9$

☐ B. $x=16$

☐ C. $x=36$

☐ D. $x=49$

11) Solve the radical inequality $\sqrt{x}>x-2$.

☐ A. $x>4$

☐ B. $0\leq x<4$

☐ C. $x\leq 0$ or $x\geq 4$

☐ D. No solution

12) Determine the solution set for the inequality $3-\sqrt[3]{2x+7}\leq 0$.

☐ A. $x\leq -10$

☐ B. $x\geq 10$

☐ C. $x\leq -\frac{13}{2}$

☐ D. $x\geq -\frac{13}{2}$

13) What is the domain of the radical function $f(x)=\sqrt[3]{5x-1}$?

☐ A. $x\geq \frac{1}{5}$

☐ B. $x\leq \frac{1}{5}$

☐ C. All real numbers

☐ D. No real numbers

14) Given the function $f(x)=\sqrt[4]{2x-8}$, which of the following represents its domain?

☐ A. $x > 4$ ☐ C. $x \leq 4$

☐ B. $x \geq 4$ ☐ D. $x < 4$

15) What is the domain of the function $f(x) = \sqrt[5]{x^3 - 5x + 1}$?

☐ A. $x > 5$ ☐ C. $x \geq 5$

☐ B. $x \leq 5$ ☐ D. All real numbers

16) Which of the following represents the graph of $f(x) = \sqrt{x} - 2$?

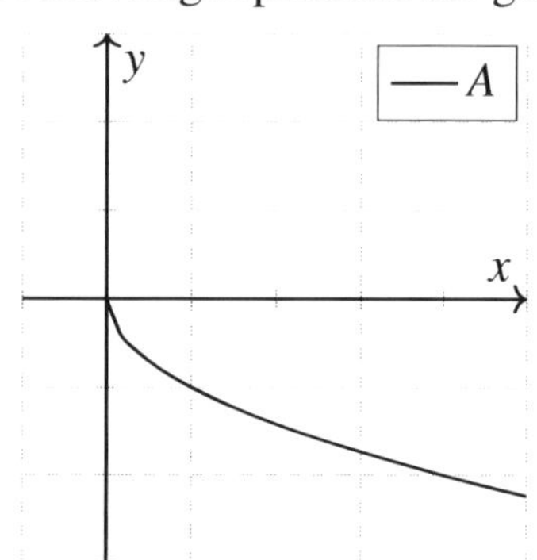

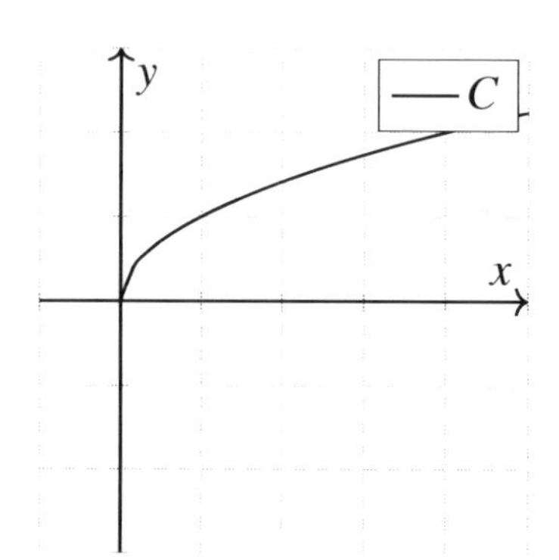

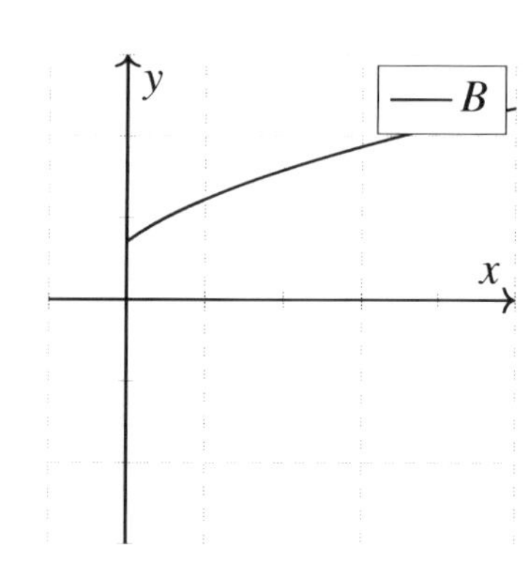

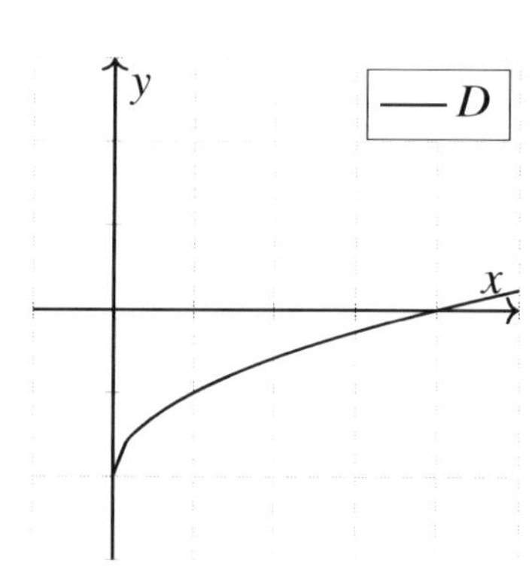

☐ A. Graph A ☐ C. Graph C

☐ B. Graph B ☐ D. Graph D

17) Which of the following represents the range of the function $f(x) = |\sqrt{x} - 3|$?

☐ A. $(-\infty, 3]$ ☐ C. $(-\infty, \infty)$

☐ B. $[0, \infty)$ ☐ D. $[3, \infty)$

18) What is the minimum value of the function $f(x) = |x^2 - 6x + 8|$?

☐ A. -8 ☐ C. 2
☐ B. 0 ☐ D. -2

19) Which of the following represents the graph of the function $f(x) = |2x - 4|$?

☐ A. A line with slope 2 and y-intercept -4
☐ B. A V-shaped graph with vertex at $(2, 0)$
☐ C. A V-shaped graph with vertex at $(0, 4)$
☐ D. A parabola opening upwards

20) For the function $h(x) = |x + 3| - 2$, what is the y-value of the vertex of its graph?

☐ A. -3 ☐ C. 0
☐ B. -2 ☐ D. 3

21) If $a = -3$ and $b = 2$, what is $|ab|$?

☐ A. -6 ☐ C. 3
☐ B. 0 ☐ D. 6

22) Which expression is equivalent to $|-x - y|$ for all real x and y?

☐ A. $-|x| - |y|$ ☐ C. $x + y$
☐ B. $|x| + |y|$ ☐ D. $|x + y|$

23) Given the floor value of 7.8 is $[7.8]$, what is $[7.8 + 3]$?

☐ A. 10 ☐ C. 7
☐ B. 11 ☐ D. 10.8

24) What is the result of $[3.2] + [2.8]$?

☐ A. 5.0 ☐ C. 5.8
☐ B. 6.0 ☐ D. 6.2

25) What is the value of the function $f(x) = 3[\frac{1}{2}x + 2] - 4$ when $x = 4$?

☐ A. 1 ☐ C. 5

☐ B. 2 ☐ D. 8

26) If $g(x) = \left[\frac{x}{3}\right]$, what is the range of $g(x)$?

☐ A. All real numbers ☐ C. All integers

☐ B. All rational numbers ☐ D. All non-negative integers

27) Identify the graph of the function $f(x) = [x+2] - 1$.

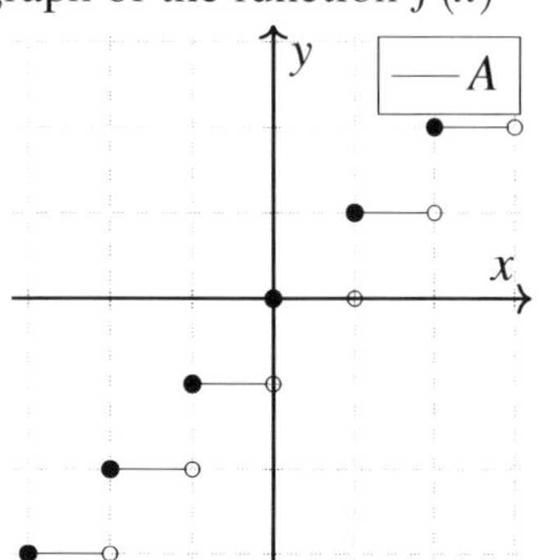

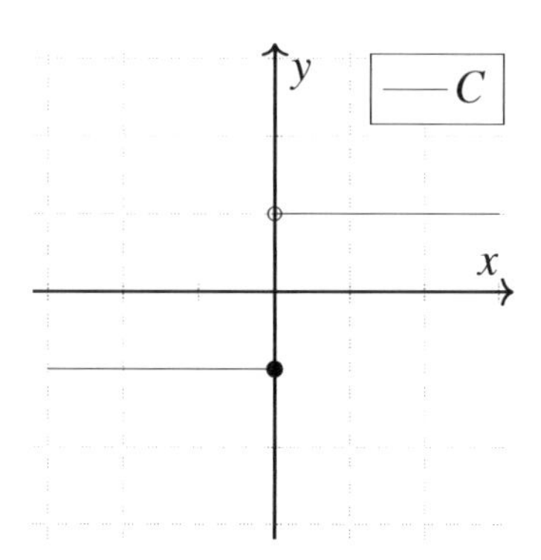

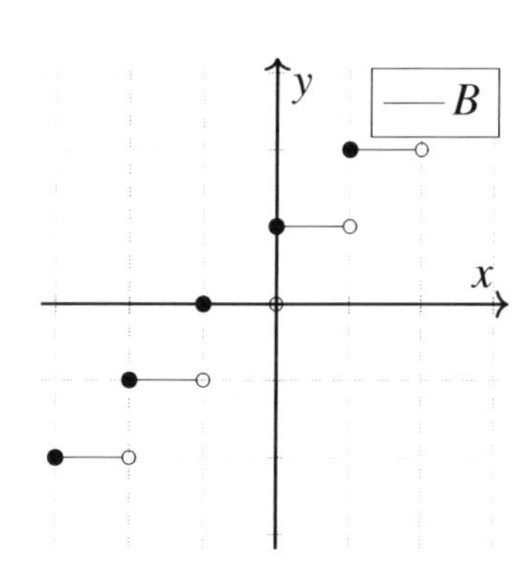

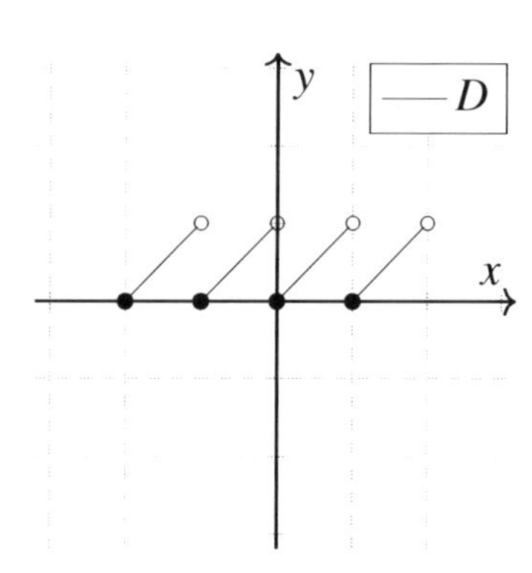

☐ A. Graph A ☐ C. Graph C

☐ B. Graph B ☐ D. Graph D

28) What is the range of the function $f(x) = -3\left[\frac{x}{2}\right] + 4$?

☐ A. $\{-3n+4 : n \in \mathbb{Z}\}$ ☐ C. $\{y \in \mathbb{Z} : y \leq -3\}$

☐ B. $\{-3n+4 : n \in \mathbb{N}\}$ ☐ D. $\{y \in \mathbb{Z} : y \geq -3\}$

Answer Keys

1) A. $5\sqrt{2}$

2) B. $2xy^2z^3\sqrt[3]{5}$

3) B. $\frac{15-5\sqrt{2}}{7}$

4) C. $\frac{2\sqrt{3}-3}{2}$

5) A. $5\sqrt{7}$

6) D. $6\sqrt[3]{5}$

7) C. $7\sqrt{14}$

8) B. $\frac{5}{3}$

9) A. $x = 0$

10) B. $x = 16$

11) B. $0 \leq x < 4$

12) B. $x \geq 10$

13) C. All real numbers

14) B. $x \geq 4$

15) D. All real numbers

16) D. Graph D

17) B. $[0, \infty)$

18) B. 0

19) B. A V-shaped graph with vertex at $(2, 0)$

20) B. -2

21) D. 6

22) D. $|x + y|$

23) A. 10

24) A. 5.0

25) D. 8

26) C. All integers

27) B. Graph B

28) A. $\{-3n + 4 : n \in \mathbb{Z}\}$

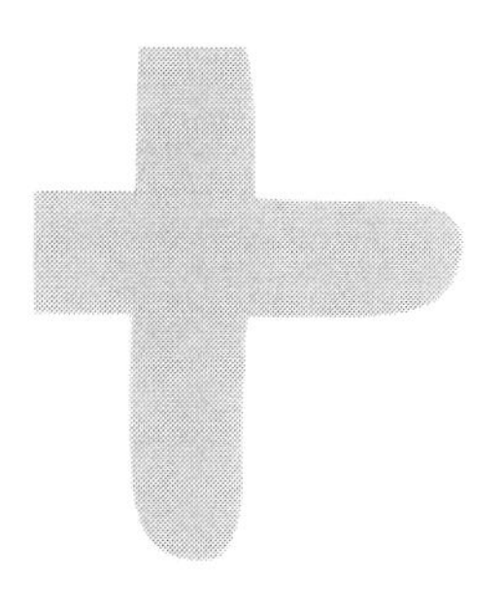

6. Functions Operations

6.1 Characteristics of Functions on Intervals

Functions $f(x)$ defined over an interval I exhibit behaviors such as being positive, negative, increasing, decreasing, or constant, each essential for understanding the function's behavior.

- **Positive Functions:** $f(x) > 0$ for all x in I, indicating the function's graph is entirely above the x-axis.
- **Negative Functions:** $f(x) < 0$ for all x in I, indicating the function's graph is entirely below the x-axis.

Key Point

Positive functions lie above the x-axis and negative functions lie below the x-axis within the given interval.

Regarding a function's growth behavior in an interval:

- **Increasing Functions:** For any $x < y$ in I, if $f(x) \leq f(y)$, the function is increasing, showing its values rise with x.
- **Decreasing Functions:** For any $x < y$ in I, if $f(x) \geq f(y)$, the function is decreasing, indicating its values decrease with x.
- **Constant Functions:** If $f(x) = f(y) = C$, for a constant C, for all $x < y$ in I, the function remains constant across the interval.

Key Point

A function is called increasing if its value does not decrease with increasing x, and decreasing if its value does not increase with increasing x.

Given the graph of function f as below,

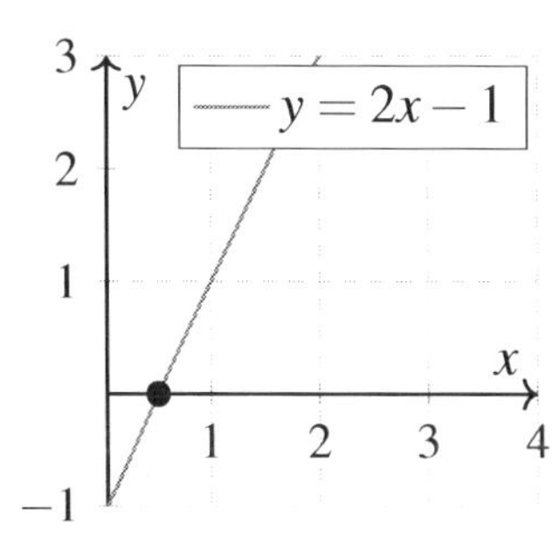

determine the intervals where the function is positive, negative, increasing, and decreasing.

Solution: From the graph:

- The function is negative on the interval $x < 0.5$, indicated by the graph residing below the x-axis.
- It becomes positive on the interval $x > 0.5$, with the graph positioned above the x-axis.
- Across its entire domain, the function is increasing, as evidenced by the continuous rise of $f(x)$ as x increases.

For the function $f(x) = x^2$, is $f(x)$ positive, negative, increasing or decreasing in the interval $I = [-1, 1]$?

Solution: In the intervals $[-1, 0)$, and $(0, 1)$ for any value of x that you pick, $f(x) = x^2$ will always be positive as the squaring operation ensures that. So, $f(x)$ is positive in the intervals $[-1, 0)$ and $(0, 1)$.
However, neither is $f(x)$ increasing nor decreasing in the entire interval because, from $x = -1$ to $x = 0$, $f(x)$ decreases, whereas from $x = 0$ to $x = 1$, $f(x)$ increases.

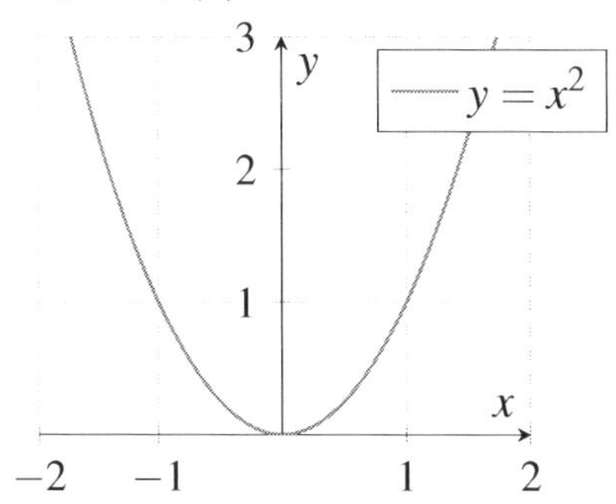

6.2 Transformations of Functions

Transformations alter a parent function $f(x)$, which represents the simplest form of a function type, into a new function. This concept is integral to precalculus and involves shifts, stretches, compressions, and reflections. These transformations can either preserve or change the properties of the parent function.

A parent function is defined as the most basic form of a type of function. Through transformations, more complex functions are derived from these parent forms.

Key Point

Key parent functions include: Constant $f(x) = c$, Linear $f(x) = x$, Absolute-value $f(x) = |x|$, Polynomial $f(x) = x^n$ (where n is a positive integer), Rational $f(x) = \frac{1}{x}$, Radical $f(x) = \sqrt{x}$, Exponential $f(x) = a^x$ (where $a > 0$ and $a \neq 1$), and Logarithmic $f(x) = \log_a x$.

For a function $f(x)$ and a constant number $k > 0$, transformations are categorized into:

Transformation Type	Description	Formula
Vertical Shift	Up/Down	$f(x) \pm k$
Horizontal Shift	Left/Right	$f(x \pm k)$
Stretching/Compressing	Horizontally	$f(kx)$
Stretching/Compressing	Vertically	$kf(x)$
Reflection	Across the x-axis	$-f(x)$
Reflection	Across the y-axis	$f(-x)$

Reflections notably can change the function's properties, such as its direction and sign.

What is the parent function of the following function and what transformations have occurred in it: $y = 2x^2 - 1$?

Solution: The given function can be rewritten as $y = 2(x^2) - 1$.

Here, we can see that $f(x) = x^2$ is our parent function, which represents a quadratic function.
Looking at the coefficient of x^2 which is 2, and this fact that if $k > 1$, the graph $y = kf(x)$ is a vertical stretch of factor k from the graph $y = f(x)$. Therefore, the function $y = x^2$ is stretching vertically by a factor of 2.

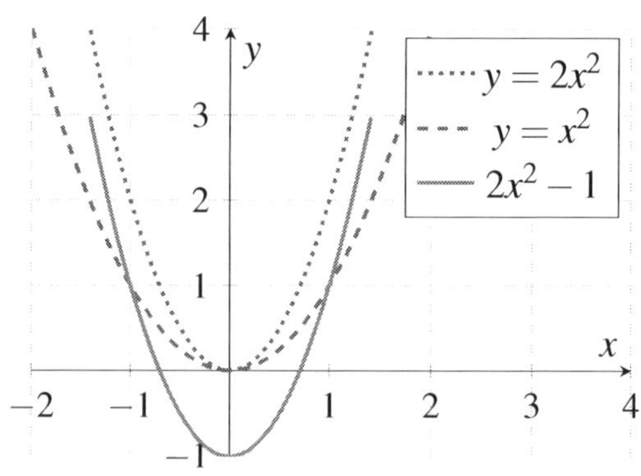

Now looking at the constant -1, we know from the fact that the graph $y = f(x) + k$; $k < 0$, is shifted k units downwards from the graph $y = f(x)$. Therefore, the function $y = 2x^2 - 1$ is shifted 1 unit downward from the graph $y = 2x^2$. So the transformations that occurred in the parent function $f(x) = x^2$ to obtain the function $y = 2x^2 - 1$ are a vertical stretch by a factor of 2 and a vertical shift of 1 unit downwards.

What is the parent function for $f(x) = 2x^2 - 4x + 3$?

Solution: The parent function for $f(x) = 2x^2 - 4x + 3$ is $f(x) = x^2$, because f can be

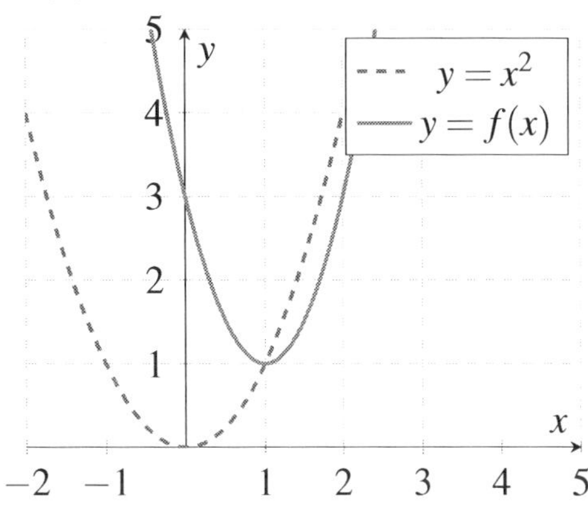

written as $f(x) = 2(x-1)^2 + 1$. Here, the parent function has been transformed by stretching vertically by a factor of 2, shifting 1 unit to the right, and then upwards by 1 unit.

6.3 Even and Odd Functions

In this section, we explore a classification of functions based on their symmetry, known as even and odd functions.

An **even function** is defined by the property $f(-x) = f(x)$ for all x in its domain, indicating symmetry across the y-axis. This means substituting $-x$ for x leaves the function unchanged.

Key Point

Even functions display symmetry with respect to the y-axis (vertical axis).

Conversely, an **odd function** satisfies $f(-x) = -f(x)$ for all x in its domain, indicating symmetry about the origin. Substituting $-x$ for x results in the negative of the original function.

Key Point

Odd functions are symmetric with respect to the origin (center of the coordinate plane).

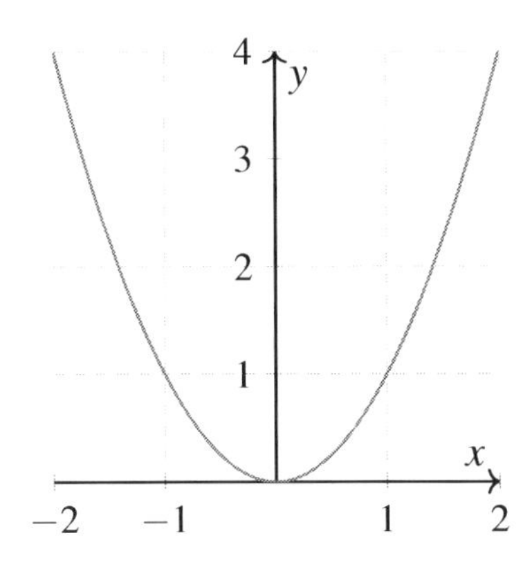

Figure 6.1: Graph of $y = x^2$, an even function.

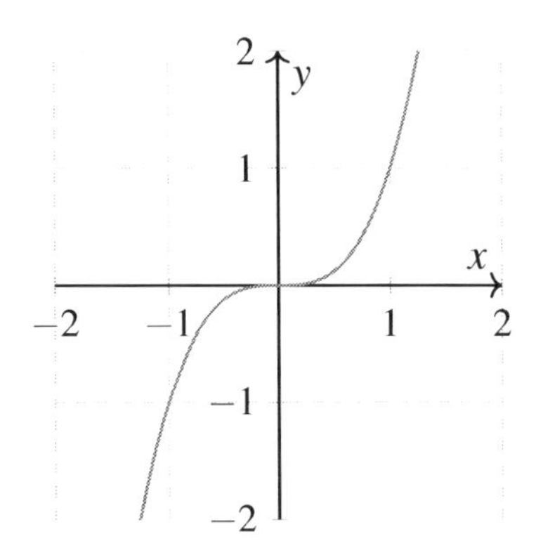

Figure 6.2: Graph of $y = x^3$, an odd function.

 Determine whether the function $f(x) = -2x^3$ is even, odd, or neither.

Solution: To find out, we substitute $-x$ for x in the function equation and simplify it.

$$f(-x) = -2(-x)^3 = -2(-x^3) = 2x^3.$$

In this case, $f(-x) = -f(x)$. Therefore, the function is odd and symmetric with respect to the origin.

 Is the function $f(x) = x^3 + x^2$ even, odd, or neither?

Solution: Let us find $f(-x)$: $f(-x) = (-x)^3 + (-x)^2 = -x^3 + x^2$.

We find that $f(-x)$ does not equal $f(x)$ nor $-f(x)$, so the function $f(x) = x^3 + x^2$ is neither even nor odd.

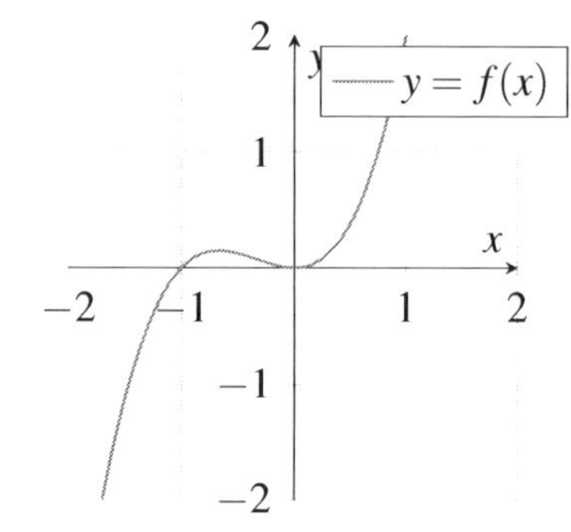

6.4 End Behavior of Polynomial Functions

Understanding the end behavior of polynomial functions enriches our grasp of how these functions behave as the input x approaches infinity ($+\infty$) or negative infinity ($-\infty$). The end behavior essentially describes the direction the graph of the function heads towards in these limits.

A polynomial function, denoted as $P(x)$, is generally expressed in the form:

$$P(x) = a_n x^n + a_{n-1} x^{n-1} + \cdots + a_1 x + a_0,$$

where $a_n, a_{n-1}, \cdots, a_1$, and a_0 are constants, with n being a non-negative integer signifying the degree of the polynomial. The coefficient a_n is the leading coefficient, and it plays a significant role in determining the function's end behavior.

Characteristics of Polynomial Functions

- The domain of $P(x)$ spans all real numbers, ensuring the function is continuous across its entire range.
- The degree of the polynomial, n, dictates the function's highest exponent of x.
- The leading coefficient, a_n, influences the function's orientation and end behavior.

Determining End Behavior

The end behavior of polynomial functions depends on two main factors: the degree (n) and the leading coefficient (a_n):

- For an **even degree** polynomial, the ends of the graph either both point up ($a_n > 0$) or down ($a_n < 0$).
- For an **odd degree** polynomial, the ends of the graph point in opposite directions, with the directionality influenced by the sign of a_n.

The following table summarizes the end behavior of polynomial functions based on the degree and leading coefficient:

If	$a_n > 0$	$a_n < 0$
n in even	$P(x) \to +\infty$, as $x \to +\infty$ $P(x) \to +\infty$, as $x \to -\infty$	$P(x) \to -\infty$, as $x \to +\infty$ $P(x) \to -\infty$, as $x \to -\infty$
n is odd	$P(x) \to +\infty$, as $x \to +\infty$ $P(x) \to -\infty$, as $x \to -\infty$	$P(x) \to -\infty$, as $x \to +\infty$ $P(x) \to +\infty$, as $x \to -\infty$

Find the end behavior of the function $f(x) = -x^4 + 3x^3 - x$.

Solution: The largest degree in the function is 4 and the coefficient of this term is -1 (which is negative). According to the table, this means the function's end behavior is:

$$f(x) \to -\infty, \text{ as } x \to +\infty,$$

$$f(x) \to -\infty, \text{ as } x \to -\infty.$$

6.5 One-to-One Function

In mathematics, a *one-to-one function*, also known as an *injective function*, plays a critical role in understanding the relationships between different sets. A function is considered one-to-one if each element in the domain maps to a unique element in the codomain, implying that no two different inputs produce the same output.

Formally, for a function f, if $f(x) = f(y)$ implies that $x = y$, then f is one-to-one. This definition ensures that each element of the domain corresponds to a distinct element in the codomain, preventing any overlap in mapping.

Key Point

A function that fails to be one-to-one has at least two distinct elements in the domain, say a and b, that map to the same element in the codomain, meaning $a \neq b$ but $f(a) = f(b)$. Such functions demonstrate that not all functions achieve a one-to-one correspondence.

The importance of one-to-one functions stretches across various fields of mathematics and its applications, including solving equations, function analysis, mathematical modeling, and establishing unique correspondences between sets. Understanding and identifying one-to-one functions is foundational in exploring more complex mathematical concepts.

Consider the function $f(x) = 2x + 3$. Is this a one-to-one function?

Solution: Suppose $f(a) = f(b)$, where a and b are elements in the domain. This gives us the equation $2a + 3 = 2b + 3$. Subtraction of 3 from both sides gives $2a = 2b$, which simplifies to $a = b$ after dividing both sides by 2. Since our assumption $f(a) = f(b)$ leads to $a = b$, we can conclude that the function is indeed a one-to-one function.

Determine whether the function $g(x) = x^2$ is a one-to-one function in its domain or not.

Solution: Let us assume $g(a) = g(b)$, where a and b are elements in the domain. Then $a^2 = b^2$. Taking the square root of both sides results in $|a| = |b|$. Since the absolute values of a and b are equal, we cannot conclude that $a = b$. Therefore, the function is not one-to-one in its domain.

6.6 Identifying The Function One-to-One From the Graph

Understanding whether a function is one-to-one (injective) from its graph is an essential skill in mathematics. Recall that a function is one-to-one if each distinct element of the domain (input values) maps to a unique element in the codomain (possible output values). To identify whether a function represented by a graph is one-to-one, we employ the *Horizontal Line Test*.

Key Point

The Horizontal Line Test involves drawing horizontal lines across the graph of the function. If any horizontal line intersects the graph more than once, the function is not one-to-one.

A function passes the test, and thus is one-to-one, if no horizontal line intersects the graph at more than one point.

Given the graph of the function $y = x^3$, is the function one-to-one?

Solution: To determine if the function is one-to-one, we apply the Horizontal Line Test.

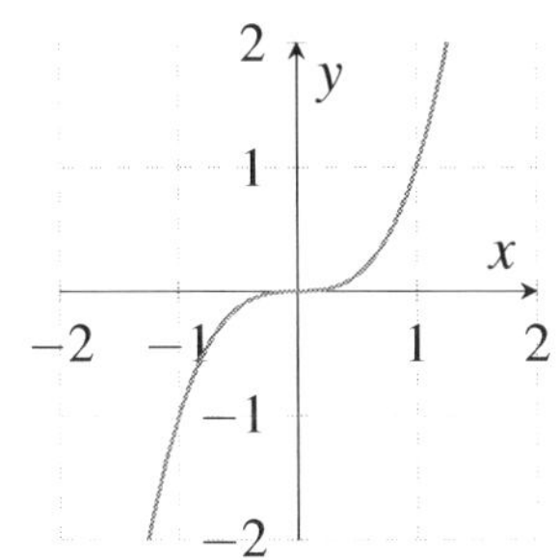

It is apparent that no horizontal line will intersect the graph of $y = x^3$ more than once. Therefore, the function $y = x^3$ is indeed one-to-one.

Is the function represented by the graph $y = x^2$ one-to-one?

Solution: Drawing a horizontal line, you can see that there will be multiple intersections with the graph of $y = x^2$. Therefore, the function fails the Horizontal Line Test and is not one-to-one.

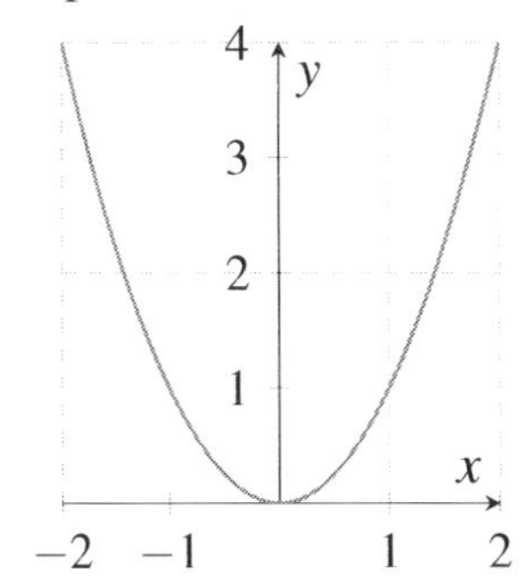

6.7 Onto (Surjective) Functions

An onto (or surjective) function $f : A \to B$ is a function where every element y in the codomain B has a pre-image in the domain A. In other words, for each y in B, there exists at least one x in A such that $f(x) = y$.

Key Point

A function is onto (surjective) if and only if its range is equal to the codomain. If there exists any element in the codomain that does not correspond to the output for any input, the function is not onto.

A function is onto within a specific interval if, for each output in the codomain of that interval, there is at least one input from the interval. To verify this, restrict the domain to the interval and then determine the range.

A function can be both one-to-one (injective) and onto (surjective). Such a function is called a bijection, meaning it establishes a one-to-one correspondence between the elements of the domain and codomain.

Example Let us check whether the function $f(x) = x^2$ is onto in the interval $[-1, 1]$ in the codomain of real numbers.

Solution: To determine this, we can take any y in the real numbers and try to solve the equation $f(x) = y$, which in this case becomes $x^2 = y$. The solution to this equation is $x = \sqrt{y}$ and $x = -\sqrt{y}$. Negative values of y have no pre-images, even though they are part of the real number codomain. So, $f(x) = x^2$ is not onto in the interval $[-1, 1]$ for the entire set real numbers. However, if we adjust our interval to $[0, 1]$, the function $f(x) = x^2$ will now be onto, since for every non-negative real number y there exists an x in the interval $[0, 1]$ with $f(x) = y$.

Example Consider the function $g(x) = x^3$ from real numbers to real numbers. Is it bijection?

Solution: Yes, for every real number y, there exists a real number x such that $g(x) = y$. For instance, for $y = 8$ we have $x = 2$ since $2^3 = 8$. So, $g(x)$ is surjective. On the other hand, as we saw before, the graph passes the Horizontal Line Test, as no horizontal line intersects the graph more than once. Therefore, the function is injective.

6.8 Function Inverses

Inverse functions reverse the effect of an initial function. Given a function f mapping an input x to an output y, its inverse f^{-1}, when applied to y, returns the original x. This relationship is captured by $f(x) = y \Leftrightarrow f^{-1}(y) = x$. For a function to have an inverse, it must be one-to-one. Graphically, a function and its inverse reflect across the line $y = x$.

Key Point

Functions f and f^{-1} undo each other, switching their domain and range:

$$f(f^{-1}(y)) = y \text{ and } f^{-1}(f(x)) = x.$$

To find an inverse, first ensure f is injective. Replace $f(x)$ with y, swap x and y, and solve for y.

Example Find the inverse of the function $f(x) = 2x - 1$.

Solution: $f(x)$ is injective. To find the inverse of the function, we will first replace $f(x)$ with y: $y = 2x - 1$. Then, we interchange x and y: $x = 2y - 1$. Now, we solve for y:

$$x = 2y - 1 \Rightarrow x + 1 = 2y \Rightarrow y = \frac{x+1}{2}.$$

Finally, we replace y with $f^{-1}(x)$: $f^{-1}(x) = \frac{x+1}{2}$.

This is the inverse function of our original function. As seen below, the function and its inverse are reflections of each other over the line $y = x$.

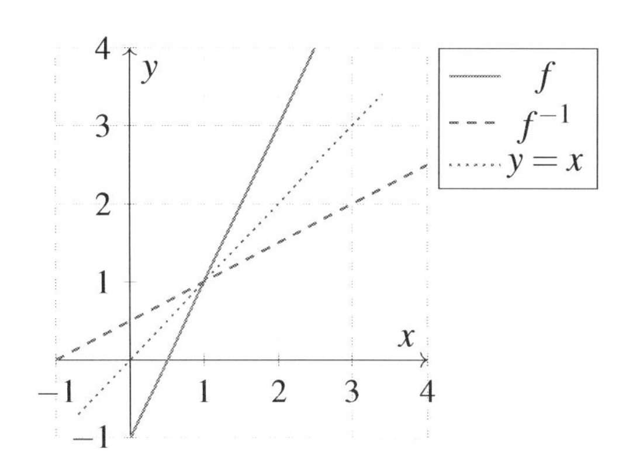

6.9 Inverse Function Graph

Expanding on inverse functions, graphically determining a function's inverse enriches comprehension. The inverse, f^{-1}, is graphically found by mirroring f across the line $y = x$, due to the role reversal of x and y coordinates. For any point (a, b) on f, the inverse f^{-1} will have a corresponding point (b, a).

To exemplify, consider the function $y = x^2$ over $[0, +\infty)$ and its inverse $y = \sqrt{x}$. The inverse graph showcases reflection across the line $y = x$, illustrating the concept visually.

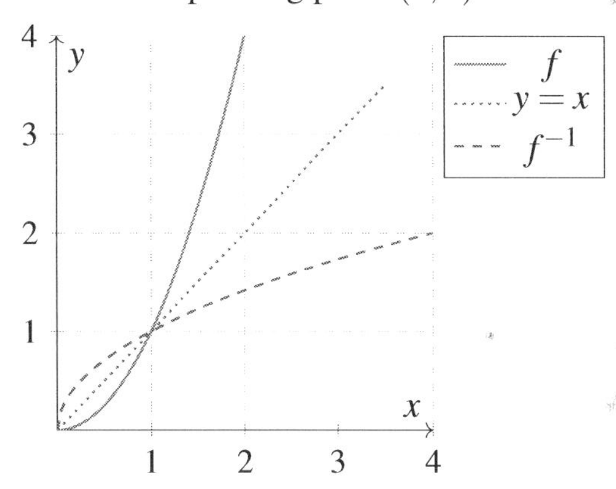

Key Point

The graph of the inverse of a function can be obtained by reflecting the graph of the original function about the line $y = x$.

We emphasize again that not all functions have an inverse. A function has an inverse if and only if it is one-to-one.

Given the graph of the function $y = 2x - 1$, find its inverse and graph it.

Solution: The function $y = 2x - 1$ can be rewritten as $x = \frac{1}{2}y + \frac{1}{2}$ to find the inverse, which is $y = \frac{1}{2}x + \frac{1}{2}$. Now, let us graph the inverse function along with the original one.

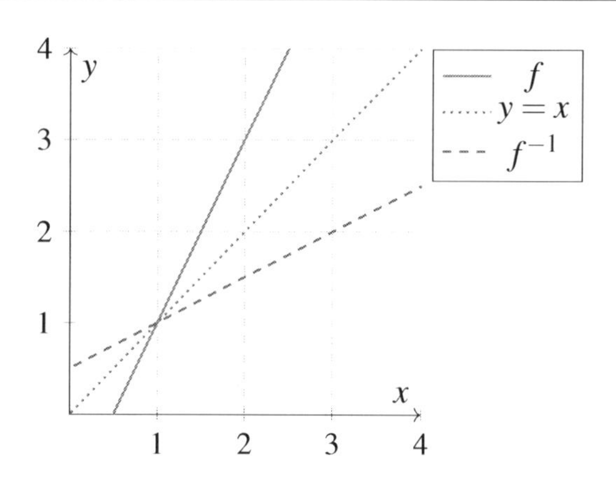

6.10 Inverse Variation

Inverse variation describes a unique relationship where one quantity's increase corresponds to another's decrease, maintaining a constant product, distinctly different from direct variation's simultaneous quantity changes.

Mathematically, this relationship is defined as $xy = k$ or $y = \frac{k}{x}$, with $x, y \neq 0$ and k, a non-zero constant, symbolizing the proportionality constant that remains unchanged in inverse variations.

Key Point

In an inverse variation, as one variable increases, the other decreases in such a way that the product of the two variables is always a constant.

Considering pairs (x_1, y_1) and (x_2, y_2) under inverse variation, both adhere to $x_1y_1 = k$ and $x_2y_2 = k$, highlighting the constant nature of k in such relationships.

Example If y varies inversely as x and $y = 12$ when $x = 5$. What is the value of y when x is 3?

Solution: To solve this, we need to find the value of k. We can do this by plugging $y = 12$ and $x = 5$ into the inverse variation formula $xy = k$. Calculating, we get $k = (5)(12) = 60$. Having obtained k, we can go on to find y, using the known value of $x = 3$. Thus, $xy = 60$, so substituting $x = 3$, we get $60 = 3y$, which we can rearrange to find $y = 20$.

Example Given $x_1 = 2$, $y_1 = 5$, and $x_2 = 4$, find the value of y_2. The given values of x_1 and y_1 follow the inverse variation rule.

Solution: $k = x_1y_1 = (2)(5) = 10$. Now, using x_2 value and k, we find y_2 by inverse variation rule. Given $x_2y_2 = k$. Substituting the value of k, we have $4y_2 = 10$. Solving for y_2, we get $y_2 = \frac{10}{4}$.

6.11 Piece-wise Functions

Piece-wise functions are defined by multiple formulas across different segments of their domain, rather than a single expression. They adapt to various conditions within their domain.

A general representation of a piece-wise function is:

$$f(x) = \begin{cases} f_1(x), & x \in Domain\ f_1(x) \\ f_2(x), & x \in Domain\ f_2(x) \\ \vdots & \\ f_n(x), & x \in Domain\ f_n(x) \end{cases}$$

where $n \geq 2$, and the domain of the function is the union of all of the smaller domains.

For example, absolute value functions are naturally piece-wise, represented as:

$$|x| = \begin{cases} x, & x \geq 0 \\ -x, & x < 0 \end{cases}$$

This format efficiently handles different function behaviors over its domain.

Key Point

Piece-wise functions use different formulas for different parts of their domain.

 Graph the piece-wise function:

$$f(x) = \begin{cases} 2, & x < 0 \\ x+1. & x \geq 1 \end{cases}$$

Solution: We first draw the line $y = 2$ for all $x < 0$. This is a horizontal line that intersects the y-axis at $y = 2$. Then, draw the graph $y = x + 1$ for all $x \geq 1$. This is a straight line with a slope of one and a y-intercept of 1.

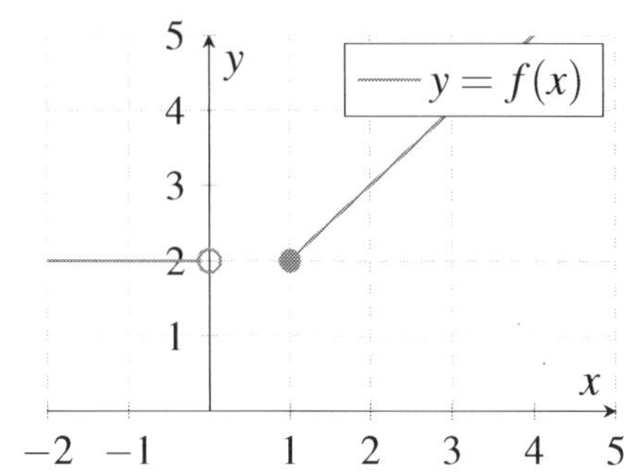

6.12 Practices

1) Consider the function $f(x) = x^3 - 3x$. Which of the following intervals is $f(x)$ decreasing?

☐ A. $(-\infty, -1]$
☐ B. $(-1, 1)$
☐ C. $[1, +\infty)$
☐ D. $(-\infty, +\infty)$

2) Given the function $f(x) = \frac{1}{x}$, for which interval(s) is the function positive and decreasing?

☐ A. $(0, +\infty)$
☐ B. $(-\infty, 0)$
☐ C. Both A and B
☐ D. Neither A nor B

3) Which of the following transformations represents a shift of the parent function $f(x) = \sqrt{x}$ to the right by 3 units?

☐ A. $g(x) = \sqrt{x+3}$
☐ B. $g(x) = \sqrt{x}+3$
☐ C. $g(x) = \sqrt{x-3}$
☐ D. $g(x) = \sqrt{3x}$

4) If the function $g(x) = -3f(x)$ is a transformation of the parent function $f(x) = x^2$, what transformation has occurred?

☐ A. Horizontal stretch by a factor of 3
☐ B. Vertical stretch by a factor of 3
☐ C. Reflection across the x-axis and vertical stretch by a factor of 3
☐ D. Reflection across the y-axis and horizontal stretch by a factor of 3

5) Which of the following functions is even?

☐ A. $f(x) = x^5 - x^3$
☐ B. $f(x) = \cos(x)$
☐ C. $f(x) = \ln x$
☐ D. $f(x) = e^x$

6) If $f(x) = x^6 + 2x^4 - x^2 + 5$ is an even function, for which of the following values will $f(-x)$ equal $f(x)$?

☐ A. $x = 2$
☐ B. $x = -2$
☐ C. $x = 1$
☐ D. All of the above

7) Determine the end behavior of the polynomial function $g(x) = 2x^5 - 6x^3 + 4$.

☐ A. $g(x) \to -\infty$ as $x \to +\infty$ and $g(x) \to -\infty$ as $x \to -\infty$
☐ B. $g(x) \to +\infty$ as $x \to +\infty$ and $g(x) \to +\infty$ as $x \to -\infty$
☐ C. $g(x) \to +\infty$ as $x \to +\infty$ and $g(x) \to -\infty$ as $x \to -\infty$
☐ D. $g(x) \to -\infty$ as $x \to +\infty$ and $g(x) \to +\infty$ as $x \to -\infty$

8) What is the end behavior of the polynomial function $f(x) = -3x^6 + 9x^2 - 1$?
☐ A. $f(x) \to -\infty$ as $x \to +\infty$ and $f(x) \to -\infty$ as $x \to -\infty$
☐ B. $f(x) \to +\infty$ as $x \to +\infty$ and $f(x) \to +\infty$ as $x \to -\infty$
☐ C. $f(x) \to +\infty$ as $x \to +\infty$ and $f(x) \to -\infty$ as $x \to -\infty$
☐ D. $f(x) \to -\infty$ as $x \to +\infty$ and $f(x) \to +\infty$ as $x \to -\infty$

9) Which of the following functions is not one-to-one?

☐ A. $f(x) = x^3$
☐ B. $g(x) = 3x + 1$
☐ C. $h(x) = \sin(x)$ for x in $[-\frac{\pi}{2}, \frac{\pi}{2}]$
☐ D. $j(x) = x^2$

10) Determine if the function $f(x) = \frac{1}{x}$ is a one-to-one function on its domain.

☐ A. Yes
☐ B. No
☐ C. Cannot be determined
☐ D. One-to-one on a limited domain

11) Which of the following functions has a one-to-one graph?

☐ A. $y = \sin(x)$
☐ B. $y = |x|$
☐ C. $y = e^x$
☐ D. $y = x^2 - 2$

12) Consider the graph of the function shown below. Determine if the function is one-to-one.

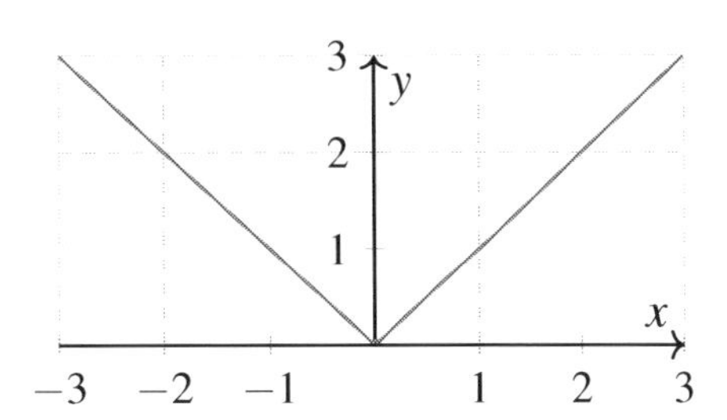

☐ A. Yes
☐ B. No
☐ C. It cannot be determined
☐ D. This is not a function

13) Which of the following functions $f : \mathbb{R} \to \mathbb{R}$ is onto?

☐ A. $f(x) = x^2$
☐ B. $f(x) = \tan(x)$
☐ C. $f(x) = e^x$
☐ D. $f(x) = \frac{1}{x}$

14) Consider the function $f : \mathbb{R} \to \mathbb{R}$ defined by $f(x) = 3x + 2$. This function is onto because:

☐ A. It passes the Vertical Line Test.
☐ B. It passes the Horizontal Line Test.
☐ C. Its graph is a parabola.
☐ D. It has an inverse function on $\mathbb{R}$.

15) Which of the following functions does not have an inverse function?

☐ A. $f(x) = x^3 - 1$
☐ B. $f(x) = 2x + 3$
☐ C. $f(x) = \frac{1}{x}$
☐ D. $f(x) = x^2 + 1$

16) What is the inverse function of $f(x) = \sqrt{x+3}$?

☐ A. $f^{-1}(x) = x^2 - 3$
☐ B. $f^{-1}(x) = x^2 + 3$
☐ C. $f^{-1}(x) = (x-3)^2$
☐ D. $f^{-1}(x) = (x+3)^2$

17) Which graph represents the inverse of the function $f(x) = -\frac{1}{3}x + 2$?

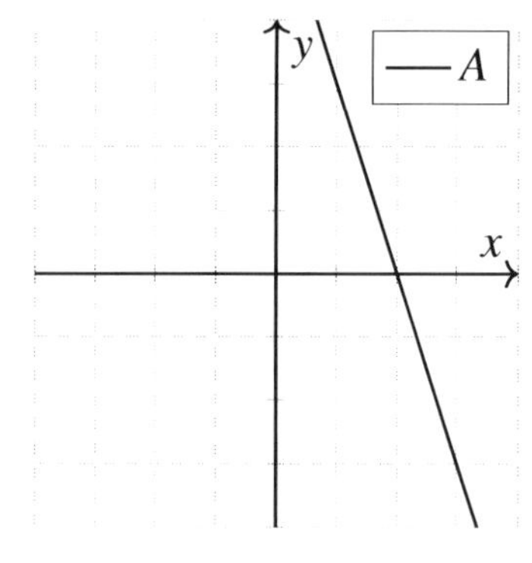

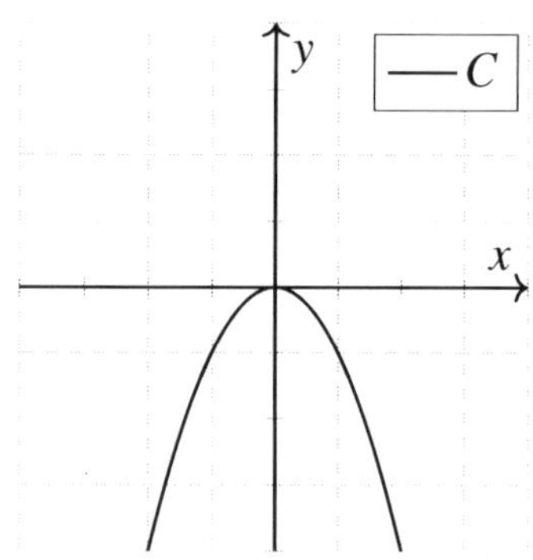

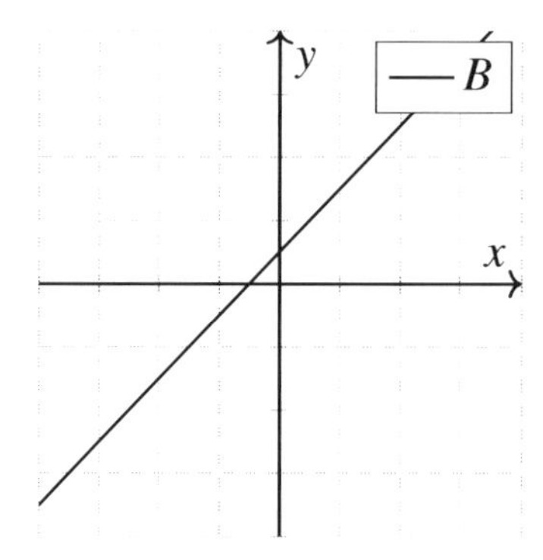

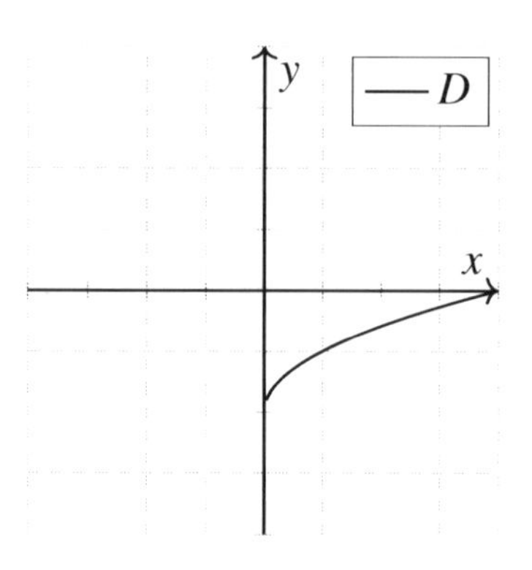

☐ A. Graph A

☐ B. Graph B

☐ C. Graph C

☐ D. Graph D

18) The graph of which function is a reflection of the graph of $y = x^3$ about the line $y = x$?

☐ A. $y = \sqrt[3]{x}$

☐ B. $y = x^2$

☐ C. $y = -x^3$

☐ D. $y = -\sqrt[3]{x}$

19) If z varies inversely as w and $z = 4$ when $w = 15$, what is the value of z when $w = 5$?

☐ A. 12

☐ B. 20

☐ C. 60

☐ D. 3

20) The quantity A varies inversely with B. If $A = \frac{2}{3}$ when $B = 9$, what is the constant of variation k?

☐ A. 6

☐ B. 8

☐ C. $\frac{3}{2}$

☐ D. 54

21) Which of the following defines the piece-wise function given by the graph below?

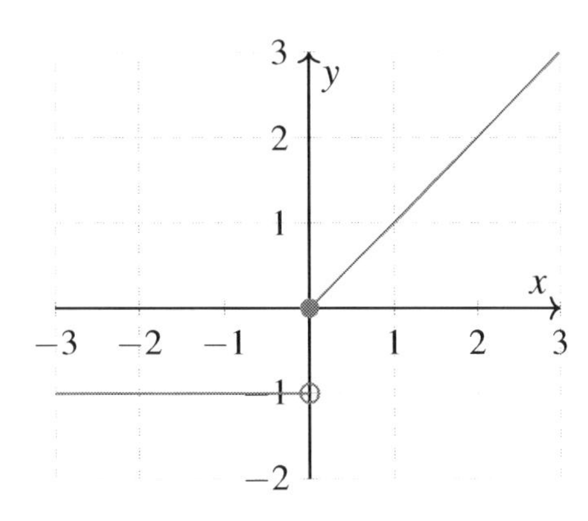

☐ A. $f(x) = \begin{cases} -1, & x < 0 \\ x, & x \geq 0 \end{cases}$

☐ B. $f(x) = \begin{cases} -1, & x \leq 0 \\ x, & x > 0 \end{cases}$

☐ C. $f(x) = \begin{cases} x, & x < 0 \\ -1, & x \geq 0 \end{cases}$

☐ D. $f(x) = \begin{cases} x, & x \leq 0 \\ -1, & x > 0 \end{cases}$

22) Evaluate the piece-wise function at $x = -3$ and $x = 2$.

$$f(x) = \begin{cases} x^2, & x \leq -2 \\ 4, & -2 < x < 3 \\ 2x - 1, & x \geq 3 \end{cases}$$

☐ A. $f(-3) = 9,\ f(2) = 4$

☐ B. $f(-3) = -9,\ f(2) = 4$

☐ C. $f(-3) = 9,\ f(2) = 3$

☐ D. $f(-3) = -9,\ f(2) = 3$

Answer Keys

1) B. $(-1,1)$

2) A. $(0,+\infty)$

3) C. $g(x)=\sqrt{x-3}$

4) C.

5) B. $f(x)=\cos(x)$

6) D. All of the above

7) C.

8) A.

9) D. $j(x)=x^2$

10) A. Yes

11) C. $y=e^x$

12) B. No

13) B. $f(x)=\tan(x)$

14) D. It has an inverse function on $\mathbb{R}$.

15) D. $f(x)=x^2+1$

16) A. $f^{-1}(x)=x^2-3$

17) A. Graph A

18) A. $y=\sqrt[3]{x}$

19) A. 12

20) A. 6

21) A.

22) A. $f(-3)=9, f(2)=4$

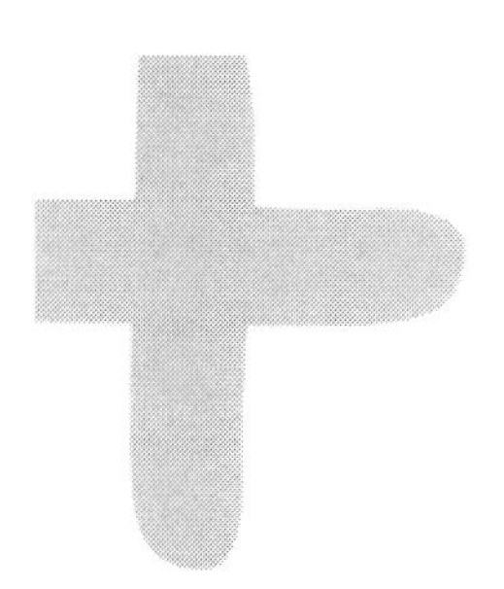

7. Sequences and Series

7.1 Sequence

A sequence is an ordered list of numbers, usually following a specific pattern, represented as $a_1, a_2, \cdots, a_i, \cdots$. Each number, or *term*, in a sequence is denoted by a_n, where n is the term's position or index. The *general term* a_n is crucial as it allows the calculation of any term's value by substituting n with the term's index.

Key Point

To find a term in a sequence, apply the formula for its general term a_n by inserting the term's position number n and computing the result.

For the sequence represented by $a_n = 4n + 5$, calculate a_6.

Solution: Insert $n = 6$ into our formula:

$$a_6 = 4(6) + 5 = 24 + 5 = 29.$$

For the sequence defined by $a_n = 3\left(2^n\right)$, determine a_4.

Solution: Substitute $n = 4$ we get:

$$a_4 = 3\left(2^4\right) = 3(16) = 48.$$

Evaluate the sequence $a_n = n^2 + 2n + 1$ for a_3.

Solution: Plug $n = 3$ in the formula $a_n = n^2 + 2n + 1$. So, we get:

$$a_3 = 3^2 + 2(3) + 1 = 9 + 6 + 1 = 16.$$

Example Find the first three terms of the sequence with the general term $a_n = (-1)^{n-1} n$, where $n \geq 1$.

Solution: Since $n \geq 1$, the sequence begins from one. To find the first term, we insert $n = 1$. Then:

$$a_1 = (-1)^{1-1} \times 1 = 1.$$

Similarly, we enter natural numbers to find other terms. Therefore:
For $n = 2$, we find $a_2 = (-1)^{2-1} \times 2 = -2$.
For $n = 3$, we find $a_3 = (-1)^{3-1} \times 3 = 3$.
Hence, the first three terms of the sequence are 1, -2, and 3.

7.2 Recursive Formula

Recursive sequences define each term T_n using its predecessor(s), such as T_{n-1}, T_{n-2}, ..., T_{n-k}, making the sequence's analysis intriguing and powerful. The sequence's order corresponds to the number of terms involved in its recursive definition. Formally, it is expressed as:

$$T_{n+1} = f(T_n, T_{n-1}, T_{n-2}, \ldots, T_{n-k}),$$

accompanied by initial conditions to uniquely determine the sequence.

Key Point

In recursion, the term T_n is calculated from the preceding k terms, with k indicating the sequence's order.

To find terms in a recursive sequence:
1. Start with initial term(s).
2. Apply the recursive formula for subsequent terms.
3. Continue for the required number of terms.

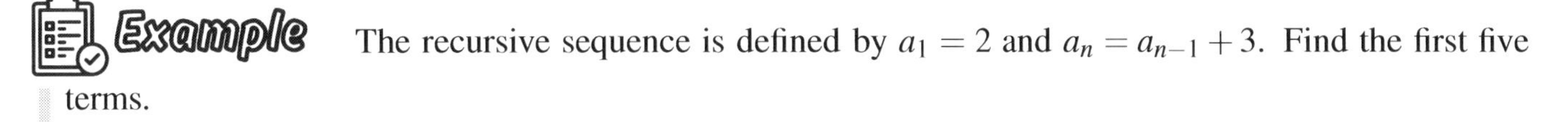

Example The recursive sequence is defined by $a_1 = 2$ and $a_n = a_{n-1} + 3$. Find the first five terms.

Solution: First, we know that $a_1 = 2$ from the initial condition. Then, by using the recursive formula, we can express a_2 in terms of a_1:

$a_2 = a_1 + 3 = 2 + 3 = 5.$

Continuing this process, we can find:

$a_3 = a_2 + 3 = 5 + 3 = 8,$

$a_4 = a_3 + 3 = 8 + 3 = 11,$

$a_5 = a_4 + 3 = 11 + 3 = 14.$

Hence, the first five terms of the sequence are $2, 5, 8, 11, 14$.

A recursive sequence is given by $a_1 = 4$ and $a_n = 2a_{n-1}$. Find a_4.

Solution: Let us find the required term using the recursive formula. As an initial condition, we are given $a_1 = 4$:

$a_2 = 2a_1 = 2(4) = 8,$

$a_3 = 2a_2 = 2(8) = 16,$

$a_4 = 2a_3 = 2(16) = 32.$

So, $a_4 = 32$.

Given the recursive formula $T_n = T_{n-1} + T_{n-2}$, $n \geq 3$, $T_1 = 1$ and $T_2 = 3$. Write the first five terms of the sequence.

Solution: To find the terms of the sequence, we need the previous two terms; since they are given, we can start by finding T_3:

$T_3 = T_2 + T_1 = 3 + 1 = 4,$

$T_4 = T_3 + T_2 = 4 + 3 = 7,$

$T_5 = T_4 + T_3 = 7 + 4 = 11.$

Therefore, the first five terms of the sequence are $1, 3, 4, 7, 11$.

7.3 Write a Formula for a Recursive Sequence

Creating a formula for a recursive sequence requires identifying its initial term(s) and the consistent numerical relationship across its terms. This involves:

1. Identifying initial term(s), denoted as a_1 (and possibly a_2).
2. Analyzing the sequence to find a consistent pattern, such as the difference or ratio between successive terms, involving a_{n-1}, a_{n-2}, etc.
3. Expressing this pattern as a function of the preceding term(s).

The formula for a recursive sequence must define a_n in terms of its previous term(s), including the initial term(s).

Find the recursive formula for the sequence $12, 17, 22, 27, 32, \cdots$.

Solution: Define the first term $a_1 = 12$. This is true for other consecutive terms. From examining the differences in the sequence, the pattern $a_n - a_{n-1} = 5$ is obtained. Consequently, $a_n = a_{n-1} + 5$ is the recurring formula, which holds for $n > 1$.

Find the recursive formula for the sequence $1, 3, 9, 27, 81, \cdots$.

Solution: Define the first term $a_1 = 1$. Checking the sequence yields $a_n = 3a_{n-1}$. The rule holds for $n \geq 2$.

Example Find the recursive formula for the sequence $1, 3, 2, -1, -3, -2, \cdots$.

Solution: This particular sequence does not readily display a difference or ratio pattern. Therefore, another relationship is considered. The difference between each term and its predecessor is identified as the sequence $2, -1, -3, -4, \cdots$. Here, the general formula $a_{n+1} = a_n - a_{n-1}$ is obtained where $a_1 = 1$ and $a_2 = 3$ and $n \geq 2$.

7.4 Arithmetic Sequences

An arithmetic sequence is a series of numbers with a constant difference, d, known as the 'common difference'. For example, starting from 6 and adding 2 to each subsequent number yields the sequence $6, 8, 10, 12, 14, \cdots$, demonstrating a common difference of 2.

Key Point

In an arithmetic sequence, each term is determined by the formula $a_n = a_1 + (n-1)d$, highlighting a uniform difference, d, between consecutive terms. Here, a_1 represents the initial term, d the common difference, and n the term's ordinal position.

Find the first five terms of the sequence. Given $a_8 = 38$ and $d = 3$.

Solution: Firstly, we need to calculate a_1. We use the arithmetic sequence formula $a_n = a_1 + (n-1)d$.

Given that $a_8 = 38$, we substitute $n = 8$ into the formula:

$$38 = a_1 + 3(8-1) = a_1 + 21.$$

Solving for a_1, we get $a_1 = 38 - 21 = 17$. So, the first five terms of the sequence are:

$$17, 20, 23, 26, 29.$$

List the first five terms of the arithmetic sequence, given that $a_1 = 18$ and $d = 2$.

Solution: We can use the arithmetic sequence formula to calculate the next four terms:
$a_2 = 18 + 2(2-1) = 20$,
$a_3 = 18 + 2(3-1) = 22$,
$a_4 = 18 + 2(4-1) = 24$,
$a_5 = 18 + 2(5-1) = 26$.
Hence, the first five terms of the sequence are:

$$18, 20, 22, 24, 26.$$

7.5 Geometric Sequences

A geometric sequence is a sequence where each term after the first is obtained by multiplying the previous one by a constant non-zero number, known as the 'common ratio' (r or q). For example, in the sequence $2, 4, 8, 16, 32, \cdots$, the common ratio is 2, with each term derived by multiplying the preceding term by 2.

Key Point

The formula for any term in a geometric sequence is:

$$a_n = a_1 r^{n-1},$$

where a_1 is the first term, r is the common ratio, and n is the term's position in the sequence.

Example Given the first term and the common ratio of a geometric sequence, find the first five terms of the sequence. $a_1 = 3$, $r = -2$.

Solution: We will use the geometric sequence formula:

$$a_n = ar^{n-1} \Rightarrow a_n = 3(-2)^{n-1}.$$

If $n = 1$, then:

$$a_1 = 3(-2)^{1-1} = 3(1) = 3.$$

Thus, the first five terms of this sequence are $3, -6, 12, -24, 48$.

Example Given two terms in a positive geometric sequence, find the 8-th term. $a_3 = 10$, and $a_5 = 40$.

Solution: We know that:

$$a_3 = ar^{3-1} = a_1r^2 = 10, \text{ and } a_5 = a_1r^{5-1} = ar^4 = 40.$$

Dividing a_5 by a_3 gives:

$$\frac{a_5}{a_3} = \frac{a_1r^4}{a_1r^2} = \frac{40}{10}.$$

Simplifying gives:

$$r^2 = 4 \Rightarrow r = \pm 2.$$

Given that the sequence is positive, we accept $r = 2$. We can find a_1 now by solving:

$$a_1r^2 = 10 \Rightarrow a_1(2^2) = 10 \Rightarrow a_1 = \frac{5}{2}.$$

Finally, using these values in the formula to find the 8th term gives us:

$$a_8 = \frac{5}{2}(2)^{8-1} = \frac{5}{2}(128) = 320.$$

7.6 Sigma Notation

Sigma notation, also known as 'summation notation', streamlines expressing the sum of sequence terms. It utilizes the Greek letter Σ for compact representation of sums.

Given a sequence with terms $a_1, a_2, \cdots, a_i, \cdots$, the sum from the k-th to the n-th term is written as:

$$\sum_{i=k}^{n} a_i,$$

where i is the summation index, a_i the i-th term, k the start, and n the end of the summation interval.

Key Point

Summation notation efficiently represents series sums, with the summation index (i) running from k to n.

Summation properties for sequences $\{a_i\}_i$, $\{b_i\}_i$, constant c, and positive integer n include:

- $\sum_{i=1}^{n} ca_i = c\sum_{i=1}^{n} a_i$.
- $\sum_{i=1}^{n}(a_i + b_i) = \sum_{i=1}^{n} a_i + \sum_{i=1}^{n} b_i$.
- $\sum_{i=1}^{n} c = nc$.
- $\sum_{i=1}^{n} i = \frac{n(n+1)}{2}$.
- $\sum_{i=1}^{n} i^2 = \frac{n(n+1)(2n+1)}{6}$.

Evaluate $\sum_{i=1}^{10}(i^2 - 1)$.

Solution: Using the property $\sum_{i=1}^{n}(a_i + b_i) = \sum_{i=1}^{n} a_i + \sum_{i=1}^{n} b_i$, we have:

$$\sum_{i=1}^{10}(i^2 - 1) = \sum_{i=1}^{10} i^2 - \sum_{i=1}^{10} 1.$$

Now, we use the formulas $\sum_{i=1}^{n} i^2 = \frac{n(n+1)(2n+1)}{6}$ and $\sum_{i=1}^{n} c = nc$. For $n = 10$, we get:

$$\sum_{i=1}^{10} i^2 = \frac{10(10+1)(2(10)+1)}{6} = 385, \text{ and } \sum_{i=1}^{10} 1 = 10(1) = 10.$$

Therefore, $\sum_{i=1}^{10}(i^2 - 1) = 385 - 10 = 375$.

7.7 Arithmetic Series

Arithmetic series result from summing the terms of an arithmetic sequence, characterized by a constant difference d between consecutive terms. This concept leverages the sigma notation to succinctly express sums of series. Given an arithmetic sequence with first term a_1 and common difference d, an arithmetic series up to the n-th term is denoted as:

$$S_n = \sum_{k=1}^{n} a_k = \sum_{k=1}^{n}[a_1 + (k-1)d].$$

Applying the identity $\sum_{k=1}^{n} k = \frac{1}{2}n(n+1)$, we obtain:

$$S_n = \frac{n}{2}[2a_1 + (n-1)d].$$

Given $a_n = a_1 + (n-1)d$, the series sum simplifies to:

$$S_n = \frac{n}{2}(a_1 + a_n).$$

This formula provides a straightforward way to calculate the sum of the first n terms of an arithmetic sequence.

Example Compute the sum of the first 10 terms in the arithmetic series $4, 11, 18, \cdots$.

Solution: In this arithmetic series, $a_1 = 4$, $d = 11 - 4 = 7$, and $n = 10$. Use the arithmetic series formula to calculate the sum:

$$S_{10} = \frac{10}{2}[2(4) + 7(10-1)] = 5(8+63) = 355.$$

So, the sum of the first 10 terms is 355.

Example Determine the sum of the first four terms given that $d = 4$ and $a_{10} = 46$.

Solution: First find a_1 using the formula $a_n = a_1 + (n-1)d$. So we have:

$$46 = a_1 + 4(10-1).$$

Solve for a_1 to get:

$$a_1 = 46 - 36 = 10.$$

Therefore, the first four terms are 10, 14, 18, and 22. The sum S_4 is:

$$S_4 = \frac{4}{2}[2(10) + 4(4-1)] = 64.$$

So, the sum of the first four terms is 64.

7.8 Finite Geometric Series

A finite geometric series sums the terms of a geometric sequence, wherein each term is obtained by multiplying the preceding term by a constant non-zero number, the common ratio r.

The formula for the sum of the first n terms of a finite geometric series is:

$$S_n = \sum_{k=1}^{n} a_n = \sum_{k=1}^{n} a_1 r^{k-1} = a_1 \left(\frac{1-r^n}{1-r} \right),$$

where n is the series' term count, S_n the sum of these terms, a_1 the initial term, and r the common ratio. This formula is crucial for calculating the sum of a geometric series up to a finite term n.

For the geometric sequence represented by $a_n = 3^{n-1}$, find the sum of the first 4 terms.

Solution: We will use the formula $S_n = a_1\left(\frac{1-r^n}{1-r}\right)$ to solve this problem. Here we have $a_1 = 1$, $r = 3$, and $n = 4$. Plugging these values into the formula, we get:

$$S_n = (1)\left(\frac{1-3^4}{1-3}\right) = 1\left(\frac{1-81}{1-3}\right) = \left(\frac{-80}{-2}\right)1 = 40.$$

Hence, the sum of the first 4 terms of $a_n = 3^{n-1}$ for $n = 4$ is 40.

For the geometric sequence represented by $a_n = -2^{n-1}$, find the sum for $n = 5$.

Solution: We use the formula $S_n = a_1\left(\frac{1-r^n}{1-r}\right)$ considering $a_1 = -1$, $r = 2$, and $n = 5$. Substituting these values into the formula, we get:

$$S_n = (-1)\left(\frac{1-2^5}{1-2}\right) = (-1)\left(\frac{-31}{-1}\right) = -31.$$

So, the sum of the first 5 terms is -31.

For the sequence represented by $\left(-\frac{1}{2}\right)^{n-1}$, find the sum for $n = 7$.

Solution: In this problem, $a_1 = 1$, $r = -\frac{1}{2}$, and $n = 7$. We use the geometric sum formula to find:

$$S_n = (1)\left(\frac{1-\left(-\frac{1}{2}\right)^7}{1-\left(-\frac{1}{2}\right)}\right) = \left(\frac{\frac{129}{128}}{\frac{3}{2}}\right) = \frac{129}{192} = \frac{43}{64}.$$

So, the sum of the first 7 terms is $\frac{43}{64}$.

7.9 Infinite Geometric Series

An infinite geometric series has an unlimited number of terms, extending indefinitely. The sum of such a series can be finite or infinite, primarily dependent on the common ratio r.

The series converges to a finite sum if $|r| < 1$, calculated by:

$$S = \frac{a_1}{1-r},$$

where a_1 represents the first term and r the common ratio. Conversely, if the absolute value of r is greater than or equal to 1, the series sum is infinite.

Key Point

An infinite geometric series demonstrates the intriguing concept that a series with an infinite number of terms can have a finite sum, provided $|r| < 1$.

Example Evaluate the infinite geometric series described by $(-\frac{2}{3})^{i-1}$.

Solution: Since the absolute value of the common ratio $-\frac{2}{3}$ is less than 1, the sum of the series is finite. Now, we can use the formula $S = \frac{a_1}{1-r}$ to find the sum:

$$S = \frac{1}{1-(-\frac{2}{3})} = \frac{1}{\frac{5}{3}} = \frac{3}{5}.$$

Example Evaluate the infinite geometric series described by $(\frac{1}{3})^{i-1}$.

Solution: The common ratio, in this case, is $\frac{1}{3}$ which is less than 1. Thus, we can use the formula $S = \frac{a_1}{1-r}$ to find the sum:

$$S = \frac{1}{1-\frac{1}{3}} = \frac{1}{\frac{2}{3}} = \frac{3}{2}.$$

Example Evaluate the infinite geometric series described by $-2\left(\frac{1}{4}\right)^{k-1}$.

Solution: The common ratio here is $\frac{1}{4}$ and $a_1 = -2$. Using the formula $S = \frac{a_1}{1-r}$, we find the sum:

$$S = \frac{-2}{1-\frac{1}{4}} = \frac{-2}{\frac{3}{4}} = \frac{-8}{3}.$$

Example Evaluate the infinite geometric series described by $(\frac{1}{4})7^{k-1}$.

Solution: The common ratio is 7 which is greater than 1. Therefore, the sum of this infinite geometric series cannot be evaluated as it does not converge to a finite value.

7.10 Convergent and Divergent Series

Series can be classified into two types based on their sum as the number of terms increases indefinitely:

- A **convergent series** has a sum that approaches a finite value.
- A **divergent series's** sum does not approach a finite value, instead it may grow without bounds, oscillate, or not settle on any specific value.

Additionally, we explore the concept of an **alternating series**, characterized by terms whose signs alternate between positive and negative, represented as:

$$\sum_{k=1}^{\infty} (-1)^k a_k,$$

where $a_k \geq 0$ for all k, and the sequence may begin with either a positive or negative term.
For an alternating series to be considered convergent, it must satisfy the following conditions:

1. $0 \leq a_{k+1} \leq a_k$ for all $k \geq 1$, indicating that each subsequent term is less than or equal to the previous term.
2. The terms of the series approach zero as k approaches infinity.

These conditions help in determining whether an alternating series converges to a finite sum.

Determine whether the following series converge or diverge:

$$\sum_{i=1}^{\infty} (-1)^i \frac{2}{i+5}.$$

Solution: Let $a_i = \frac{2}{i+5}$. Then, we know that a_i approaches 0 when i goes to infinity since the numerator is constant while the denominator tends to infinity.

In addition, $0 \leq a_{k+1} \leq a_k$ must hold:

$$\frac{2}{(i+1)+5} \leq \frac{2}{i+5} \rightarrow \frac{2}{i+6} \leq \frac{2}{i+5} \rightarrow i+6 \geq i+5 \rightarrow 6 \geq 5.$$

This is true. Consequently, the alternating series in question is convergent.

Example
Determine whether the following series converge or diverge:

$$\frac{(-1)^k k}{2k+1}.$$

Solution: Let $a_k = \frac{k}{2k+1}$. Then, as k goes to infinity, $\frac{k}{2k+1}$ approaches $\frac{1}{2}$. Therefore, the alternating series in question does not satisfy all the convergence conditions, hence it is divergent.

7.11 The n-th Term Test for Divergence

The n-th term test for divergence provides a simple criterion to determine the divergence of a series. This test focuses on the behavior of the series' terms as the series progresses towards infinity.

Key Point

If the terms of a series (a_n) do not tend towards zero as the index, n, approaches infinity, then the series $\sum a_n$ is divergent.

This principle underlines that a series cannot converge if its terms do not asymptotically approach zero. Conversely, if the terms of a series do approach zero as n approaches infinity, the series may either converge or diverge, making the test inconclusive in this scenario.

Determine whether the series $\sum_{n=1}^{\infty} \frac{1}{n}$ is convergent or divergent.

Solution: Here, the n-th term of the series is given by $a_n = \frac{1}{n}$. As n gets larger and larger, a_n gets smaller and smaller and approaches 0. However, the n-th term test is inconclusive in this case, and we need more than the n-th term test to decide if the series converges or diverges.

Determine whether the series $\sum_{n=1}^{\infty} n$ is convergent or divergent.

Solution: Here, the n-th term of the series is given by $a_n = n$. As n gets larger and larger, a_n gets larger and larger and does not approach 0. Therefore, by the n-th term test for divergence, the series must diverge.

Determine whether the series $1 - 1 + 1 - 1 + 1 - 1 + \cdots$ is convergent or divergent.

Solution: Here, the n-th term of the series is given by $a_n = (-1)^n$. As n gets larger and larger, a_n gets larger and larger and does not approach 0. It alternates between -1 and 1. Therefore, by the n-th term test for divergence, the series must diverge.

7.12 Comparison, Ratio, and Root Tests

We introduce three significant tests to determine the convergence or divergence of series: the Comparison Test, the Ratio Test, and the Root Test. These tests utilize comparisons to known series, ratios of consecutive terms, and n-th term roots to study the series' behavior, assuming $a_n \geq 0$ for all n.

- **Comparison Test:** This test compares a series term-by-term with a known series. If our series is less than a known convergent series, it converges. If it is greater than a known divergent series, it diverges.
- **Ratio Test:** It checks convergence or divergence by the ratio of consecutive terms. The series converges

if this ratio is consistently less than 1, diverges if more than 1, and the test is inconclusive if the ratio equals 1.

- **Root Test:** By taking the n-th root of the n-th term, the series converges if this root approaches a number smaller than 1, diverges if it approaches a number greater than 1, and is inconclusive if it approaches 1.

Key Point

The p-series $\sum_{n=1}^{\infty} \frac{1}{n^p}$, converging for $p > 1$, is often used for comparison.

Key Point

While the Comparison Test evaluates a series against a benchmark, the Ratio and Root Tests examine intrinsic series properties—term ratios and term roots—to ascertain convergence or divergence.

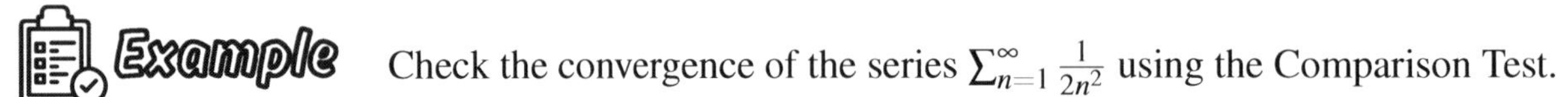

Example Check the convergence of the series $\sum_{n=1}^{\infty} \frac{1}{2n^2}$ using the Comparison Test.

Solution: We are already aware that $\sum_{n=1}^{\infty} \frac{1}{n^2}$ is a convergent series. Now, since $\frac{1}{2n^2} \leq \frac{1}{n^2}$ for all n, by the Comparison Test, the series $\sum_{n=1}^{\infty} \frac{1}{2n^2}$ also converges.

Example Check the convergence of the series $\sum_{n=1}^{\infty} \frac{3^n}{n!}$ using the Ratio Test.

Solution: Using the Ratio Test, we need to find the ratio of the $(n+1)$-th term and the n-th term:

$$\frac{a_{n+1}}{a_n} = \frac{\frac{3^{n+1}}{(n+1)!}}{\frac{3^n}{n!}} = \frac{3}{n+1}.$$

As n grows large, $\frac{3}{n+1}$ approaches 0, which is less than 1. Since the result of the ratio test is less than 1, the series converges.

Example Check the convergence of the series $\sum_{n=1}^{\infty} \frac{2^n}{n^3}$ using the Root Test.

Solution: Utilizing the Root Test, we need to determine the n-th root of the n-th term:

$$\sqrt[n]{a_n} = \sqrt[n]{\frac{2^n}{n^3}} = \frac{2}{n^{\frac{3}{n}}}.$$

As n grows large, $n^{\frac{3}{n}}$ approaches 1, hence $\sqrt[n]{a_n}$ approaches 2. Since the result of the root test equals 2, being more than 1, the series diverges.

7.13 Pascal's Triangle

Pascal's Triangle, a significant concept in algebra, is a triangular array where each entry is the sum of the two directly above it. Starting with row $n = 0$ at the top, the triangle progresses with entries in each row labeled from $k = 0$ from the left, staggered across rows. This structure assists in determining coefficients in binomial expansions. See the Figure below for a depiction

1
1 1
1 2 1
1 3 3 1
1 4 6 4 1
1 5 10 10 5 1
1 6 15 20 15 6 1
1 7 21 35 35 21 7 1
1 8 28 56 70 56 28 8 1

Pascal's Triangle with 8 rows.

Key Point

In Pascal's Triangle, the n-th row (starting from 0) has $n+1$ entries. Sum of entries in the n-th row is equal to 2^n.

The k-th entry in the n-th row is called a "Binomial Coefficient" and given by:

$$\binom{n}{k} = \frac{n!}{k!(n-k)!}.$$

Example What is the fifth entry (starting from 0) of the 7-th row of Pascal's Triangle?

Solution: According to the key point above, the k-th entry of the n-th row corresponds to the binomial coefficient $\binom{n}{k}$. So, the fifth entry of the 7-th row corresponds to $\binom{7}{5} = \frac{7!}{5!2!}$, which equals 21.

7.14 The Binomial Theorem

The binomial theorem describes the algebraic expansion of powers of a binomial. According to the theorem, it is possible to expand the polynomial $(x+y)^n$ into a sum involving terms of the form $x^a y^b$, where the exponents a and b are nonnegative integers with $a+b=n$, and the coefficient of each term is a specific integer known as a binomial coefficient.

Key Point

The formula of the binomial theorem for a positive integer n is:

$$(x+y)^n = \sum_{k=0}^{n} \binom{n}{k} x^{n-k} y^k, \quad \text{where} \quad \binom{n}{k} = \frac{n!}{k!(n-k)!}.$$

In the expansion of $(x+y)^n$, there are $n+1$ terms and the k-th term is given by:

$$\binom{n}{k-1} x^{n-k+1} y^{k-1}.$$

Key Point

In the expansion of $(x+y)^n$, the exponents on x start with n and decrease, while the exponents on y start with 0 and increase. The powers on x and y always add up to n in each term.

Example Find the expansion of $(x+y)^4$.

Solution: By directly applying the binomial theorem, we have:

$$(x+y)^4 = \binom{4}{0} x^4 + \binom{4}{1} x^3 y + \binom{4}{2} x^2 y^2 + \binom{4}{3} x y^3 + \binom{4}{4} y^4.$$

This simplifies to:

$$x^4 + 4x^3 y + 6x^2 y^2 + 4xy^3 + y^4.$$

Example Find the 3rd term of the expansion of $(a-1)^5$.

Solution: By again applying the formula for the k-th term, we get:

$$T_3 = \binom{5}{3-1} a^{5-3+1} (-1)^{3-1} = \binom{5}{2} a^3 (-1)^2.$$

This simplifies to $10a^3$.

Example Find the expansion of $(2b+2)^3$.

Solution: By directly applying the binomial theorem, we get:

$$(2b+2)^3 = \binom{3}{0}(2b)^3 + \binom{3}{1}(2b)^2(2) + \binom{3}{2}(2b)(2)^2 + \binom{3}{3}(2)^3.$$

This simplifies to $8b^3 + 24b^2 + 24b + 8$.

7.15 Practices

1) Determine the 8th term of the sequence defined by the general term $a_n = 2n - 3$.

☐ A. 13
☐ B. 14
☐ C. 15
☐ D. 16

2) Consider the sequence defined by the general term $a_n = (-1)^{n-1}n$. Calculate the value of a_5.

☐ A. -2
☐ B. 5
☐ C. -5
☐ D. 9

3) Consider the recursive sequence $b_n = 3b_{n-1} + 2$, with the initial term $b_1 = 1$. What is the value of the third term b_3?

☐ A. 5
☐ B. 17
☐ C. 9
☐ D. 11

4) Given the recursive sequence where $d_1 = 6$, $d_2 = 12$, and $d_n = d_{n-1} + d_{n-2}$, what is the value of the fourth term, d_4?

☐ A. 30
☐ B. 24
☐ C. 18
☐ D. 33

5) Which of the following represents the recursive formula for the sequence $16, 21, 26, 31, \ldots$?

☐ A. $a_n = 5n + 11$
☐ B. $a_n = a_{n-1} + 5$
☐ C. $a_n = a_{n-1} + 6$
☐ D. $a_n = 5a_{n-1}$

6) Which of the following represents the recursive formula for the sequence $2, 4, 8, 16, \ldots$?

☐ A. $a_n = \frac{1}{2}a_{n-1}$
☐ B. $a_n = 2a_{n-1}$
☐ C. $a_n = 2a_{n-2}$
☐ D. $a_n = \frac{1}{2}a_{n-2}$

7) What is the 10th term of the arithmetic sequence that starts with 5 and has a common difference of 3?

☐ A. 32
☐ B. 35
☐ C. 28
☐ D. 29

8) If the common difference of an arithmetic sequence is -4 and the 5-th term is 3, what is the 1-st term?

☐ A. 19
☐ B. 21
☐ C. 23
☐ D. -13

9) If the first three terms of a geometric sequence are $27, 9, 3$, what is the fourth term?

☐ A. 1
☐ B. $\frac{3}{2}$
☐ C. $\frac{1}{3}$
☐ D. -1

10) Given the fifth term of a geometric sequence is 32 and the common ratio is $\frac{1}{2}$, what is the first term?

☐ A. 256
☐ B. 64
☐ C. 1024
☐ D. 512

11) Evaluate the sum $\sum_{i=3}^{8}(2i+3)$.

☐ A. 60
☐ B. 84
☐ C. 80
☐ D. 94

12) Simplify $\sum_{i=1}^{n}(3i-2)$ to an expression involving n.

☐ A. $2n^2+n$
☐ B. $3n^2+n-2$
☐ C. $\frac{3}{2}n^2-\frac{1}{2}n$
☐ D. $\frac{3}{2}n^2+\frac{1}{2}n-2$

13) An arithmetic series starts with $a_1 = 6$ and has a common difference $d = 3$. What is the sum of the first 15 terms?

☐ A. 540 ☐ C. 675
☐ B. 315 ☐ D. 405

14) Find the value of n in the arithmetic series $3, 8, 13, \ldots, n$ terms, so that the sum of the first n terms equals 164.

☐ A. $n = 10$ ☐ C. $n = 8$
☐ B. $n = 15$ ☐ D. $n = 20$

15) A geometric sequence has a first term of $a_1 = 5$ and a common ratio of $r = -3$. What is the sum of the first 3 terms of this series?

☐ A. -35 ☐ C. 40
☐ B. -40 ☐ D. 35

16) A finite geometric series has a sum of $S_5 = -121$ and common ratio $r = 3$. If the first term $a_1 = -1$, determine S_6.

☐ A. -242 ☐ C. 363
☐ B. -364 ☐ D. -121

17) What is the sum of the infinite geometric series $\sum_{n=1}^{\infty} \left(\frac{1}{2}\right)^n$?

☐ A. 2 ☐ C. $\frac{1}{2}$
☐ B. 1 ☐ D. The series does not converge.

18) The third term of an infinite geometric series is 4 and the sum of the series is 32. What is the first term if the common ratio is rational?

☐ A. 12 ☐ C. 24
☐ B. 18 ☐ D. 16

19) Which of the following series converges?

☐ A. $\sum_{n=1}^{\infty} \frac{(-1)^n}{n^2}$

☐ B. $\sum_{n=1}^{\infty} \frac{(-1)^n \cdot n}{\sqrt{n}}$

☐ C. $\sum_{n=1}^{\infty} \frac{(-1)^n \cdot n^2}{n+1}$

☐ D. $\sum_{n=1}^{\infty} n$

20) Determine whether the series $\sum_{k=0}^{\infty} \frac{(-1)^{k+1}(k+1)}{3^k}$ converges or diverges.

☐ A. Converges

☐ B. Diverges

☐ C. Cannot be determined

☐ D. Converges conditionally

21) Determine whether the series $\sum_{n=2}^{\infty} \frac{1}{n-1}$ is convergent or divergent using the 'n-th term test'.

☐ A. Convergent to 0

☐ B. Convergent to 1

☐ C. Inconclusive without further tests

☐ D. Convergent to $\frac{1}{2}$

22) Assess the convergence or divergence of the series $\sum_{n=1}^{\infty} n^2$ using the 'n-th term test'.

☐ A. Convergent

☐ B. Divergent

☐ C. Inconclusive based on the n-th term test

☐ D. Depends on the value of the first term

23) Determine whether the series $\sum_{n=1}^{\infty} \frac{1}{n^{\frac{3}{2}}}$ converges or diverges using the Comparison Test.

☐ A. Converges

☐ B. Diverges

☐ C. The Comparison Test is inconclusive

☐ D. Not enough information

24) Use the Ratio Test to determine if the series $\sum_{n=1}^{\infty} \left(\frac{5}{6}\right)^n$ converges or diverges.

☐ A. Converges

☐ B. Diverges

☐ C. The Ratio Test is inconclusive

☐ D. Not enough information

25) How many entries are in the 9-th row of Pascal's Triangle?

☐ A. 9

☐ B. 10

☐ C. 45

☐ D. 11

26) What is the sum of the entries in the 6-th row of Pascal's Triangle?

☐ A. 32 ☐ C. 16

☐ B. 64 ☐ D. 128

27) What is the coefficient of the term x^2y^2 in the expansion of $(x+y)^4$?

☐ A. 4 ☐ C. 8

☐ B. 6 ☐ D. 16

28) If the constant term in the expansion of $(1-2x)^5$ is K, what is the value of K?

☐ A. -1 ☐ C. -32

☐ B. 1 ☐ D. 32

Answer Keys

1) A. 13
2) B. 5
3) B. 17
4) A. 30
5) B. $a_n = a_{n-1} + 5$
6) B. $a_n = 2a_{n-1}$
7) A. 32
8) A. 19
9) A. 1
10) D. 512
11) B. 84
12) C. $\frac{3}{2}n^2 - \frac{1}{2}n$
13) D. 405
14) C. $n = 8$
15) D. 35
16) B. -364
17) B. 1
18) D. 16
19) A. $\sum_{n=1}^{\infty} \frac{(-1)^n}{n^2}$
20) A. Converges
21) C. Inconclusive without further tests
22) B. Divergent
23) A. Converges
24) A. Converges
25) B. 10
26) B. 64
27) B. 6
28) B. 1

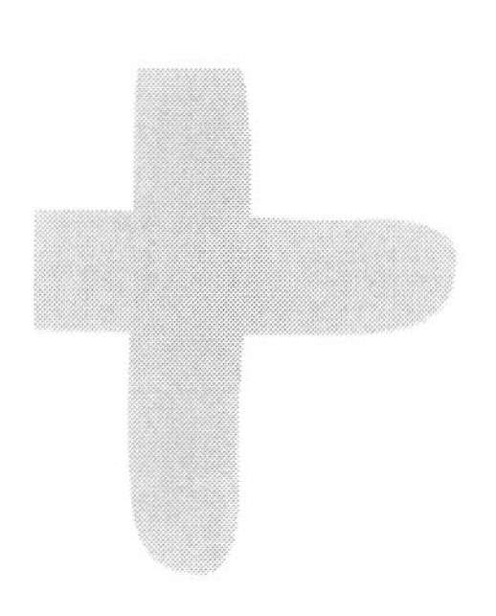

8. Trigonometry

8.1 Degrees, Radians and Angle Conversions

Trigonometry defines angles with a revolving ray around a vertex, often represented by θ. Angles are measured in degrees or radians, both inter-convertible. A degree divides a circle into 360 equal parts, with 1° symbolizing one part. For example, 45° and 90° represent 45 parts and a quarter of a circle, respectively.

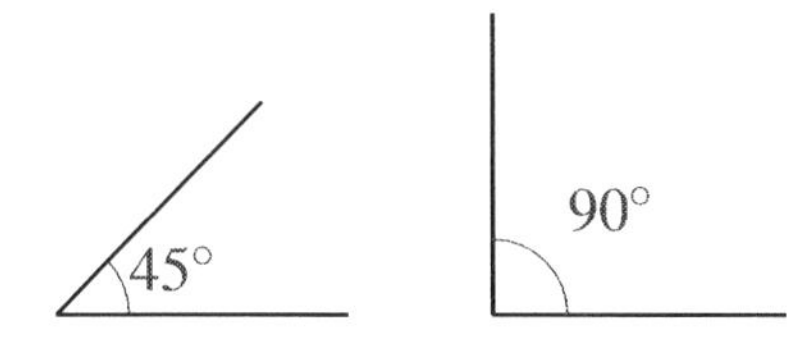

Radians measure angles based on the radius of a circle, where 2π radians equal a full circle ($\pi \approx 3.14159$). $\frac{\pi}{4}$ and $\frac{\pi}{2}$ radians denote $\frac{1}{8}$ and a quarter of a circle, respectively.

Key Point

To shift between these two angle measurement systems, you would use the following conversion formulas:

Convert degrees to radians using the formula: $Radians = Degrees \times \frac{\pi}{180}$.

Convert radians to degrees with: $Degrees = Radians \times \frac{180}{\pi}$.

Example Convert 120° to radians.

Solution: We will use the formula from the key point above:

$$Radians = Degrees \times \frac{\pi}{180}.$$

Substitute 120° into the formula:

$$Radians = 120 \times \frac{\pi}{180} = \frac{120\pi}{180} = \frac{2\pi}{3}.$$

So, 120° is equal to $\frac{2\pi}{3}$ radians.

Convert $\frac{\pi}{3}$ to degrees.

Solution: We will use the formula from the key point above:

$$Degrees = Radians \times \frac{180}{\pi}.$$

Substitute $\frac{\pi}{3}$ into the formula: $Degrees = \frac{\pi}{3} \times \frac{180}{\pi} = 60°$.

8.2 Trigonometric Ratios

Trigonometric ratios, essential in fields like engineering, architecture, and physics, are based on the relationships within a right triangle.

A right triangle features a 90° angle, with the sides named hypotenuse (opposite the right angle), and legs (adjacent and opposite to a given angle θ).

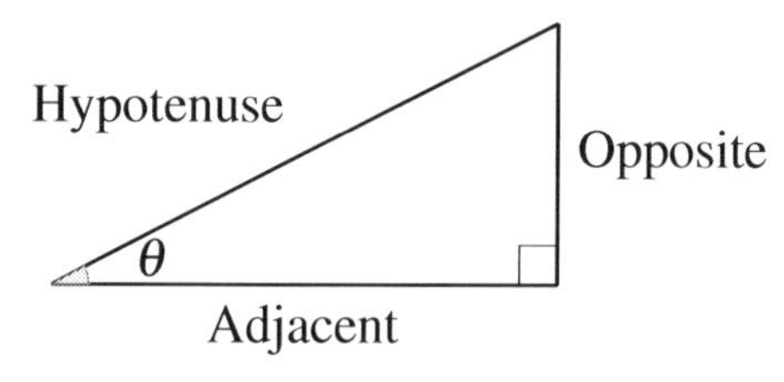

We define sine, cosine, and tangent as follows:

- Sine (sin): $\frac{\text{opposite}}{\text{hypotenuse}}$
- Cosine (cos): $\frac{\text{adjacent}}{\text{hypotenuse}}$
- Tangent (tan): $\frac{\text{opposite}}{\text{adjacent}}$

Special angles include:

θ	0°	30°	45°	60°	90°
sin	0	$\frac{1}{2}$	$\frac{\sqrt{2}}{2}$	$\frac{\sqrt{3}}{2}$	1
cos	1	$\frac{\sqrt{3}}{2}$	$\frac{\sqrt{2}}{2}$	$\frac{1}{2}$	0
tan	0	$\frac{\sqrt{3}}{3}$	1	$\sqrt{3}$	undefined

Key Point

The Pythagorean theorem, stating that in a right triangle, the square of the hypotenuse length is equal to the sum of the squares of the legs' lengths, underpins these ratios.

Example In a right triangle with a right angle at vertex B, the hypotenuse measures 10 cm, the base 8 cm, and the height 6 cm. If $\angle ACB = \theta$, find $\sin(\theta)$, $\cos(\theta)$ and $\tan(\theta)$.

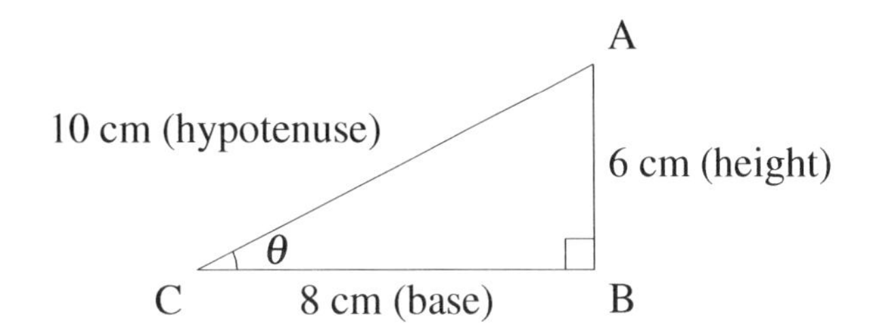

Solution: We employ the formulas $\sin(\theta)$, $\cos(\theta)$ and $\tan(\theta)$. Inserting the given side lengths, we find: $\sin(\theta) = \frac{6}{10}$, $\cos(\theta) = \frac{8}{10}$, $\tan(\theta) = \frac{6}{8}$.

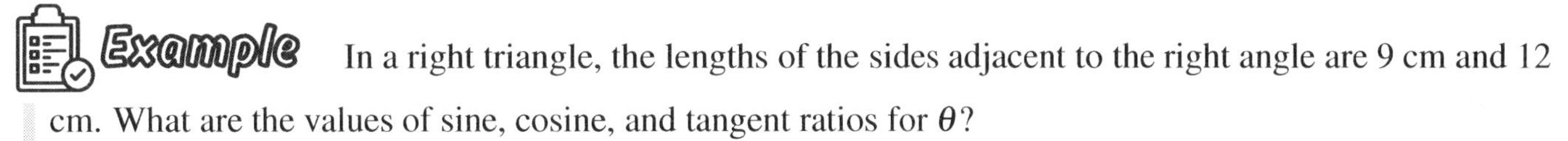

Example In a right triangle, the lengths of the sides adjacent to the right angle are 9 cm and 12 cm. What are the values of sine, cosine, and tangent ratios for θ?

Solution: First, using the Pythagorean theorem, calculate the length of the hypotenuse:

$$\sqrt{9^2 + 12^2} = 15.$$

Using the formulas $\sin(\theta)$, $\cos(\theta)$ and $\tan(\theta)$, we find: $\sin(\theta) = \frac{12}{15}$, $\cos(\theta) = \frac{9}{15}$, $\tan(\theta) = \frac{12}{9}$.

8.3 Coterminal Angles and Reference Angles

Coterminal angles and reference angles play vital roles in trigonometry, particularly in analyzing the unit circle. They facilitate simplifying problems and identifying multiple angle solutions. Angle addition and subtraction are key operations that extend to coterminal and reference angle calculations.

For degrees, angle addition and subtraction are straightforward numerical operations. For instance, given angles $A = 30^\circ$ and $B = 45^\circ$:

- Angle addition: $A + B = 75^\circ$.
- Angle subtraction: $A - B = -15^\circ$ (a negative angle indicates an opposite direction measurement).

In radians, these operations incorporate π, for example, with $C = \frac{\pi}{6}$ and $D = \frac{\pi}{4}$:

- Angle addition: $C + D = \frac{5\pi}{12}$.
- Angle subtraction: $C - D = -\frac{\pi}{12}$.

Coterminal angles are angles with the same terminal side in standard position. Adding or subtracting 360° (or 2π radians) from any angle yields a coterminal angle.

Reference angles are acute angles ($\leq 90^\circ$) formed with the x-axis by an angle's terminal side. To find a reference angle, consider:

- **First Quadrant** (0° to 90°): The reference angle is the angle itself.
- **Second Quadrant** (90° to 180°): Reference angle = 180° – angle.
- **Third Quadrant** (180° to 270°): Reference angle = angle – 180°.
- **Fourth Quadrant** (270° to 360°): Reference angle = 360° – angle.

For negative angles or those beyond 360°, adjust the angle to the 0° to 360° range before calculating the reference angle, e.g., for -1000°, add 360° multiple times to find a coterminal angle within the desired range.

Key Point

An angle with a negative sign indicates that the angle is measured in the opposite direction.

Key Point

To find an angle coterminal to a given angle, you can add or subtract 360 degrees (or 2π radians for radian measurements) from the given angle.

Find an angle coterminal to 135° and -45°.

Solution: For 135°, we can subtract 360° to get a coterminal angle: $135^\circ - 360^\circ = -225^\circ$. For -45°, we can add 360° to get a coterminal angle:

$$-45^\circ + 360^\circ = 315^\circ.$$

Find the reference angle of 34°, 120°, 213°, 333°

Solution:

- For 34°, its reference angle is also 34°.
- For 120°, the terminal side lies in the second quadrant. So we subtract the given angle from 180° to get the positive acute reference angle that is 60°
- For 213°, its reference angle is $213^\circ - 180^\circ = 33^\circ$.
- For 333°, its reference angle is $360^\circ - 333^\circ = 27^\circ$.

8.4 Angles of Rotation and Unit Circle

An angle of rotation, denoted by θ, is formed by a ray's rotation from its initial position to a terminal position, with the initial side along the positive x-axis and the vertex at the origin. Positive rotations are counterclockwise, while negative rotations are clockwise.

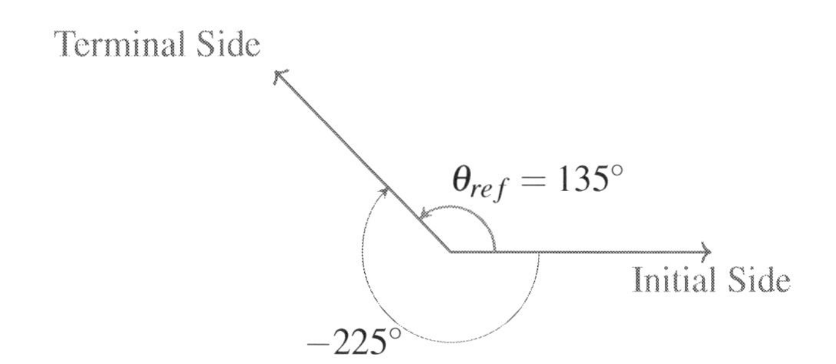

Key Point

An angle is in standard position if its initial side lies along the positive x-axis and its vertex is at the origin. Counterclockwise rotations from the initial side are positive, while clockwise rotations are negative.

The unit circle, with its center at the origin and a radius of one unit, is described by $x^2 + y^2 = 1$. It plays a crucial role in understanding sine and cosine:

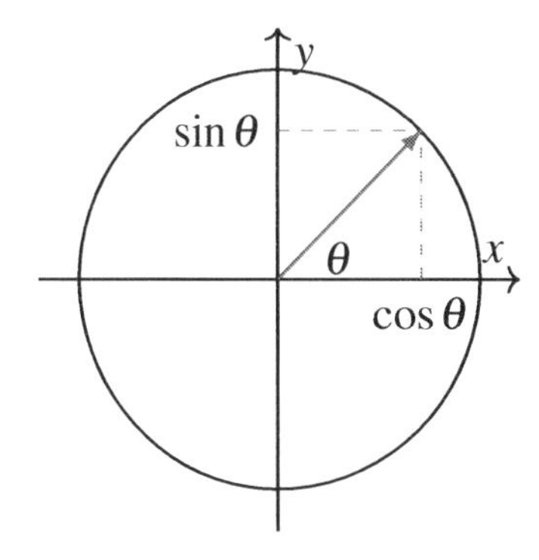

Key Point

The unit circle is crucial for grasping sine and cosine ratios. The y coordinate of a point on the unit circle represents $\sin(\theta)$, while the x coordinate represents $\cos(\theta)$.

The unit circle divides into four quadrants:

1. *Quadrant 1*: $0°$ to $90°$, where both x and y values are positive.
2. *Quadrant 2*: $90°$ to $180°$, where x values are negative, and y values are positive.
3. *Quadrant 3*: $180°$ to $270°$, where both x and y values are negative.
4. *Quadrant 4*: $270°$ to $360°$, where x values are positive, and y values are negative.

This quadrant system is essential for understanding the signs of trigonometric functions across different angles.

Example Find the sine and cosine of the angle of rotation that forms with the positive x-axis an angle of $\frac{\pi}{3}$ in counterclockwise direction.

Solution: Using the unit circle, P is on the circle at approximately $60°$ counterclockwise from the positive x-axis. Thus, P coordinates are $\left(\frac{1}{2}, \frac{\sqrt{3}}{2}\right)$. Hence, the sine and the cosine of the angle of rotation are $\frac{\sqrt{3}}{2}$, and $\frac{1}{2}$ respectively.

Example Let $P = (\frac{\sqrt{3}}{2}, -\frac{1}{2})$ be a point on the unit circle corresponding to an angle θ in standard position. Find $\sin(\theta)$ and $\cos(\theta)$.

Solution: A point $P(x,y)$ on the unit circle corresponds to $\sin(\theta) = y$ and $\cos(\theta) = x$. So, for $P = (\frac{\sqrt{3}}{2}, -\frac{1}{2})$:

$$\sin(\theta) = y - \text{coordinate of P} = -\frac{1}{2},$$

$$\cos(\theta) = x - \text{coordinate of P} = \frac{\sqrt{3}}{2}.$$

8.5 Arc Length and Sector Area

A sector is a part of a circle defined by a central angle θ (in degrees) and the two radii extending to the circle's edge. The area of this sector relates to the circle's total area, proportionate to the angle θ.

Given a circle's radius r and a sector's angle θ (in degrees), the sector's area A is a fraction of the circle's total area πr^2, calculated as:

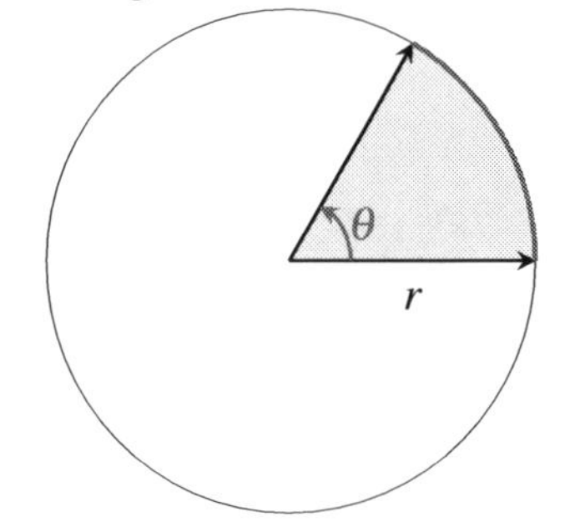

A circle with a sector marked.

Key Point

The area of a sector is calculated by: $A_{sector} = \frac{\theta}{360} \times \pi r^2$, where θ is the sector's angle in degrees, and r is the circle's radius.

Arc length refers to the distance along the circle's edge that the sector spans. With a circle's radius r and the sector's angle θ, the arc's length l is given by:

Key Point

The arc length is calculated by: $l_{arc} = \frac{\theta}{180} \times \pi r$, where θ is the angle in degrees, and r is the radius of the circle. This formula determines the portion of the circle's circumference that the arc encompasses.

 Consider a sector with a radius $r = 10$ cm, and central angle $\theta = 90$ degrees. Find the

area of the sector and the length of the arc.

Solution: Use the formula for the area of the sector:

$$A_{sector} = \frac{\theta}{360} \times \pi r^2 = \frac{90}{360} \times \pi \times 10^2 = \frac{\pi \times 100}{4} = 25\pi.$$

Hence, the area of the sector is 25π cm^2. Also, using the formula for the Arc length:

$$l_{arc} = \frac{\theta}{180} \times \pi r = \frac{90}{180} \times \pi \times 10 = \frac{\pi \times 10}{2} = 5\pi.$$

Therefore, the length of the arc is 5π cm.

8.6 Cosecant, Secant, and Cotangent

Understanding sine and cosine is crucial before exploring their reciprocal functions, which are pivotal in algebra and trigonometry for simplifying expressions. Reciprocal functions are defined as the inverse of the basic trigonometric functions $\sin\theta$, $\cos\theta$, and $\tan\theta$.

Key Point

The secant function ($\sec\theta$), reciprocal to $\cos\theta$, is defined as: $\sec\theta = \frac{1}{\cos\theta}$. It exists for all θ except where $\cos\theta = 0$, which occurs at $\theta = \frac{(2n+1)\pi}{2}$ for any integer n. The range of $\sec\theta$ is all real numbers ≥ 1 or ≤ -1.

Key Point

The cosecant function ($\csc\theta$), reciprocal to $\sin\theta$, is: $\csc\theta = \frac{1}{\sin\theta}$. Undefined for $\sin\theta = 0$ at $\theta = n\pi$, where n is any integer. Its range includes all real numbers ≥ 1 or ≤ -1.

Key Point

The cotangent function ($\cot\theta$), the reciprocal of $\tan\theta$, is given by: $\cot\theta = \frac{1}{\tan\theta}$. Undefined at $\theta = n\pi$ for any integer n, its values span all real numbers.

Key Point

Reciprocal functions (secant, cosecant, and cotangent) are undefined where their base functions (cosine, sine, and tangent, respectively) are zero.

Example Given that $\sin\theta = -\frac{3}{5}$ (with θ in the third quadrant) and $\cos\phi = \frac{4}{5}$ (with ϕ in the first quadrant), find $\csc\theta$ and $\sec\phi$.

Solution: Since the cosecant function is the reciprocal of the sine function, $\csc\theta = -\frac{5}{3}$. Likewise, as the secant function is the reciprocal of the cosine function, $\sec\phi = \frac{5}{4}$.

8.7 Domain and Range of Trigonometric Functions

The domain encompasses all possible inputs a function can accept, while the range includes all potential outputs.

For *sine* and *cosine* functions, the domain is all real numbers, $\mathbb{R}$, and their range is limited between -1 and 1, inclusive:

$$\text{Domain}(sine) = \text{Domain}(cosine) = \mathbb{R},$$

$$\text{Range}(sine) = \text{Range}(cosine) = [-1, 1].$$

Tangent and *cotangent* are not defined at $\frac{\pi}{2} + n\pi$ and $n\pi$, respectively, where n is any integer. Thus, their domains exclude these values, but their ranges include all real numbers:

$$\text{Domain}(tangent) = \mathbb{R} - \left\{\frac{\pi}{2} + n\pi \mid n \in \mathbb{Z}\right\},$$

$$\text{Domain}(cotangent) = \mathbb{R} - \{n\pi \mid n \in \mathbb{Z}\},$$

$$\text{Range}(tangent) = \text{Range}(cotangent) = \mathbb{R}.$$

Secant and *cosecant*, being reciprocals of cosine and sine, respectively, exclude the same values from their domains as tangent and cotangent. However, their ranges do not include the interval between -1 and 1:

$$\text{Domain}(secant) = \mathbb{R} - \left\{\frac{\pi}{2} + n\pi \mid n \in \mathbb{Z}\right\},$$

$$\text{Domain}(cosecant) = \mathbb{R} - \{n\pi \mid n \in \mathbb{Z}\},$$

$$\text{Range}(secant) = \text{Range}(cosecant) = (-\infty, -1] \cup [1, \infty).$$

Secant and cosecant functions' ranges include all real numbers except those between -1 and 1.

Example Find the range of the functions $f(x) = 2\sin(x) + 3$ and $g(x) = 5\tan(x)$.

Solution: Since the range of the sine function is $[-1, 1]$, for $f(x) = 2\sin(x) + 3$, the function will simply stretch vertically and shift up. Thus, the range of $f(x)$ is:

$$2[-1,1]+3=[-2,2]+3=[1,5].$$

The range of the tangent function is all real numbers. Thus, irrespective of the constant multiplier, for $g(x)=5\tan(x)$ the range remains all real numbers.

Find the domain of the function $f(x)=2\cos(x)+3$.

Solution: The domain of the function $f(x)=2\cos(x)+3$ is $(-\infty,+\infty)$

8.8 Arcsine, Arccosine, and Arctangent

Arcsine, arccosine, and arctangent are inverse trigonometric functions. These functions reverse the action of their respective trigonometric counterparts: sine, cosine, and tangent. In mathematics, they are used to determine the angle of a right-angled triangle when the lengths of two sides are known.

Key Point

The **arcsine** function ($\sin^{-1}(x)$ or $\arcsin(x)$) determines the angle whose sine is x.

$$\text{Domain}:[-1,1] \quad \text{and} \quad \text{Range}:\left[-\frac{\pi}{2},\frac{\pi}{2}\right].$$

Key Point

The **arccosine** function ($\cos^{-1}(x)$ or $\arccos(x)$) determines the angle whose cosine is x.

$$\text{Domain}:[-1,1] \quad \text{and} \quad \text{Range}:[0,\pi].$$

Key Point

The **arctangent** function ($\tan^{-1}(x)$ or $\arctan(x)$) determines the angle whose tangent is x.

$$\text{Domain}:(-\infty,\infty) \quad \text{and} \quad \text{Range}:\left(-\frac{\pi}{2},\frac{\pi}{2}\right).$$

Key Point

For trigonometric functions and their inverses, remember that:

$$\sin(\sin^{-1}(x))=x, \quad \cos(\cos^{-1}(x))=x, \quad \tan(\tan^{-1}(x))=x.$$

 Find the angle θ if $\cos(\theta) = -0.5$ and θ is in the second quadrant.

Solution: Since the cosine of the angle is negative and the angle is in the second quadrant, we first calculate the reference angle (in the first quadrant) using, $\cos^{-1}(0.5)$ which is $\frac{\pi}{3}$ (or 60 degrees). Now, to get the angle in the second quadrant, we subtract from π which gives the result as $\theta = \pi - \frac{\pi}{3} = \frac{2\pi}{3}$.

 Find $\tan(\tan^{-1}(3))$.

Solution: Since $\tan(\tan^{-1}(x)) = x$ for all x in the domain of $\tan^{-1}(x)$, then: $\tan(\tan^{-1}(3)) = 3$.

8.9 Applications of Inverse Trigonometric Function

Inverse trigonometric functions find extensive applications in various disciplines, from geometry to physics, demonstrating their importance across numerous fields.

The following sections will delve into the applications in more detail.

Geometry: In scenarios where the side lengths of a triangle are known and an angle needs to be determined, inverse trigonometric functions are invaluable. They are particularly useful in solving triangles and finding unknown angles.

Trigonometric Identities: These functions help derive and establish various trigonometric identities, facilitating mathematical calculations and problem-solving.

Calculus: In calculus, inverse trigonometric functions are fundamental in solving derivatives and integrals involving trigonometric functions, enabling effective problem-solving in integration and differentiation.

Engineering: Engineers rely on inverse trigonometric functions for analyzing waveforms, oscillations, and for structural design in various engineering fields, including electronics, mechanical, and civil engineering.

Physics: Physics extensively utilizes inverse trigonometric functions in wave mechanics, kinematics, and dynamics, especially in studying oscillatory systems.

Navigation: These functions are essential in navigation and cartography for determining distances, bearings, and angles.

Computer Graphics: In computer science, inverse trigonometric functions are used for transforming and rendering 2D and 3D models, image processing, and computational geometry.

Key Point

Inverse trigonometric functions are used to find angles in various applications, including geometry, calculus, engineering, physics, navigation, and computer graphics.

Example Find the angle θ in a right-angled triangle with opposite side length 3 units and hypotenuse length 5 units.

Solution: We know that the sine of an angle in a right-angled triangle is the ratio of the length of the opposite side to the hypotenuse. Hence, $\sin(\theta) = \frac{3}{5}$. To find the angle θ, we use the arcsine function, which is the inverse of the sine function. Equating $\sin(\theta) = \frac{3}{5}$ to $\theta = \arcsin\left(\frac{3}{5}\right) \approx 36.87^\circ$.

8.10 Fundamental Trigonometric Identities

Trigonometric identities are critical in trigonometry, derived directly from the definitions of trigonometric functions.

Key Point

Trigonometric identities include:

1. $\tan\theta = \frac{\sin\theta}{\cos\theta}$,
2. $\cot\theta = \frac{\cos\theta}{\sin\theta}$,
3. $\csc\theta = \frac{1}{\sin\theta}$,
4. $\sec\theta = \frac{1}{\cos\theta}$,
5. $\cot\theta = \frac{1}{\tan\theta}$,
6. $\cos^2\theta + \sin^2\theta = 1$, (Pythagorean Identity)
7. $\sin^2\theta = 1 - \cos^2\theta$,
8. $\cos^2\theta = 1 - \sin^2\theta$,
9. $\tan^2\theta + 1 = \sec^2\theta$,
10. $1 + \cot^2\theta = \csc^2\theta$.

Verify that $\sin^2\frac{\pi}{4} + \cos^2\frac{\pi}{4} = 1$.

Solution: The value of $\cos\frac{\pi}{4}$ and $\sin\frac{\pi}{4}$ is $\frac{\sqrt{2}}{2}$. So, we get:

$$\sin^2\frac{\pi}{4} + \cos^2\frac{\pi}{4} = \left(\frac{\sqrt{2}}{2}\right)^2 + \left(\frac{\sqrt{2}}{2}\right)^2 = \frac{1}{2} + \frac{1}{2} = 1.$$

So, we have verified the Pythagorean identity as promised.

Given that $\sin\theta = -\frac{3}{5}$ (with θ in the third quadrant) find $\cos\theta$.

Solution: We can find the cosine using the Pythagorean identity $\cos^2\theta = 1 - \sin^2\theta$, yielding $\cos\theta = -\frac{4}{5}$.

Example Simplify the following expression using fundamental identities:

$$\tan\theta(\sec^2\theta - 1).$$

Solution: We start by replacing $\sec^2\theta$ with $\tan^2\theta + 1$ from the identities list:

$$\tan\theta(\sec^2\theta - 1) = \tan\theta(\tan^2\theta + 1 - 1) = \tan\theta\tan^2\theta = \tan^3\theta.$$

8.11 Co-Function, Even-Odd, and Periodicity Identities

Co-Function identities describe the relationship between sine, cosine, and other trigonometric functions. Even-Odd identities describe the behavior of trigonometric functions with respect to the sign of the input. Periodicity identities describe the behavior of trigonometric functions when their input is increased by a specific value called the period.

Key Point

Co-function identities arise from complementary angles, totaling 90°. These identities include:

1. $\sin(90° - x) = \cos(x)$,
2. $\cos(90° - x) = \sin(x)$,
3. $\tan(90° - x) = \cot(x)$,
4. $\cot(90° - x) = \tan(x)$,
5. $\sec(90° - x) = \csc(x)$,
6. $\csc(90° - x) = \sec(x)$.

Key Point

Even-odd identities distinguish trigonometric functions' symmetry. They are:

1. $\sin(-x) = -\sin(x)$,
2. $\cos(-x) = \cos(x)$,
3. $\tan(-x) = -\tan(x)$,
4. $\cot(-x) = -\cot(x)$,
5. $\sec(-x) = \sec(x)$,
6. $\csc(-x) = -\csc(x)$.

A function is even if $f(x) = f(-x)$, and it is odd if $f(x) = -f(x)$.

Key Point

Sine, tangent, cotangent, and cosecant are odd functions. Cosine and secant are even functions.

A function is deemed 'periodic' if $f(x+P) = f(x)$ for all x, where P is referred to as the period.

Key Point

Periodicity identities indicate how trigonometric functions repeat over intervals. For any integer n:

1. $\sin(x+2\pi n) = \sin(x)$,
2. $\cos(x+2\pi n) = \cos(x)$,
3. $\tan(x+\pi n) = \tan(x)$,
4. $\cot(x+\pi n) = \cot(x)$,
5. $\sec(x+2\pi n) = \sec(x)$,
6. $\csc(x+2\pi n) = \csc(x)$.

Example Find the value of $\tan 120^\circ$ using Co-Function identities.

Solution: By using the identities $\cot(90^\circ - x) = \tan(x)$ and $\cot(-x) = -\cot(x)$, the value of $\tan 120^\circ$ can be determined.

$$\tan 120^\circ = \cot(90^\circ - 120^\circ) = \cot(-30^\circ) = -\cot 30^\circ = -\sqrt{3}.$$

Example Find the exact value of $\cos\left(\frac{15\pi}{6}\right)$ using the periodicity identity of the cosine function.

Solution: Since $\frac{15\pi}{6} = 2\pi + \frac{\pi}{2}$, we can write:

$$\cos\left(\frac{15\pi}{6}\right) = \cos\left(2\pi + \frac{\pi}{2}\right).$$

Using the periodicity identity $\cos(x+2\pi) = \cos(x)$, we get:

$$\cos\left(2\pi + \frac{\pi}{2}\right) = \cos\left(\frac{\pi}{2}\right) = 0.$$

8.12 Double Angle and Half-Angle Formulas

Double-Angle and Half-Angle Formulas significantly simplify the process of solving trigonometric equations, especially when angles are doubled or halved.

Key Point

The double-angle formulas for an angle A are:

1. $\sin(2A) = 2\sin(A)\cos(A)$,
2. $\cos(2A) = \cos^2(A) - \sin^2(A)$,
3. $\tan(2A) = \frac{2\tan(A)}{1-\tan^2(A)}$.

Key Point

The half-angle identities include:

1. $\sin\left(\frac{A}{2}\right) = \pm\sqrt{\frac{1-\cos(A)}{2}}$,
2. $\cos\left(\frac{A}{2}\right) = \pm\sqrt{\frac{1+\cos(A)}{2}}$,
3. $\tan\left(\frac{A}{2}\right) = \pm\sqrt{\frac{1-\cos(A)}{1+\cos(A)}} = \pm\frac{\sin(A)}{1+\cos(A)} = \pm\frac{1-\cos(A)}{\sin(A)}$.

The sign ($\pm$) is determined by the quadrant in which the half-angle resides, reflecting the sign behavior of trigonometric functions across different quadrants.

Example Evaluate $\sin(2A)$ if $\sin(A) = \frac{3}{5}$ and A lies in the second quadrant.

Solution: Since $\sin(A)$ is given, we can use the double-angle sine identity:

$$\sin(2A) = 2\sin(A)\cos(A).$$

First, we need to calculate $\cos(A)$. We know that $\cos^2(A) = 1 - \sin^2(A)$. Substituting the value of given $\sin(A)$ into the above equation, we get $\cos(A) = -\frac{4}{5}$. The negative sign indicates that cosine is negative in the second quadrant. Substituting these values in $\sin(2A) = 2\sin(A)\cos(A)$, we find:

$$\sin(2A) = 2\left(\frac{3}{5}\right)\left(-\frac{4}{5}\right) = -\frac{24}{25}.$$

Example Calculate $\tan\left(\frac{A}{2}\right)$ if $\cos(A) = \frac{1}{2}$ and A lies in the fourth quadrant.

Solution: To evaluate this, we use the half-angle identity for tangent:

$$\tan\left(\frac{A}{2}\right) = \pm\sqrt{\frac{1-\cos(A)}{1+\cos(A)}}.$$

Substitute the value $\cos(A) = \frac{1}{2}$ into the equation, which gives:

$$\tan\left(\frac{A}{2}\right) = \pm\sqrt{\frac{1-\frac{1}{2}}{1+\frac{1}{2}}} = \pm\sqrt{\frac{1}{3}}.$$

Since A is in the fourth quadrant where $\tan(\text{angle})$ is negative, the answer is $-\sqrt{\frac{1}{3}}$.

8.13 Sum and Difference Formulas

The sum and difference formulas allow us to evaluate the value of trigonometric functions at angles that can be expressed as the sum or difference of certain angles-namely, 0°, 30°, 45°, 60°, 90°, and 180°.

Key Point

The sum and difference formulas for trigonometric functions are:

1. $\sin(A+B) = \sin A \cos B + \cos A \sin B$,
2. $\sin(A-B) = \sin A \cos B - \cos A \sin B$,
3. $\cos(A+B) = \cos A \cos B - \sin A \sin B$,
4. $\cos(A-B) = \cos A \cos B + \sin A \sin B$,
5. $\tan(A+B) = \frac{\tan A + \tan B}{1 - \tan A \tan B}$,
6. $\tan(A-B) = \frac{\tan A - \tan B}{1 + \tan A \tan B}$.

These formulas facilitate the manipulation and simplification of expressions involving the sum or difference of angles.

Find the value of $\sin(105^\circ)$.

Solution: To solve this problem, we can apply the sum formula for the sine function, since we can write $105 = 60 + 45$:

$$\sin(60^\circ + 45^\circ) = \sin(60^\circ)\cos(45^\circ) + \cos(60^\circ)\sin(45^\circ).$$

By using the known values of sine and cosine at these standard angles, we have:

$$\sin(60^\circ + 45^\circ) = \left(\frac{\sqrt{3}}{2}\right)\left(\frac{\sqrt{2}}{2}\right) + \left(\frac{1}{2}\right)\left(\frac{\sqrt{2}}{2}\right).$$

Finally, we simplify to get:

$$\sin(60^\circ + 45^\circ) = \frac{\sqrt{6}}{4} + \frac{\sqrt{2}}{4} = \frac{\sqrt{6}+\sqrt{2}}{4}.$$

So, the value of $\sin(105^\circ)$ is $\frac{\sqrt{6}+\sqrt{2}}{4}$.

8.14 Product-to-Sum and Sum-to-Product Formulas

The Product-to-Sum formulas expand on the Sum and Difference Formulas, transforming products of sines and cosines into sums or differences to simplify expressions.

Key Point

Product-to-Sum formulas transition from products to sums or differences:

1. $\sin(A)\sin(B) = \frac{1}{2}[\cos(A-B) - \cos(A+B)]$,
2. $\cos(A)\cos(B) = \frac{1}{2}[\cos(A-B) + \cos(A+B)]$,
3. $\sin(A)\cos(B) = \frac{1}{2}[\sin(A+B) + \sin(A-B)]$.

Key Point

Sum-to-Product formulas consolidate sums or differences into products:

1. $\sin(A) + \sin(B) = 2\sin(\frac{A+B}{2})\cos(\frac{A-B}{2})$,
2. $\sin(A) - \sin(B) = 2\cos(\frac{A+B}{2})\sin(\frac{A-B}{2})$,
3. $\cos(A) + \cos(B) = 2\cos(\frac{A+B}{2})\cos(\frac{A-B}{2})$,
4. $\cos(A) - \cos(B) = -2\sin(\frac{A+B}{2})\sin(\frac{A-B}{2})$.

Example Verify the identity $\sin(A)\sin(B) = \frac{1}{2}[\cos(A-B) - \cos(A+B)]$ for $A = 60^\circ$ and $B = 30^\circ$.

Solution: First, we calculate the sine values:

$$\sin(60^\circ) = \frac{\sqrt{3}}{2}, \quad \sin(30^\circ) = \frac{1}{2}.$$

The product of the sines is:

$$\sin(60^\circ)\sin(30^\circ) = \frac{\sqrt{3}}{2} \cdot \frac{1}{2} = \frac{\sqrt{3}}{4}.$$

Next, we find the cosine values for the difference and sum of A and B:

$$\cos(60^\circ - 30^\circ) = \cos(30^\circ), \quad \cos(60^\circ + 30^\circ) = \cos(90^\circ).$$

These values are known to be:

$$\cos(30^\circ) = \frac{\sqrt{3}}{2}, \quad \cos(90^\circ) = 0.$$

We then compute the right-hand side of the identity:

$$\frac{1}{2}[\cos(30^\circ) - \cos(90^\circ)] = \frac{1}{2}\left[\frac{\sqrt{3}}{2} - 0\right] = \frac{\sqrt{3}}{4}.$$

Comparing both sides, we find:

$$\sin(60^\circ)\sin(30^\circ) = \frac{\sqrt{3}}{4} = \frac{1}{2}[\cos(30^\circ) - \cos(90^\circ)],$$

confirming that the identity holds true for $A = 60^\circ$ and $B = 30^\circ$.

8.15 Practices

1) What is the radian measure of a 30° angle?

☐ A. $\frac{\pi}{6}$
☐ B. $\frac{\pi}{4}$
☐ C. $\frac{\pi}{3}$
☐ D. $\frac{\pi}{2}$

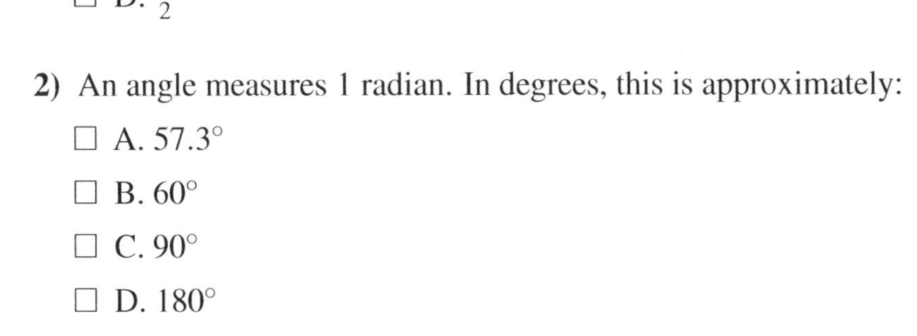

2) An angle measures 1 radian. In degrees, this is approximately:

☐ A. 57.3°
☐ B. 60°
☐ C. 90°
☐ D. 180°

3) Given the lengths of a right triangle's legs are 7 cm and 24 cm, which of the following options is correct for the angle θ opposite to the 7 cm side?

☐ A. $\cos(\theta) = \frac{7}{25}$
☐ B. $\sin(\theta) = \frac{7}{25}$
☐ C. $\tan(\theta) = \frac{17}{24}$
☐ D. $\sin(\theta) = \frac{4}{25}$

4) Which of the following is the value of $\cos(45^\circ)$?

☐ A. 1
☐ B. 0
☐ C. $\frac{\sqrt{2}}{2}$
☐ D. $\frac{\sqrt{3}}{2}$

5) Which of the following angles is NOT coterminal with 165°?

☐ A. 525°
☐ B. -195°
☐ C. 705°

☐ D. -555°

6) An angle of $\frac{7\pi}{6}$ radians is coterminal with which of the following angles?

☐ A. $\frac{19\pi}{6}$ radians

☐ B. $\frac{8\pi}{6}$ radians

☐ C. $\frac{11\pi}{6}$ radians

☐ D. $\frac{5\pi}{6}$ radians

7) When the terminal side of an angle in standard position rests in Quadrant II, which of the following is true for the trigonometric ratios of θ?

☐ A. $\sin(\theta) < 0$ and $\cos(\theta) < 0$

☐ B. $\sin(\theta) > 0$ and $\cos(\theta) < 0$

☐ C. $\sin(\theta) < 0$ and $\cos(\theta) > 0$

☐ D. $\sin(\theta) > 0$ and $\cos(\theta) > 0$

8) What is the value of $\tan(\theta)$ if $\theta = 270^\circ$?

☐ A. 0

☐ B. 1

☐ C. Undefined

☐ D. -1

9) A circle has a radius of 6 cm. What is the area of a sector formed by a central angle of 30 degrees?

☐ A. $3\pi\,\text{cm}^2$

☐ B. $6\pi\,\text{cm}^2$

☐ C. $9\pi\,\text{cm}^2$

☐ D. $13\pi\,\text{cm}^2$

10) Which length is the arc length of a circle with a radius of 12 cm associated with a 45° central angle?

☐ A. 6π cm

☐ B. 7π cm

☐ C. 3π cm

☐ D. 12π cm

11) If $\tan\theta = -2$, where θ is in the second quadrant, which of the following is equal to $\cot\theta$?

☐ A. 2

☐ B. -2

☐ C. -0.5

☐ D. 0.5

12) If $\cos\theta = \frac{2}{3}$, and θ is in the fourth quadrant, which of the following represents $\sec\theta$?

☐ A. 1

☐ B. $\frac{2}{3}$

☐ C. $-\frac{3}{2}$

☐ D. $\frac{3}{2}$

13) Which of the following represents the domain of the secant function, $\sec(x)$?

☐ A. $\mathbb{R}$

☐ B. $\mathbb{R} - \{n\pi \mid n \in \mathbb{Z}\}$

☐ C. $\mathbb{R} - \{\frac{\pi}{2} + n\pi \mid n \in \mathbb{Z}\}$

☐ D. $[-1, 1]$

14) For $f(x) = \frac{1}{3}\csc(x - \pi) + 2$, what is the range of $f(x)$?

☐ A. $\mathbb{R}$

☐ B. $(-\infty, \frac{5}{3}] \cup [\frac{7}{3}, \infty)$

☐ C. $(-\infty, 1] \cup [3, \infty)$

☐ D. $\mathbb{R} - \{2, \frac{1}{3}\}$

15) If $\sin(\alpha) = 0.6$ and $-\frac{\pi}{2} \leq \alpha \leq \frac{\pi}{2}$, what is α?

☐ A. $\sin^{-1}(0.1)$

☐ B. $\sin^{-1}(0.6)$

☐ C. $\cos^{-1}(0.6)$

☐ D. $\tan^{-1}(0.6)$

16) If $\sin(\theta) = -0.5$ and $\pi < \theta < \frac{3\pi}{2}$, what is the value of θ?

☐ A. $\sin^{-1}(0.5)$

☐ B. $\pi - \sin^{-1}(0.5)$

☐ C. $1 + \sin^{-1}(0.5)$

☐ D. $\pi + \sin^{-1}(0.5)$

17) In a right-angled triangle, if the adjacent side to the angle θ is 4 units and the hypotenuse is 8 units, what is the value of θ?

☐ A. $\arccos\left(\frac{1}{2}\right)$

☐ B. $\arccos\left(\frac{3}{8}\right)$

☐ C. $\arcsin\left(\frac{4}{8}\right)$

☐ D. $\arctan\left(\frac{4}{8}\right)$

18) Which of the following expressions is equivalent to $\sin^{-1}\left(\frac{\sqrt{3}}{2}\right)$?

☐ A. $\frac{\pi}{6}$

☐ B. $\frac{\pi}{3}$

☐ C. $\frac{\pi}{4}$

☐ D. $\frac{\pi}{2}$

19) If $\tan\theta = \frac{3}{4}$ and $0 < \theta < \frac{\pi}{2}$, what is the value of $\csc\theta$?

☐ A. $\frac{4}{3}$

☐ B. $\frac{4}{5}$

☐ C. $\frac{5}{3}$

☐ D. $\frac{5}{4}$

20) Simplify the expression $\frac{\sec^2\theta - \tan^2\theta}{\cot\theta}$ given that $\cot\theta \neq 0$.

☐ A. $\sin\theta$

☐ B. $\cos\theta$

☐ C. $\tan\theta$

☐ D. $\csc\theta$

21) Using the co-function identities, which of the following is equal to $\cos 60^\circ$?

☐ A. $\sin 30^\circ$

☐ B. $\sin 60^\circ$

☐ C. $\tan 30^\circ$

☐ D. $\cot 30^\circ$

22) What is $\csc(-\frac{\pi}{3})$? Select the correct answer using the even-odd identities.

☐ A. $\csc(\pi)$

☐ B. $-\csc(\frac{\pi}{3})$

☐ C. $-\csc(\pi)$

☐ D. $\csc(\frac{\pi}{3})$

23) If $\cos(2\theta) = \frac{1}{3}$ and θ is in the first quadrant then what is $\sin(\theta)\cos(\theta)$?

☐ A. $\frac{1}{6}$

☐ B. $\frac{1}{3}$

☐ C. $-\frac{\sqrt{2}}{3}$

☐ D. $\frac{\sqrt{2}}{3}$

24) Using the half-angle identities, which expression equals $\sin\left(\frac{\beta}{2}\right)$ if $\sin(\beta) = \frac{5}{13}$ and $\cos(\beta) < 0$?

☐ A. $\frac{-2}{\sqrt{26}}$
☐ B. $\frac{2}{\sqrt{13}}$
☐ C. $\frac{5}{\sqrt{26}}$
☐ D. $\frac{3}{\sqrt{13}}$

25) Evaluate $\cos(75°)$ using sum or difference formulas.

☐ A. $\frac{\sqrt{6}-\sqrt{2}}{4}$
☐ B. $\frac{\sqrt{6}+\sqrt{2}}{4}$
☐ C. $\frac{\sqrt{3}+\sqrt{2}}{4}$
☐ D. $\frac{\sqrt{2}-\sqrt{3}}{4}$

26) Which of the following is the value of $\tan(15°)$?

☐ A. $\frac{\sqrt{3}-1}{\sqrt{3}+1}$
☐ B. $\frac{\sqrt{3}-1}{\sqrt{3}-1}$
☐ C. $2-\sqrt{3}$
☐ D. $\sqrt{3}-1$

27) Use the product-to-sum formulas to express $\cos(3x)\cos(5x)$ as a sum.

☐ A. $\frac{1}{2}[\cos(8x)+\cos(2x)]$
☐ B. $\frac{1}{2}[\cos(8x)-\cos(2x)]$
☐ C. $\frac{1}{2}[\sin(8x)+\sin(2x)]$
☐ D. $\sin(8x)\sin(2x)$

28) Use the sum-to-product formulas to rewrite $\sin(x)-\sin(3x)$ as a product.

☐ A. $2\cos(2x)\sin(x)$
☐ B. $-2\cos(2x)\sin(x)$
☐ C. $2\sin(2x)\cos(x)$
☐ D. $-2\sin(2x)\cos(x)$

Answer Keys

1) A. $\frac{\pi}{6}$

2) A. 57.3°

3) B. $\sin(\theta) = \frac{7}{25}$

4) C. $\frac{\sqrt{2}}{2}$

5) C. 705°

6) A. $\frac{19\pi}{6}$ radians

7) B. $\sin(\theta) > 0$ and $\cos(\theta) < 0$

8) C. Undefined

9) A. $3\pi\,\text{cm}^2$

10) C. 3π cm

11) C. -0.5

12) D. $\frac{3}{2}$

13) C. $\mathbb{R} - \left\{\frac{\pi}{2} + n\pi \mid n \in \mathbb{Z}\right\}$

14) B. $(-\infty, \frac{5}{3}] \cup [\frac{7}{3}, \infty)$

15) B. $\sin^{-1}(0.6)$

16) D. $\pi + \sin^{-1}(0.5)$

17) A. $\arccos\left(\frac{1}{2}\right)$

18) B. $\frac{\pi}{3}$

19) C. $\frac{5}{3}$

20) C. $\tan\theta$

21) A. $\sin 30^\circ$

22) B. $-\csc(\frac{\pi}{3})$

23) D. $\frac{\sqrt{2}}{3}$

24) C. $\frac{5}{\sqrt{26}}$

25) A. $\frac{\sqrt{6}-\sqrt{2}}{4}$

26) C. $2-\sqrt{3}$

27) A. $\frac{1}{2}[\cos(8x) + \cos(2x)]$

28) B. $-2\cos(2x)\sin(x)$

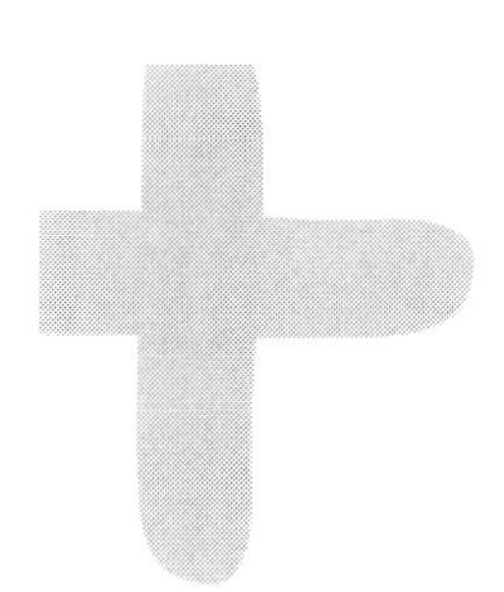

9. Trigonometric Functions and Graphs

9.1 Graph of the Sine Function

The sine function can be represented as an infinite set of ordered pairs of real numbers (x, y), where each ordered pair represents a point on the Cartesian coordinate system. Using the interval $0 \leq x \leq 2\pi$, we can generate a portion of the complete graph of the sine function.

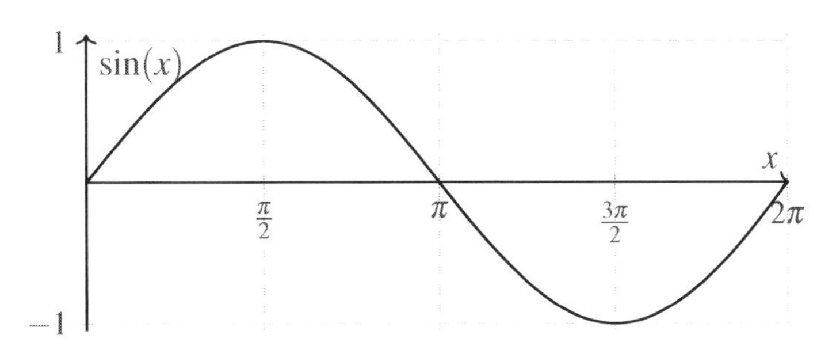

Graph of the sine function in the interval $0 \leq x \leq 2\pi$.

Key Point

For sine function ($y = \sin(x)$) note how y changes as x increases in the interval:

- By increasing x from 0 to $\frac{\pi}{2}$, y increases from 0 to 1.
- By increasing x from $\frac{\pi}{2}$ to π, y decreases from 1 to 0.
- By increasing x from π to $\frac{3\pi}{2}$, y continues to decreases from 0 to -1.
- By increasing x from $\frac{3\pi}{2}$ to 2π, y increases from -1 to 0.

Example In the interval $-2\pi \leq x \leq 0$, for what values of x does $y = \sin(x)$ increase?

Solution: Referring to the graph of the sine function, examine the portion of the graph in the negative x-axis. As depicted in the following figure, $y = \sin(x)$ increases in the intervals $-2\pi \leq x \leq -\frac{3\pi}{2}$, and

$-\frac{\pi}{2} \leq x \leq 0$.

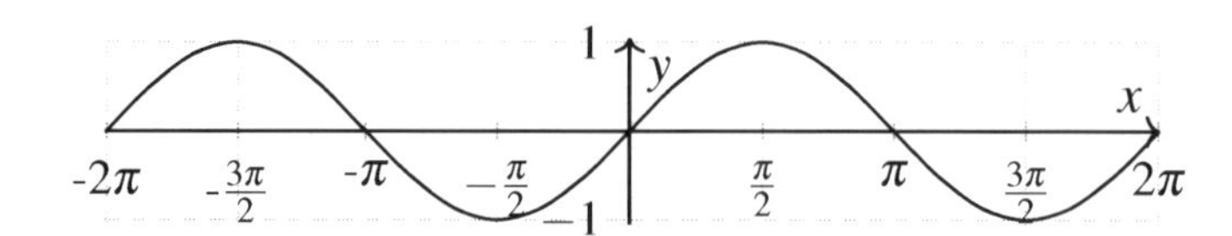

9.2 Graph of the Cosine Function

To graph the cosine function, focus on the interval $0 \leq x \leq 2\pi$, where its behavior repeats every 2π unit along the x-axis. See the next figure for the typical look of a cosine graph on this interval.

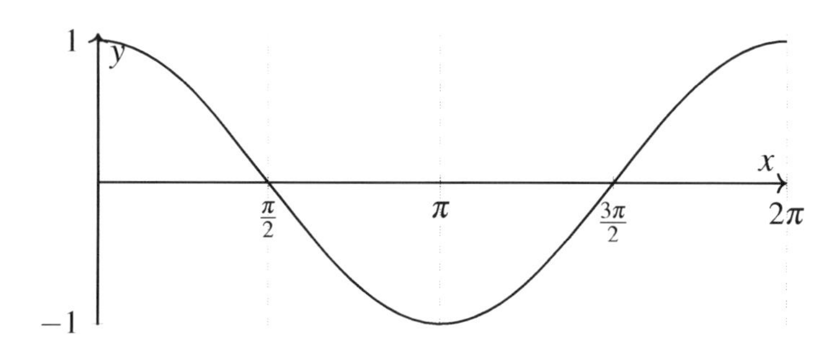

Graph of the cosine function in the interval $0 \leq x \leq 2\pi$.

Key Point

For cosine function ($y = \cos(x)$) note how y changes as x increases in the interval:

- As x increases from 0 to $\frac{\pi}{2}$, y decreases from 1 to 0.
- Increasing x from $\frac{\pi}{2}$ to π leads to y decreasing from 0 to -1.
- When x ascends from π to $\frac{3\pi}{2}$, y climbs from -1 to 0.
- Finally, as x grows from $\frac{3\pi}{2}$ to 2π, y rises from 0 to 1.

Example In the interval $-2\pi \leq x \leq 0$, for what values of x does $y = \cos(x)$ increase?

Solution: Referring to the graph of the cosine function, examine the portion of the graph on the negative x-axis. As depicted in the following figure, $y = \cos(x)$ increases in the interval $-\pi \leq x \leq 0$.

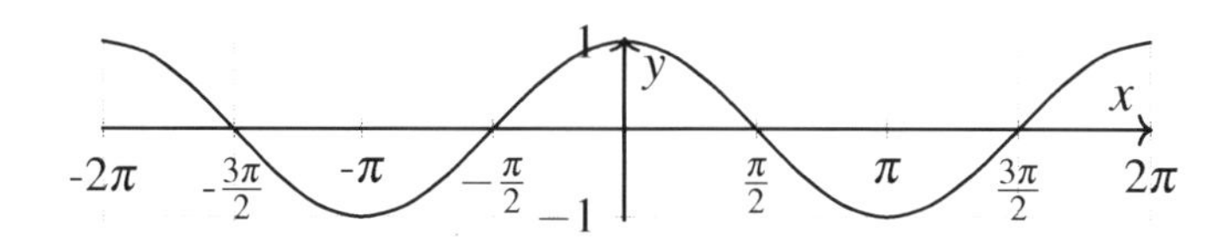

9.3 Amplitude, Period, and Phase Shift

The amplitude, period, and phase shift are key properties of the functions such as $y = a\cos(bx+c)+d$ and $y = a\sin(bx+c)+d$. Understanding these properties is essential for analyzing and graphing trigonometric functions.

Key Point

For $y = a\cos(bx+c)+d$ or $y = a\sin(bx+c)+d$, the amplitude, representing half the distance between the function's maximum and minimum, is $|a|$. This measures the peak height or depth the graph reaches, indicating the function's vertical stretch or compression.

The period, indicating one full cycle's length, shows how long before the function's graph repeats.

Key Point

The period of $y = a\cos(bx+c)+d$ or $y = a\sin(bx+c)+d$ is $|\frac{2\pi}{b}|$.

Key Point

The phase shift of a function $y = a\cos(bx+c)+d$ or $y = a\sin(bx+c)+d$ is $-\frac{c}{b}$. This corresponds to a horizontal shift or displacement of the function on the graph.

Example Determine the amplitude, the period, and the phase shift of $y = -9\cos(8x+\pi)-8$.

Solution: Here, $a = -9$, $b = 8$ and $c = \pi$. The minus sign on a refers to the reflection in the x-axis, the value gives the amplitude of the graph, which is $|a| = 9$. The period is computed as $\frac{2\pi}{|b|} = \frac{2\pi}{8} = \frac{\pi}{4}$. The phase shift is $-\frac{\pi}{8}$.

Example Determine the amplitude, the period, and the phase shift of $y = \sin(3x-4)+5$.

Solution: Here, $a = 1$, $b = 3$ and $c = -4$. The value of a gives the amplitude which is $|a| = 1$. The period is computed as $\frac{2\pi}{|b|} = \frac{2\pi}{3}$. The phase shift is $-\frac{c}{b} = \frac{4}{3}$.

9.4 Writing the Equation of a Sine Graph

The general form of the sine function equation is $y = a\sin(bx+c)+d$. We can find a, b, c and d using its graph.

Key Point

The amplitude a is obtained by finding the maximum and minimum y values and taking half of their difference: $a = \frac{y_{\max} - y_{\min}}{2}$.

We define a basic cycle of the sine or cosine graph as an interval $[x_0, x_1]$, within which the entire graph pattern is captured. For instance, $[0, 2\pi]$ and $[-\pi, \pi]$ are both basic cycles for $y = \sin(x)$.

Key Point

The period of the function can be determined by $x_1 - x_0$ for any basic cycle $[x_0, x_1]$. The coefficient b in our function is related to the period as $b = \frac{2\pi}{x_1 - x_0}$.

Key Point

The value of d is the average of the maximum and minimum: $d = \frac{y_{\max} + y_{\min}}{2}$.

Key Point

Using the general equation, substitute the amplitude a, the vertical shift d, and the period coefficient b. Determine the phase shift c by using a known point on the graph and solving for c in the equation.

By finding a, b, c, and d, the sine graph is fully described by $y = a\sin(bx + c) + d$.

Example Determine the equation of the sine graph shown below.

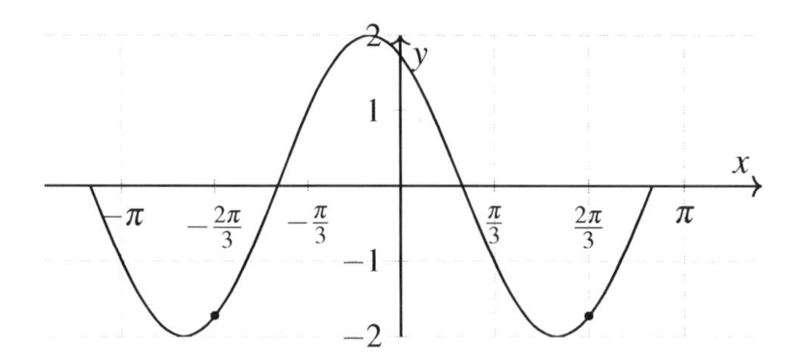

Solution: The maximum and minimum values on the y-axis are 2 and -2 respectively. So, the amplitude is $a = \frac{2-(-2)}{2} = 2$. Observing a basic cycle of the curve from $-\frac{2\pi}{3} \le x \le \frac{2\pi}{3}$, we determine the period to be $\frac{4\pi}{3}$. Hence, $b = \frac{2\pi}{\frac{4\pi}{3}} = \frac{3}{2}$. Also, for d, we have: $d = \frac{y_{\max}+y_{\min}}{2} = \frac{2+(-2)}{2} = 0$. With these, we get the equation $y = 2\sin(\frac{3}{2}x + c)$. To find c, we insert the point $(\frac{\pi}{3}, -1)$ into the equation, yielding $c = \frac{2\pi}{3}$. Thus, the equation for this sine graph is $y = 2\sin\left(\frac{3}{2}x + \frac{2\pi}{3}\right)$.

9.5 Writing the Equation of a Cosine Graph

The general form of a cosine function is $y = a\cos(bx+c)+d$. We can find a, b, c and d using its graph.

Key Point

The amplitude a is found by calculating half the difference between the maximum and minimum y values: $a = \frac{y_{\max}-y_{\min}}{2}$.

A basic cycle of the cosine graph is an interval $[x_0, x_1]$ that captures the entire pattern. Examples include $[0, 2\pi]$ and $[-\pi, \pi]$ for $y = \cos(x)$.

Key Point

The period of the function is $x_1 - x_0$. The coefficient b is related to the period by the formula $b = \frac{2\pi}{x_1 - x_0}$.

Key Point

The vertical shift d is the average of the maximum and minimum values: $d = \frac{y_{\max}+y_{\min}}{2}$.

Key Point

Substitute the amplitude a, the vertical shift d, and the period coefficient b into the general equation. Determine the phase shift c by using a known point on the graph and solving for c.

By finding a, b, c, and d, the cosine graph is fully described by $y = a\cos(bx+c)+d$.

Determine the equation of the graph below in the form $y = a\cos(bx+c)+d$.

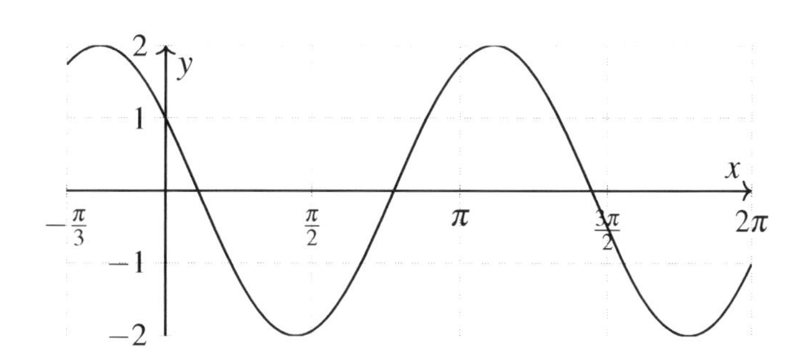

Solution: The maximum and minimum values on the y-axis are 2 and -2 respectively. So, the amplitude is: $a = \frac{2-(-2)}{2} = 2$. Observing a basic cycle of the curve from $-\frac{\pi}{3} \le x \le \pi$, we determine the period to be: $\pi - \left(-\frac{\pi}{3}\right) = \frac{4\pi}{3}$. Hence, $b = \frac{2\pi}{\frac{4\pi}{3}} = \frac{3}{2}$. Also, for d, we have: $d = \frac{y_{\max}+y_{\min}}{2} = \frac{2+(-2)}{2} = 0$. With these, we get the equation: $y = 2\cos(\frac{3}{2}x + c)$. To find c, we use a known point on the graph. For instance, if at $x = 0$, $y = 1$:

$$1 = 2\cos(0 + c) \Rightarrow \cos(c) = \frac{1}{2} \Rightarrow c = \frac{\pi}{3}.$$

Thus, the equation for this cosine graph is: $y = 2\cos(\frac{3}{2}x + \frac{\pi}{3})$.

9.6 Graph of the Tangent Function

The tangent graph displays a curve that starts from negative infinity, passes through zero, and then rises to positive infinity. This pattern repeats after every π interval along the x-axis. The formal definition of the tangent function is given by the ratio of the sine over the cosine functions, i.e., $\tan(x) = \frac{\sin(x)}{\cos(x)}$.

Key Point

The graph of the tangent function is undefined at odd multiples of $\frac{\pi}{2}$. These points are represented by vertical asymptotes on the graph.

The graph shows a complete cycle of the $\tan(x)$ function in the interval from $x = -\frac{\pi}{2}$ to $x = \frac{\pi}{2}$. The period of this function is π, and it has vertical asymptotes at $x = -\frac{\pi}{2}$ and $x = \frac{\pi}{2}$. These vertical asymptotes represent discontinuities in the function where it is undefinable because the cosine function in the denominator equals zero.

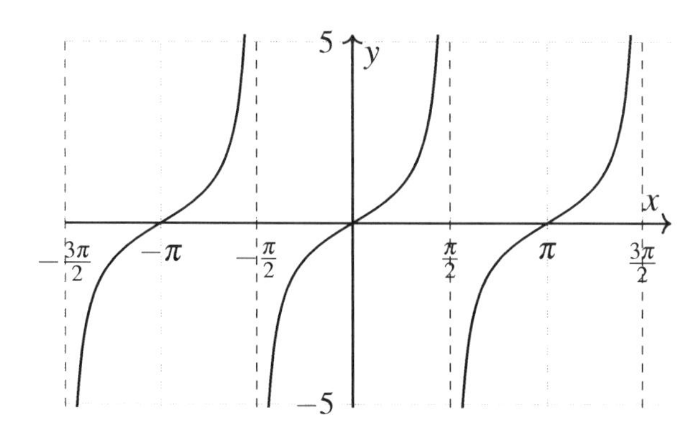

Find the graph of $y = \tan\left(x - \frac{\pi}{4}\right)$ in the interval of $-\frac{\pi}{4} < x < \frac{3\pi}{4}$.

Solution: The graph of $y = \tan\left(x - \frac{\pi}{4}\right)$ is the same as the graph of $y = \tan(x)$, but shifted to the right by $\frac{\pi}{4}$. As you can see in the graph, shifts $\frac{\pi}{4}$ units to the right due to the phase shift $\frac{\pi}{4}$, resulting in the vertical asymptotes occurring at these shifted positions.

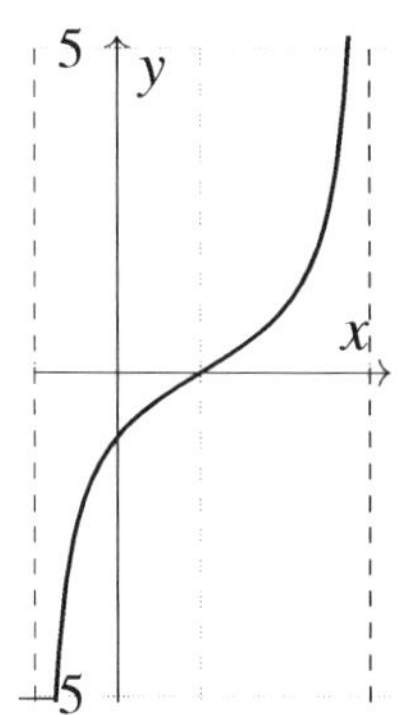

9.7 Graph of the Cosecant Function

The cosecant function, denoted as $\csc(x)$, is basically the reciprocal of the sine function, i.e., $\csc(x) = \frac{1}{\sin(x)}$. The sine function gives values between -1 and 1 inclusively, and it is defined for all x in $\mathbb{R}$. Consequently, the cosecant function gives values for $-\infty < \csc x \leq -1$ and for $1 \leq \csc x < \infty$.

Key Point

The cosecant function, $\csc(x)$, is undefined for values of x that are multiples of π since the sine value at those points is zero, making the cosecant undefined.

When graphing the cosecant function, we first draw the sine function and then take the reciprocals of the sine function values. The resulting curve has vertical asymptotes where the values of x are multiples of π. The vertical lines shown in the plot are asymptotes to the graph at $x = n\pi$ where n can be any integer i.e., the lines $x = 0, \pm\pi, \pm 2\pi, \ldots$ are asymptotes to the cosecant function.

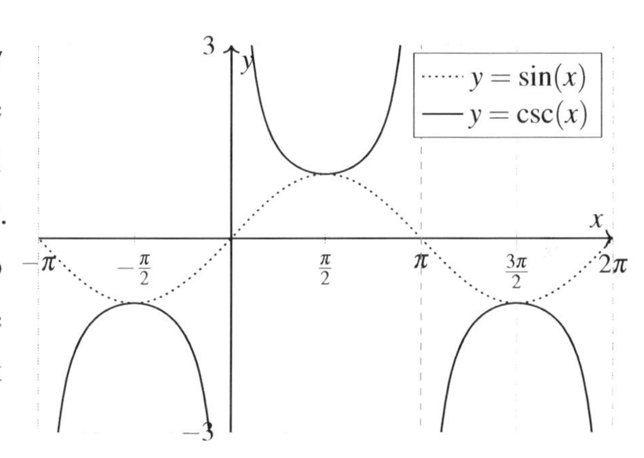

Draw the sine graph for one period of $y = -3\csc(4x)$.

Solution: To draw this graph, we first create the graph for the function $y = -3\sin(4x)$.

Then we sketch the asymptotes and fill in the cosecant curve. As you can see, the vertical asymptotes divide the curve into several sections. The curve itself drops below the x-axis where the sine function is negative and rises above where the sine function is positive.

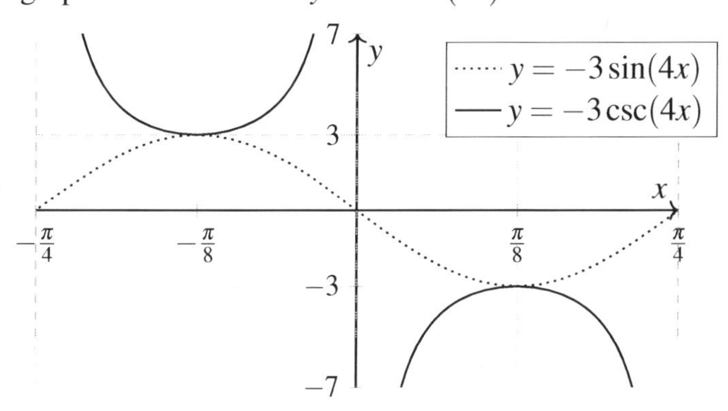

9.8 Graph of the Secant Function

The secant function, denoted as $\sec(x)$, is the reciprocal of the cosine function, represented mathematically as $\sec(x) = \frac{1}{\cos(x)}$. It is important to note that the secant function is not defined for the values of x at which the cosine function is zero.

To construct the graph of the secant function, we rely heavily on the cosine function due to their reciprocal relationship. Equivalently, for each x with $-1 \leq \cos x < 0$ or $0 < \cos x \leq 1$, we have $-\infty < \sec x \leq -1$ or $1 \leq \sec x < \infty$ respectively.

Key Point

The secant function is undefined for x values that are odd multiples of $\frac{\pi}{2}$ since $\cos x = 0$ at these points.

The graph consists of an infinite number of hyperbolas and the x values of their asymptotes can be determined using the formula $x = \frac{\pi}{2} + n\pi$, where n is an integer. These are the points where $\cos x = 0$, hence $\sec x$ is not defined.

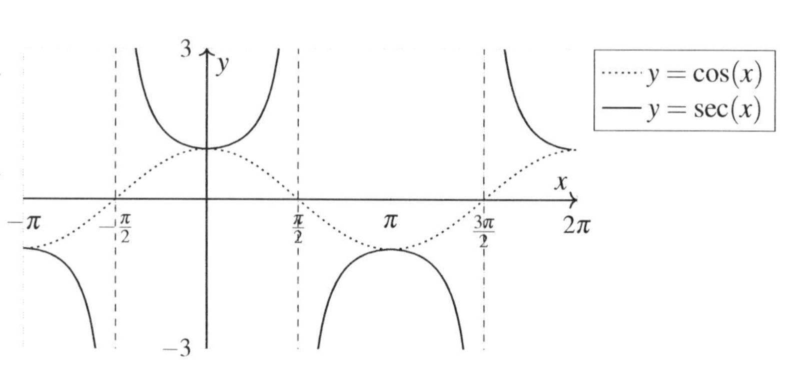

Example Draw one period of $y = \sec\left(2x - \frac{\pi}{2}\right) + 3$.

Solution: To begin the graphing process, first graph $y = \cos\left(2x - \frac{\pi}{2}\right) + 3$.

Next, identify the points where the cosine function reaches zero; these are the points where you will have asymptotes for the secant function. Finally, sketch in the hyperbolas of the secant function in between these asymptotes.

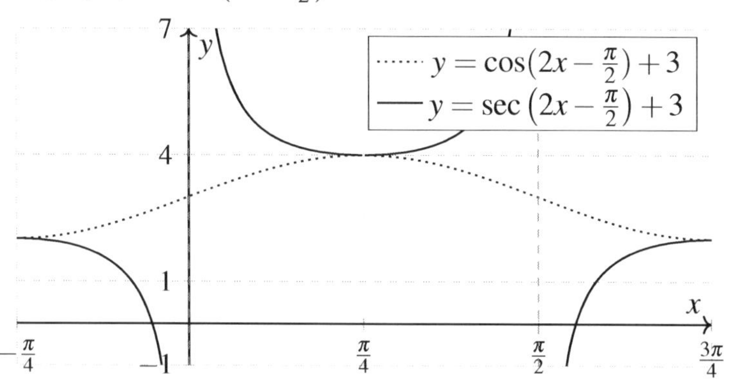

9.9 Graph of the Cotangent Function

The cotangent function can be identified in terms of the tangent function. Mathematically, it is represented as: $\cot x = \dfrac{1}{\tan x}$.

The graph of the cotangent function can be drawn using reciprocals of the tangent function values. It is important to observe patterns and make connections between the tangent and cotangent function.

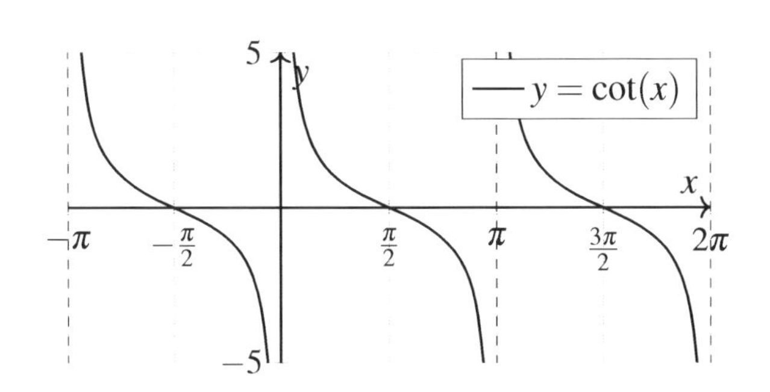

A notable feature of the graph of the cotangent function is that it is not defined where $\tan x = 0$, that is, for multiples of π ($x = n\pi$, where n is an integer). Conversely, the cotangent function equals zero where the tangent function is not defined. As a result, when we graph the cotangent function, we observe vertical asymptotes at integral multiples of π.

If x is a multiple of π, then $\tan x = 0$ and $\cot x$ is undefined. On the other hand, if the $\tan x$ is not defined, then $\cot x = 0$. For integral values of n, the vertical lines on the graph are asymptote at $x = n\pi$.

Example Draw one period of the function $y = \cot\left(x + \frac{\pi}{2}\right)$.

Solution: The cotangent function, $\cot(x)$, has a period of π.

For $\cot\left(x + \frac{\pi}{2}\right)$, the graph of the function will be shifted to the left by $\frac{\pi}{2}$.

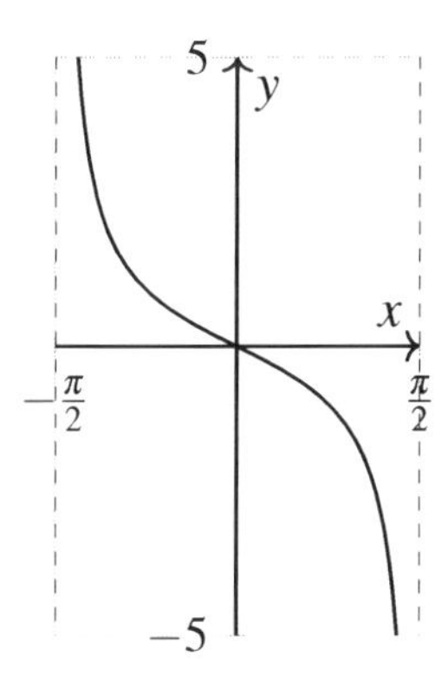

9.10 Graph of Inverse of the Sine Function

The sine function, $\sin(x)$, is not one-to-one over its entire range due to its periodic nature, repeating every 2π interval.

Key Point

Only a one-to-one function can have an inverse. To graph the inverse sine function (arcsine), we restrict the domain of the sine function to $-\frac{\pi}{2} \leq x \leq \frac{\pi}{2}$.

Reflecting the domain-restricted sine function across the line $y = x$ yields its inverse, depicted as $y = \sin^{-1}(x)$. This reflection produces a graph where $y = \sin^{-1}(x)$ serves as the mirror image of $y = \sin(x)$ within the restricted domain, across the line $y = x$.

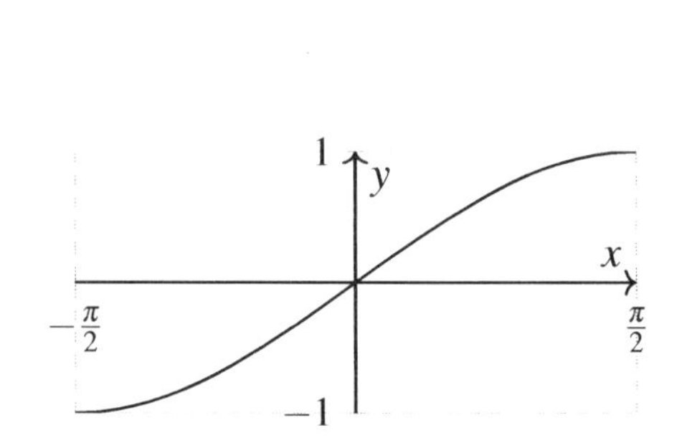

Sine function with restricted domain.

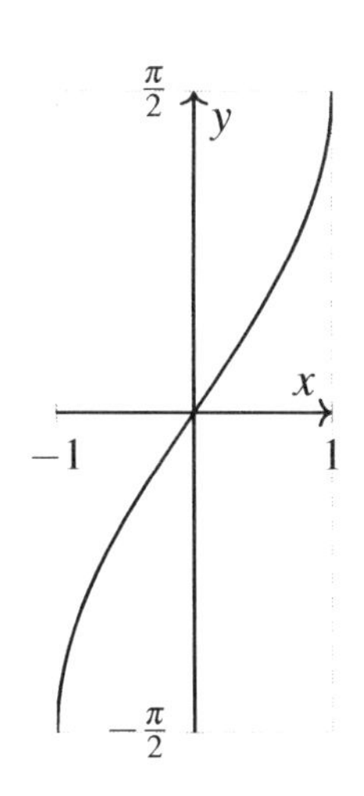

The graph of $y = \sin^{-1}(x)$.

Example Graph $f(x) = \sin^{-1}(x-1)$.

Solution: The graph of $f(x) = \sin^{-1}(x-1)$ is the graph of $y = \sin^{-1}(x)$, but with a phase shift of 1 unit to the right. The graph can be produced as follows:

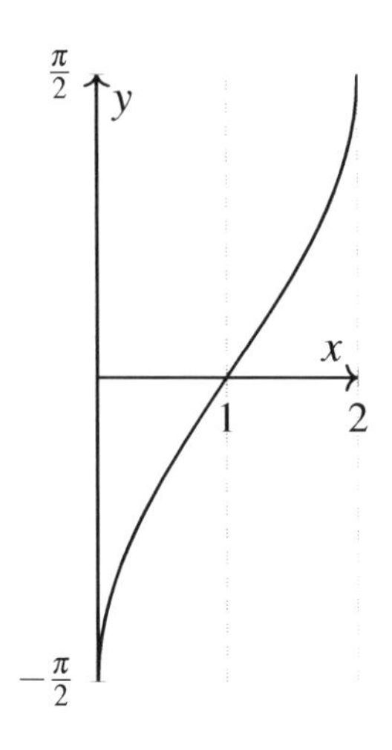

9.11 Graph of Inverse of the Cosine Function

The cosine function, $\cos(x)$, does not maintain a one-to-one relationship throughout its entire range due to its periodicity, repeating every 2π interval.

Key Point

Restricting the domain of the cosine function to $[0, \pi]$ transforms it into a one-to-one function, allowing us to efficiently calculate its inverse.

Reflecting the domain-restricted cosine function across the line $y = x$ gives us its inverse, represented as $y = \cos^{-1}(x)$. This action generates a graph where $y = \cos^{-1}(x)$ mirrors the restricted $y = \cos(x)$ function across the line $y = x$.

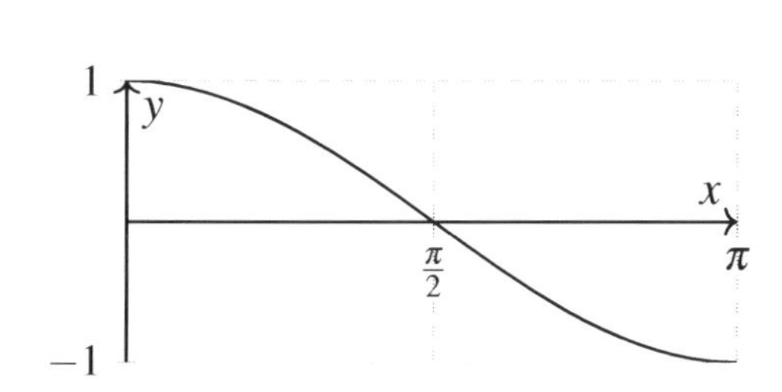

Cosine function with restricted domain.

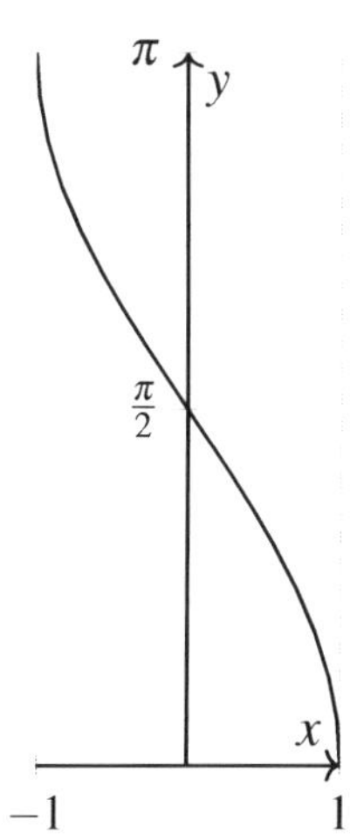

The graph of $y = \arccos(x)$.

Graph the function $y = 2\cos^{-1}(x-1)$.

Solution: The graph of $y = 2\cos^{-1}(x-1)$ results from the graph of $y = \cos^{-1}(x)$, but with a phase shift of 1 and a vertical stretch by a factor of 2. By observing the below graph, it is clear that $y = 2\cos^{-1}(x-1)$ is in fact a vertical stretch by a factor of 2 and phase shift of 1 from the $\cos^{-1}(x)$ function.

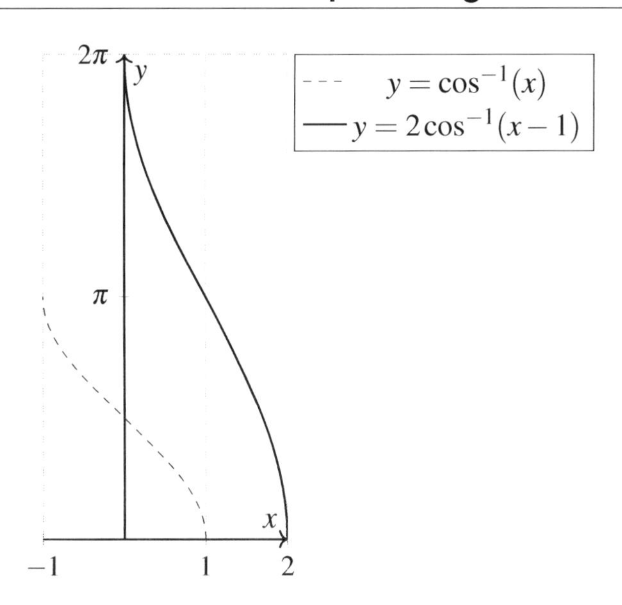

9.12 Graph of Inverse of the Tangent Function

The Tangent function, which is not a one-to-one function in its entire domain, can have an inverse if its domain is restricted.

Key Point

The inverse of the tangent function can be obtained when we limit the domain of the tangent function to $-\frac{\pi}{2} < x < \frac{\pi}{2}$.

This restriction is necessary for the function to pass the horizontal line test. When we reflect this subset on the line $y = x$, we get the image of the function which represents the graph of the inverse tangent function.

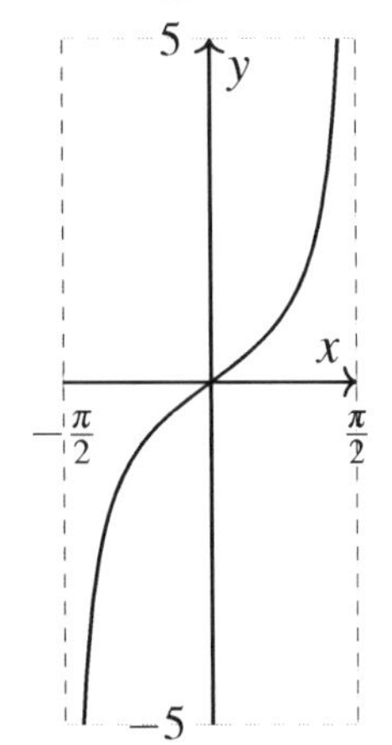

The function $y = \tan(x)$ with restricted domain.

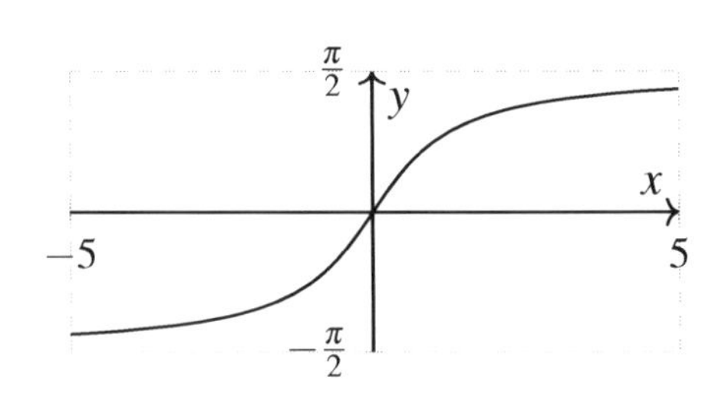

The graph of $y = \arctan(x)$.

Draw the function $y = -2\tan^{-1}(x)$.

Solution: The graph of $y=-2\tan^{-1}(x)$ is essentially the graph of $y=\tan^{-1}(x)$ with two main transformations: scaling the output variable y by a factor of 2, and a reflection across the x-axis. This means that the magnitude of y values in $y=\tan^{-1}(x)$ is doubled in $y=-2\tan^{-1}(x)$, and the graph is inverted. The horizontal asymptotes of $y=\tan^{-1}(x)$ at $y=\frac{\pi}{2}$ and $y=-\frac{\pi}{2}$ become $y=-\pi$ and $y=\pi$ in $y=-2\tan^{-1}(x)$, respectively.

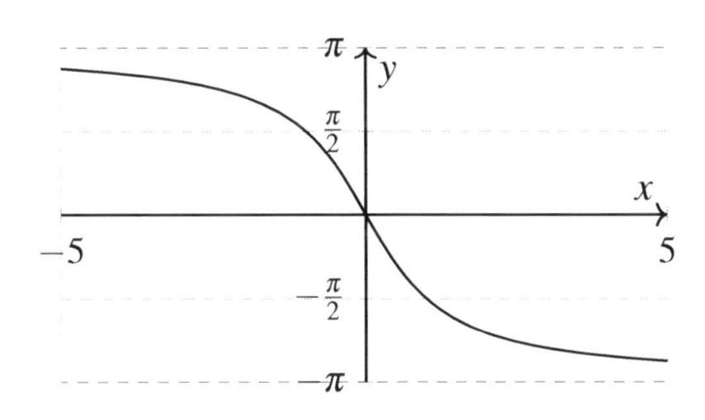

9.13 Practices

1) What is the amplitude of the sine function $y=\sin(x)$ graphed from 0 to 2π?

☐ A. 0
☐ B. $\frac{1}{2}$
☐ C. 1
☐ D. 2

2) At what value of x in the interval $0\le x\le 2\pi$ does the graph of $y=\sin(x)$ first cross the x-axis?

☐ A. 0
☐ B. $\frac{\pi}{2}$
☐ C. π
☐ D. $\frac{3\pi}{2}$

3) Which of the following values of x results in $y=\cos(x)$ being equal to 0?

☐ A. 0
☐ B. $\frac{\pi}{2}$
☐ C. π
☐ D. $\frac{2\pi}{3}$

4) At which intervals is the cosine function increasing on $0\le x\le 2\pi$?

☐ A. $\pi\le x\le\frac{3\pi}{2}$
☐ B. $0\le x\le\frac{\pi}{2}$ and $\frac{3\pi}{2}\le x\le 2\pi$
☐ C. $\frac{\pi}{2}\le x\le\frac{3\pi}{2}$
☐ D. $\pi\le x\le 2\pi$

5) Given the function $y=3\cos\left(2\left(x-\frac{\pi}{3}\right)\right)$, what is its amplitude?

☐ A. 2
☐ B. 3
☐ C. 6
☐ D. $\frac{\pi}{3}$

6) For the function $y=2\sin\left(\frac{\pi}{6}x+\frac{\pi}{2}\right)$, what is the period?

☐ A. 3
☐ B. 6
☐ C. 12
☐ D. $\frac{\pi}{2}$

7) The graph of a sine function has an amplitude of 3, a period of π, and is phase-shifted to the right by $\frac{\pi}{2}$. What is the equation of this sine function?

☐ A. $y=3\sin(2x-\pi)$
☐ B. $y=3\sin(x-\frac{\pi}{2})$
☐ C. $y=3\sin(x+\frac{\pi}{2})$
☐ D. $y=3\sin(2x+\pi)$

8) Given the sine graph that starts at origin 0 and has its first maximum at $(\frac{\pi}{4},2)$, what is the equation of the sine function?

☐ A. $y=2\sin(2x)$
☐ B. $y=\sin(4x)$
☐ C. $y=\sin(2x)$
☐ D. $y=2\sin(4x)$

9) Consider the graph of a cosine function with an amplitude of 3 and a period of π. Which of the following options can be the equation for the cosine function?

☐ A. $y=3\cos(2x)$
☐ B. $y=3\cos\left(\frac{x}{2}\right)$
☐ C. $y=3\cos(\pi x)$
☐ D. $y=3\cos\left(\frac{x}{\pi}\right)$

10) A cosine graph has a maximum at $y=5$ and a minimum at $y=-5$. If the graph completes one cycle from $x=0$ to $x=3\pi$, which of the following options can be the equation of the graph?

☐ A. $y=5\cos\left(\frac{2x}{3}\right)$
☐ B. $y=10\cos\left(\frac{x}{3}\right)$
☐ C. $y=10\cos(3x)$
☐ D. $y=5\cos\left(\frac{3x}{2}\right)$

11) What is the period of the function $y=\tan(2x)$?

☐ A. 2π
☐ B. π
☐ C. $\frac{\pi}{2}$
☐ D. $\frac{\pi}{4}$

12) Which of the following describes the graph of $y = \tan\left(x + \frac{3\pi}{4}\right)$?

☐ A. A shift to the left by $\frac{3\pi}{4}$ units.
☐ B. A shift to the right by $\frac{3\pi}{4}$ units.
☐ C. A reflection over the y-axis.
☐ D. A reflection over the x-axis.

13) What is the period of the cosecant function $y = \csc(2x)$?

☐ A. π
☐ B. 2π
☐ C. $\frac{\pi}{2}$
☐ D. 4π

14) Which of the following represents the graph with vertical asymptotes of the function $y = \csc(x + \frac{\pi}{2})$?

☐ A. $x = k\pi;\ k \in \mathbb{Z}$
☐ B. $x = k\pi - \frac{\pi}{2};\ k \in \mathbb{Z}$
☐ C. $x = k\pi + \frac{3\pi}{2};\ k \in \mathbb{Z}$
☐ D. $x = (2k+1)\pi;\ k \in \mathbb{Z}$

15) Which of the following values is not in the range of $y = \sec(x)$?

☐ A. -2
☐ B. 0
☐ C. 1
☐ D. 2

16) Suppose $y = \sec(x)$ has a vertical asymptote at $x = \frac{3\pi}{2}$. Which of the following is the next larger value of x where another vertical asymptote will occur?

☐ A. 2π
☐ B. $\frac{5\pi}{2}$
☐ C. 3π
☐ D. 4π

17) Which of the following statements is true for the graph of $y = \cot(x)$?

☐ A. The function has horizontal asymptotes at $y = 0$.
☐ B. The function has vertical asymptotes at $x = k\pi$, where k is an integer.
☐ C. The function has a period of 2π.
☐ D. The function is defined for all real numbers.

18) What is the period of the function $y = 3\cot(2x)$?

☐ A. π

☐ B. $\frac{\pi}{2}$

☐ C. 2π

☐ D. 3π

19) What is the range of $y = \sin^{-1}(x)$?

☐ A. $-1 \leq y \leq 1$

☐ B. $-\pi \leq y \leq \pi$

☐ C. $-\frac{\pi}{2} \leq y \leq \frac{\pi}{2}$

☐ D. All real values of y

20) If $y = \sin^{-1}(\frac{1}{2})$, what is the value of y?

☐ A. $\frac{\pi}{4}$

☐ B. $\frac{\pi}{3}$

☐ C. $\frac{\pi}{2}$

☐ D. $\frac{\pi}{6}$

21) What is the range of the function $y = \arccos(x)$?

☐ A. $[-1, 1]$

☐ B. $[0, \pi]$

☐ C. $(-\infty, \infty)$

☐ D. $[-\pi, 0]$

22) Which of the following graphs represents the function $y = \arccos(x)$?

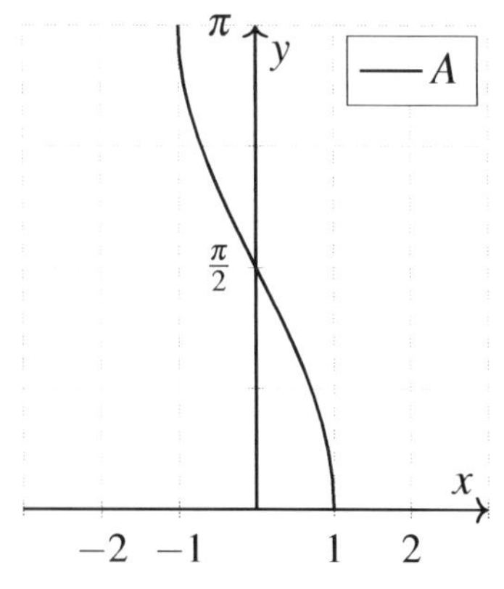

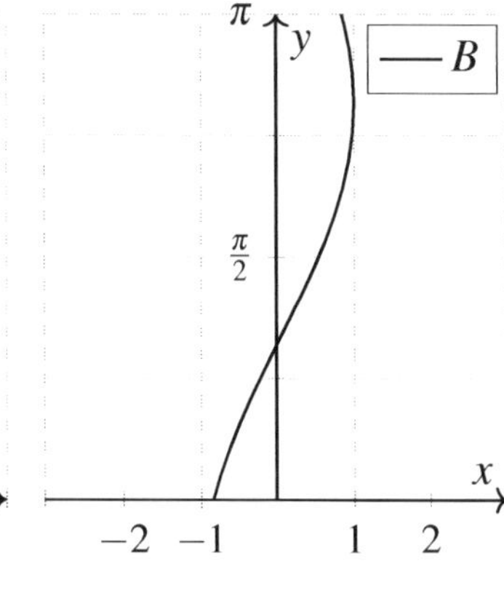

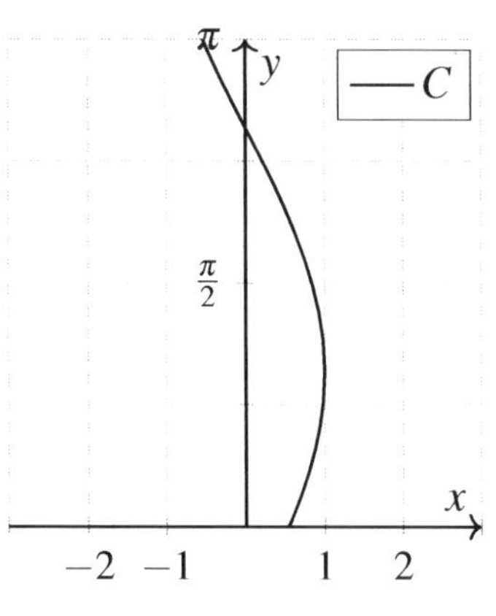

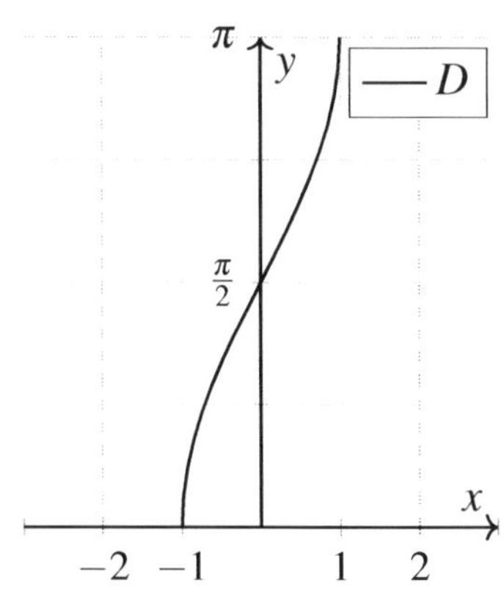

☐ A. Graph A

☐ B. Graph B

☐ C. Graph C

☐ D. Graph D

23) Which of the following is the correct range of the function $y = \arctan(x)$?

☐ A. $(-\infty, \infty)$

☐ B. $[0, \pi]$

☐ C. $(-\frac{\pi}{2}, \frac{\pi}{2})$

☐ D. $[-\frac{\pi}{2}, \frac{\pi}{2}]$

24) What is the graph of the function $y = 3\arctan(x)$?

☐ A. The graph is compressed vertically relative to $y = \arctan(x)$.

☐ B. The graph is stretched vertically by a factor of 3 and the horizontal asymptotes are at $y = \pm\frac{\pi}{2}$.

☐ C. The graph is stretched vertically by a factor of 3 and the horizontal asymptotes are at $y = \pm\frac{3\pi}{2}$.

☐ D. The graph is the same as $y = \arctan(x)$, just moved up by 3 units.

Answer Keys

1) C. 1
2) A. 0
3) B. $\frac{\pi}{2}$
4) D. $\pi \leq x \leq 2\pi$
5) B. 3
6) C. 12
7) A. $y = 3\sin(2x - \pi)$
8) A. $y = 2\sin(2x)$
9) A. $y = 3\cos(2x)$
10) A. $y = 5\cos\left(\frac{2x}{3}\right)$
11) C. $\frac{\pi}{2}$
12) A. A shift to the left by $\frac{3\pi}{4}$ units.
13) A. π
14) B. $x = k\pi - \frac{\pi}{2}; k \in \mathbb{Z}$
15) B. 0
16) B. $\frac{5\pi}{2}$
17) B.
18) B. $\frac{\pi}{2}$
19) C. $-\frac{\pi}{2} \leq y \leq \frac{\pi}{2}$
20) D. $\frac{\pi}{6}$
21) B. $[0, \pi]$
22) A. Graph A
23) C. $(-\frac{\pi}{2}, \frac{\pi}{2})$
24) C.

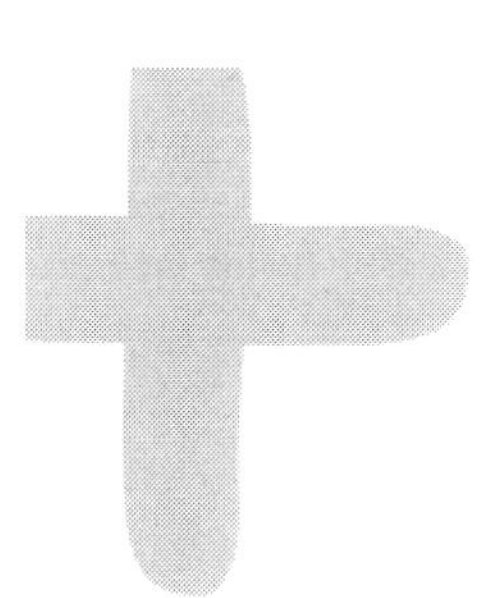

10. Trigonometric Equations

10.1 Basic Techniques for Solving Trigonometric Equations

A trigonometric equation involves an unknown variable within a trigonometric function. Solving these equations typically requires using fundamental trigonometric identities and algebraic methods.

1. Simplification: Apply trigonometric identities (reciprocal, quotient, Pythagorean, co-function, double-angle) to simplify the equation.

2. Factoring: Simplify complex equations by factoring trigonometric expressions.

3. Isolating the variable: Rearrange the equation to isolate the trigonometric function containing the variable.

4. Using inverse functions: Use inverse trigonometric functions (sine, cosine, tangent) to find the variable's value by cancelling out the trigonometric functions.

5. Multiple angles: For equations with multiple solutions, employ the periodicity and symmetries of trigonometric functions to determine all solutions.

6. Graphing: Visualize the function's behavior through graphing to find approximate solutions, especially when algebraic methods are cumbersome.

Key Point

To solve trigonometric equations, always consider all solutions within the given interval, not just the first solution found.

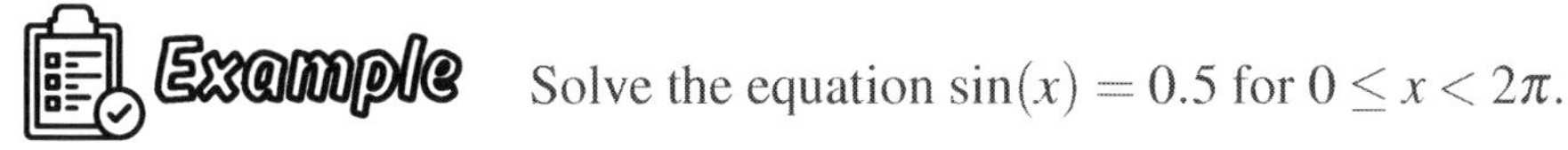

Example Solve the equation $\sin(x) = 0.5$ for $0 \leq x < 2\pi$.

Solution: Here we do not need a simplification or factoring step. Isolating the variable is also not needed because the original equation is isolated. Now we use the inverse trigonometric functions method $x = \sin^{-1}(0.5)$. We know that $\sin^{-1}(0.5) = \frac{\pi}{6}$. However, the sine function is positive in both the first and second quadrants. To find all solutions in the interval $0 \leq x < 2\pi$, we must also consider the second quadrant. The general solution for $\sin(x) = 0.5$ is:

$$x = \frac{\pi}{6} \quad \text{or} \quad x = \pi - \frac{\pi}{6}.$$

Therefore, the second solution is: $x = \pi - \frac{\pi}{6} = \frac{5\pi}{6}$. Thus, the solutions to $\sin(x) = 0.5$ in the interval $0 \leq x < 2\pi$ are:

$$x = \frac{\pi}{6} \quad \text{and} \quad x = \frac{5\pi}{6}.$$

10.2 Factoring and Simplifying Trigonometric Expressions

Mastering the simplification and factoring of trigonometric expressions is essential for transforming complex problems into simpler ones. This process typically involves a combination of algebraic and trigonometric techniques.

1. Factoring Techniques: Employ algebraic methods like grouping, difference of squares, and sum/difference of cubes to simplify trigonometric expressions.

2. Trigonometric Identities: Utilize fundamental identities (reciprocal, quotient, Pythagorean, co-function, double-angle) to either simplify expressions or rewrite them in alternative forms.

3. Rationalizing Denominators: For expressions with trigonometric functions in the denominator, multiply both numerator and denominator by the conjugate to simplify the expression by eliminating complex trigonometric terms.

These strategies are pivotal for easing the solution of trigonometric equations and understanding the mathematical principles involved.

Key Point

Simplifying trigonometric expressions often requires the application of trigonometric identities and algebraic factoring to rewrite and simplify the expressions.

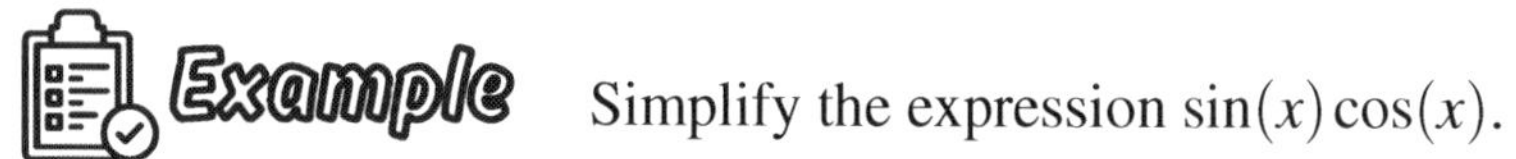

Example Simplify the expression $\sin(x)\cos(x)$.

Solution: This problem is a straightforward application of the double angle identity for sine:

$$\sin(2x) = 2\sin(x)\cos(x).$$

So, $\sin(x)\cos(x)$ simplifies to $\frac{1}{2}\sin(2x)$.

Example Factor the expression $1 - 2\cos^2(x)$.

Solution: The expression does not seem to factor easily at first glance. For factorizing, firstly, we write the Pythagorean identity:

$$\sin^2(x) + \cos^2(x) = 1,$$

thus transforming $\cos^2(x)$ to the right of equation:

$$\sin^2(x) = 1 - \cos^2(x).$$

Subtracting $\cos^2(x)$ from both sides of the equation gives: $\sin^2(x) - \cos^2(x) = 1 - 2\cos^2(x)$. Now, we recognize the left-hand of the above equation as a difference of squares, which can be factored into:

$$(\sin(x) - \cos(x))(\sin(x) + \cos(x)).$$

So, we can rewrite our expression as:

$$1 - 2\cos^2(x) = (\sin(x) - \cos(x))(\sin(x) + \cos(x)),$$

Simplify the expression $\sec^2(x) - \tan^2(x)$.

Solution: This problem involves the use of the identity $1 + \tan^2(x) = \sec^2(x)$. Instead of $\sec^2(x)$, we put $1 + \tan^2(x)$. Rearranging this gives $\sec^2(x) - \tan^2(x) = 1$. So, we prove that $\sec^2(x) - \tan^2(x)$ simplifies to 1.

10.3 Solving Equations with Multiple Angles

Trigonometric equations involving multiple angles (e.g., $2x$, $3x$) require specific strategies for effective solving:

1. Simplification using identities: Utilize double-angle, half-angle, and triple-angle identities to rewrite multi-angle expressions into simpler forms. For example, the double-angle identity $\sin(2x) = 2\sin(x)\cos(x)$ simplifies equations with $\sin(2x)$.

2. Substitution: Temporarily replace the multi-angle term (e.g., $2x$) with a new variable (e.g., t), converting the equation into a single-angle format. Solve for the new variable, then revert back to the original to find the solution for the initial variable.

3. Solving for one angle at a time: In equations with different multiples of angles across trigonometric functions, it is sometimes efficient to solve for one angle first, ensuring all functions are expressed in terms of a single angle.

Key Point

To solve trigonometric equations with multi-angle expressions:

1. Simplify using identities (e.g., $\sin(2x) = 2\sin(x)\cos(x)$).
2. Substitute a new variable for multi-angle terms, solve, and revert.
3. Solve for one angle at a time by expressing all functions in terms of a single angle.

Key Point

Since trigonometric functions are periodic, multiple-angle equations may have multiple solutions within a given range.

Find the values of x between $0°$ and $360°$ in the following equation:

$$2\sin(2x) + \cos(x) = 0.$$

Solution: Rewrite the double-angle identity for the sine as: $2\sin(2x) = 2(2\sin(x)\cos(x))$. Substituting this into the equation gives: $2(2\sin(x)\cos(x)) + \cos(x) = 0$. Expanding and rearranging the terms, we get: $4\sin(x)\cos(x) + \cos(x) = 0$. This implies that: $\cos(x)(4\sin(x) + 1) = 0$. Now we have two possible cases to consider: $\cos(x) = 0$ and $4\sin(x) + 1 = 0$.

For $\cos(x) = 0$, x equals $90°$ and $270°$.

For case $4\sin(x) + 1 = 0$: $4\sin(x) = -1 \Rightarrow \sin(x) = -\frac{1}{4}$. Applying the inverse sine function delivers two solutions: $x = \arcsin\left(-\frac{1}{4}\right) \approx -14.48°$, and $x = 180 - \arcsin\left(-\frac{1}{4}\right) \approx 194.48°$.

We find that the solutions are $x = 90°$, $x = 270°$, $x \approx -14.48°$, and $x \approx 194.48°$. By considering the given range of x which is between $0°$ and $360°$, we discard the negative value to obtain the final solutions as $x = 90°$, $x = 270°$, and $x \approx 194.48°$.

10.4 Practices

1) Solve the equation $2\cos^2(x) - 3\cos(x) + 1 = 0$ for $0 \leq x < 2\pi$.

☐ A. $x = 0, \frac{\pi}{3}, \frac{5\pi}{3}$

☐ B. $x = 0, \frac{2\pi}{3}, \pi, \frac{4\pi}{3}$

☐ C. $x = \frac{\pi}{3}, \frac{2\pi}{3}, \frac{4\pi}{3}, \frac{5\pi}{3}$

☐ D. $x = \frac{\pi}{3}, \pi, \frac{5\pi}{3}$

2) Solve $\tan(x) = \sqrt{3}$ for $-\pi < x < \pi$.

☐ A. $x = -\frac{\pi}{6}, \frac{\pi}{3}$

- ☐ B. $x = \frac{\pi}{3}, \frac{4\pi}{3}$
- ☐ C. $x = -\frac{\pi}{3}, \frac{\pi}{3}$
- ☐ D. $x = -\frac{2\pi}{3}, \frac{\pi}{3}$

3) Simplify the trigonometric expression $\cos^2(x) - \sin^2(x)$.

- ☐ A. $\cos(2x)$
- ☐ B. $\sin(2x)$
- ☐ C. 1
- ☐ D. $-\sin(2x)$

4) What is the simplified form of the expression $1 - \sin^2(x) - \cos^2(x)$?

- ☐ A. 1
- ☐ B. 0
- ☐ C. -1
- ☐ D. $\sin^2(x) - \cos^2(x)$

5) Solve the trigonometric equation $\cos(3x) = \frac{1}{2}$ for x where $0 \leq x < \pi$.

- ☐ A. $x = \frac{\pi}{9}, \frac{5\pi}{9}, \frac{7\pi}{9}$
- ☐ B. $x = \frac{\pi}{9}, \frac{7\pi}{9}, \frac{11\pi}{9}$
- ☐ C. $x = \frac{2\pi}{9}, \frac{4\pi}{9}, \frac{8\pi}{9}$
- ☐ D. $x = \frac{\pi}{6}, \frac{\pi}{2}, \frac{5\pi}{3}$

6) Which of the following is a solution to the equation $\tan(2x) = \sqrt{3}$ in the interval $[0, 2\pi)$?

- ☐ A. $x = \frac{\pi}{6}$
- ☐ B. $x = \frac{\pi}{3}$
- ☐ C. $x = \frac{4\pi}{3}$
- ☐ D. $x = \frac{\pi}{2}$

Answer Keys

1) A. $x=0, \frac{\pi}{3}, \frac{5\pi}{3}$

2) D. $x=-\frac{2\pi}{3}, \frac{\pi}{3}$

3) A. $\cos(2x)$

4) B. 0

5) A. $x=\frac{\pi}{9}, \frac{5\pi}{9}, \frac{7\pi}{9}$

6) A. $x=\frac{\pi}{6}$

11. Trigonometric Applications

11.1 The Law of Sines: Definition and Applications

The Law of Sines asserts that the ratio of a side length to the sine of its opposite angle is the same for all three sides and angles of a triangle.

Key Point

Representing the sides as a, b, and c, and their opposite angles as A, B, and C respectively, the Law of Sines formula is given by:

$$\frac{a}{\sin A} = \frac{b}{\sin B} = \frac{c}{\sin C}.$$

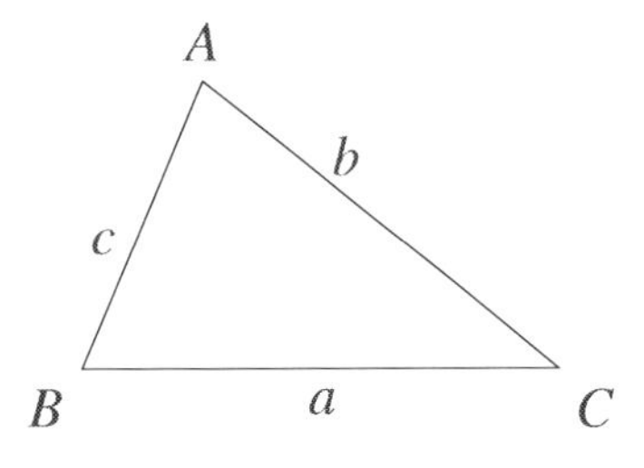

Example For a triangle, with given measures $A = 62°$, $B = 55°$ and $c = 5\ cm$, determine the length of side b.

Solution: First, we need to find the measure of angle C. Since the sum of the measures of the angles of a triangle is $180°$, we arrive at: $A+B+C = 180° \Rightarrow C = 63°$. Now, we can use the Law of Sines to find the length of side b:

$$\frac{b}{\sin B} = \frac{c}{\sin C}.$$

Knowing that $\sin 55^\circ \approx 0.82$, and $\sin 63^\circ \approx 0.89$ and solving for b we have:

$$b = \frac{5\sin 55^\circ}{\sin 63^\circ} \approx \frac{5\times 0.82}{0.89} \approx 4.61\ cm.$$

11.2 Law of Cosines: Definition and Applications

The Law of Cosines is an essential theorem in trigonometry that links the lengths of a triangle's sides with the cosines of its angles. This law is particularly useful for determining unknown sides or angles in a triangle when two sides and the included angle, or three sides (SAS or SSS), are known.

It states that the square of a side is equal to the sum of the squares of the other two sides minus twice the product of those sides and the cosine of their included angle. Representing the sides as a, b, and c, and the angle opposite to side c as C, the Law of Cosines is formulated as:

$$a^2 = b^2 + c^2 - 2bc\cos A$$

$$b^2 = a^2 + c^2 - 2ac\cos B$$

$$c^2 = a^2 + b^2 - 2ab\cos C$$

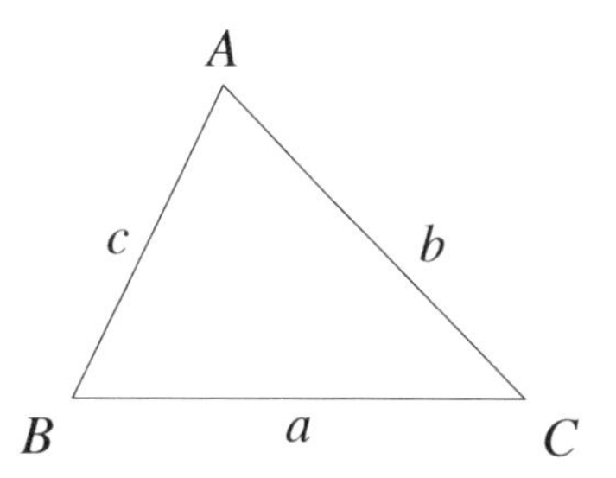

Key Point

The Law of Cosines facilitates the calculation of unknown side lengths and angles in triangles, serving as a cornerstone for solving many trigonometry problems.

Example In triangle ABC, sides a, b, and c are given as 8, 14, and 10 units respectively. Find angle B in degrees.

Solution: We can apply the Law of Cosines to find angle B:

$$\cos B = \frac{a^2 + c^2 - b^2}{2ac}.$$

Let us put the given lengths into the equation:

$$\cos B = \frac{8^2 + 10^2 - 14^2}{2\times 8\times 10} = \frac{64 + 100 - 196}{160} = -0.2.$$

Thus, $B = \cos^{-1}(-0.2) = 101.54^\circ$.

11.3 Area of a Triangle Using Trigonometry

Trigonometry offers a powerful tool for calculating the area of triangles, which is handy when we have information about two sides and the included angle. This method is invaluable for figures in two-dimensional spaces and for applications involving vectors. To calculate the area of a triangle when you know two sides and the angle between them, use the formula:

$$A = \frac{1}{2} a \times b \times \sin C,$$

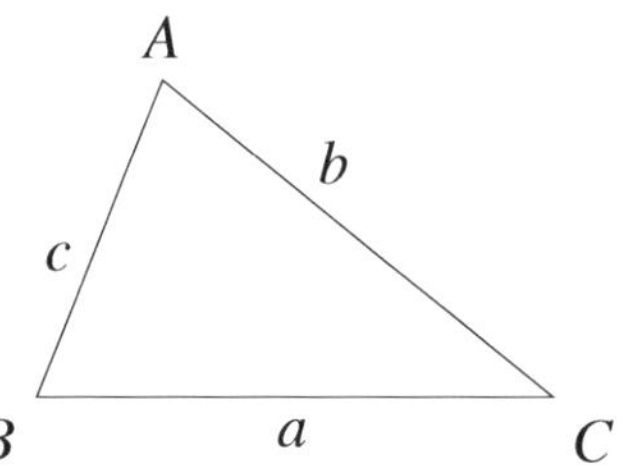

where in this formula a and b are the lengths of two sides of the triangle and $\sin C$ is the sine of the angle C enclosed by sides a and b.

Key Point

The area of any triangle, whether acute, obtuse, or right-angled, is given by $A = \frac{1}{2} ab \sin C$, where a and b are sides with their included angle C. This demonstrates the wide applicability of trigonometry in geometry.

Example Consider a triangle with sides $a = 5$ units and $b = 7$ units, and the angle $C = 60^\circ$ between them. Find the area of this triangle.

Solution: To solve this, we make use of the formula for finding the area of a triangle using trigonometric principles discussed earlier. Hence, the area A of the triangle can be calculated as follows:

$$A = \frac{1}{2} \times 5 \times 7 \times \sin 60^\circ = \frac{1}{2} \times 5 \times 7 \times \frac{\sqrt{3}}{2} = \frac{35\sqrt{3}}{4} \text{ square units.}$$

So, the area of the triangle is $\frac{35\sqrt{3}}{4}$ square units.

11.4 Practices

1) In triangle XYZ, side $x = 8\ cm$, side $z = 10\ cm$, and angle $X = 45^\circ$. What is the measure of angle Z?

☐ A. 53.27°
☐ B. 58.12°
☐ C. 61.64°
☐ D. 67.8°

2) In a triangle with sides $a = 7\ cm$, $b = 12\ cm$ and angle $A = 35°$, what is the nearest value to the measure of angle B?

☐ A. 75°
☐ B. 80°
☐ C. 79°
☐ D. 90°

3) In triangle PQR, side $p = 7$ units, side $q = 24$ units, and angle $R = 45°$. What is the length of side r?

☐ A. 13.5 units
☐ B. 17.29 units
☐ C. 22.45 units
☐ D. 19.68 units

4) Triangle XYZ has sides $x = 15$ units, $y = 9$ units, and $z = 12$ units. What is the measure of angle X?

☐ A. 90°
☐ B. 68°
☐ C. 92°
☐ D. 118°

5) A triangle has sides of lengths 6 units and 8 units with an angle of 45° between them. What is the area of the triangle?

☐ A. 12 square units
☐ B. 24 square units
☐ C. $12\sqrt{2}$ square units
☐ D. $24\sqrt{2}$ square units

6) Given a triangle with sides 4 units and 9 units and an angle of 30° between them, which of the following is the area of the triangle?

☐ A. 6 square units
☐ B. 9 square units
☐ C. 18 square units
☐ D. 36 square units

Answer Keys

1) C. 61.64°

2) C. 79°

3) D. 19.68 units

4) A. 90°

5) C. $12\sqrt{2}$ square units

6) B. 9 square units

12. Complex Numbers

12.1 Graphs of Polar Equations

Graphing polar equations is essential for visualizing complex numbers and understanding their geometric representations. Polar coordinates (r,θ) offer a different perspective than Cartesian coordinates (x,y), emphasizing the relationship between angles and distances from a central point.

Here is a brief step-by-step process for graphing polar equations:

1. **Understand the Polar Coordinate System:** Defined by distance r from the origin and angle θ from the positive x-axis.
2. **Identify the Polar Equation:** Recognize the form of the polar equation (e.g., circles, spirals, roses).
3. **Create a Table of Values:** For various θ, calculate corresponding r values to determine points that satisfy the equation.
4. **Plot Points:** On a polar graph, mark the points (r,θ), with θ measured counterclockwise from the x-axis and r indicating distance from the origin.
5. **Connect the Dots:** Draw a smooth curve to complete the graph.
6. **Analyze the Graph:** Identify the graph's properties, such as symmetry or intercepts, and determine the maximum and minimum values.

Example Consider the polar equation $r=2+2\cos(\theta)$. Sketch the graph of this polar equation.

Solution: To create the graph, we begin by computing values of r for several values of θ.

1. For $\theta=0$, $r=2+2\cos(0)=4$. So the coordinate is $(4,0)$.
2. For $\theta=\frac{\pi}{4}$, $r=2+2\cos(\frac{\pi}{4})=2+\sqrt{2}\approx 3.41$. Here, the coordinate is $(3.41,\frac{\pi}{4})$.
3. For $\theta=\frac{\pi}{2}$, $r=2+2\cos(\frac{\pi}{2})=2$. This gives us the coordinate $(2,\frac{\pi}{2})$.
4. For $\theta=\pi$, $r=2+2\cos(\pi)=0$. The coordinate is $(0,\pi)$.

5. For $\theta = \frac{3\pi}{2}$, $r = 2 + 2\cos(\frac{3\pi}{2}) = 2$. The coordinate is $(2, \frac{3\pi}{2})$.
6. For $\theta = \frac{7\pi}{4}$, $r = 2 + 2\cos(\frac{7\pi}{4}) = 2 + \sqrt{2} \approx 3.41$. The coordinate is $(3.41, \frac{7\pi}{4})$.
7. For $\theta = 2\pi$, $r = 2 + 2\cos(2\pi) = 4$. The coordinate is $(4, 2\pi)$.

Following this process for several more values of θ will give us a set of points that we can plot on a polar plane. Observe that for the coordinates we have so far, the angle increases as we move counter-clockwise, which is the positive direction in polar coordinates. After calculating and plotting these points, we connect the points to construct our graph.

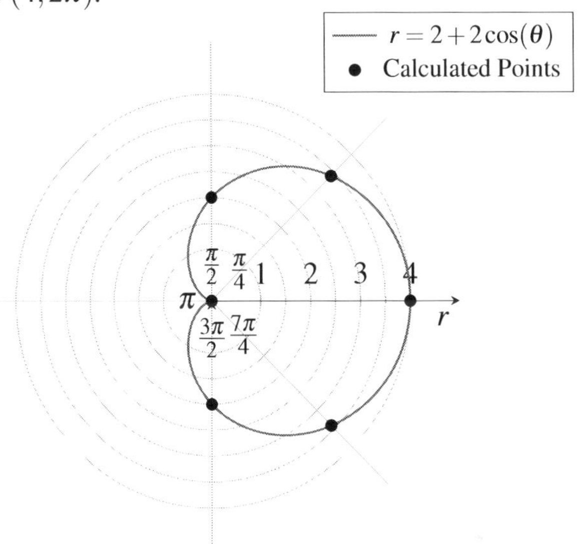

12.2 Converting Between Polar and Rectangular Coordinates

Learning how to switch between polar and rectangular coordinates is essential, as it allows us to apply the appropriate coordinate system to simplify calculations and analysis of various problems.

Key Point

Convert Cartesian to polar coordinates using $r = \sqrt{x^2 + y^2}$ for radius. For angle θ, use $\arctan\left(\frac{y}{x}\right)$, adjusting for the correct quadrant:

- Quadrant I: $\theta = \arctan\left(\frac{y}{x}\right)$,
- Quadrant II and III: $\theta = \pi + \arctan\left(\frac{y}{x}\right)$,
- Quadrant IV: $\theta = 2\pi + \arctan\left(\frac{y}{x}\right)$.

Key Point

Given polar coordinates (r, θ), the coordinates are found using the cosine and sine functions:

$$x = r\cos(\theta) \quad \text{and} \quad y = r\sin(\theta).$$

Key Point

Use radians for θ when applying trigonometric functions for conversion between polar and rectangular coordinates.

Example Convert the polar coordinates $(3, \frac{\pi}{4})$ to rectangular coordinates.

Solution: By using the key point equations to convert from polar to rectangular coordinates, we get:

$$x = r\cos(\theta) = 3\cos(\frac{\pi}{4}) = 3(\frac{\sqrt{2}}{2}) = \frac{3\sqrt{2}}{2},$$

$$y = r\sin(\theta) = 3\sin(\frac{\pi}{4}) = 3(\frac{\sqrt{2}}{2}) = \frac{3\sqrt{2}}{2}.$$

Hence, the polar coordinates $(3, \frac{\pi}{4})$ are equal to the rectangular coordinates $(\frac{3\sqrt{2}}{2}, \frac{3\sqrt{2}}{2})$.

Example Convert the rectangular coordinates $\left(1, \sqrt{3}\right)$ to polar coordinates.

Solution: We use the equations to convert from rectangular to polar coordinates:

$$r = \sqrt{x^2 + y^2} = \sqrt{1^2 + \sqrt{3}^2} = \sqrt{1+3} = \sqrt{4} = 2.$$

We find θ in radians:

$$\theta = \arctan\left(\frac{\sqrt{3}}{1}\right) = \arctan\left(\sqrt{3}\right) = \frac{\pi}{3}.$$

Therefore, the rectangular coordinates $\left(1, \sqrt{3}\right)$ are equivalent to the polar coordinates $\left(2, \frac{\pi}{3}\right)$.

12.3 Introduction to Complex Numbers

Complex numbers introduce an ‘imaginary’ dimension to the real number system, crucial for Precalculus. Defined by $z = a + bi$ where $a, b \in \mathbb{R}$ and i satisfies $i^2 = -1$, they encompass both a real part (a) and an imaginary part (b). Originating from solutions to equations like $x^2 + 1 = 0$, which lacks real solutions but finds resolution in complex numbers as i and $-i$, they form an essential numerical set represented by $\mathbb{C}$.

For visualization, complex numbers are plotted on the Argand Plane, akin to Cartesian coordinates, with the real and imaginary parts aligning with the x-axis and y-axis, respectively. Additionally, a complex number’s magnitude, or absolute value, is given by $|z| = \sqrt{a^2 + b^2}$, reflecting its ‘distance’ in the plane.

Key Point

A complex number comprises both a real and an imaginary part. The set of all complex numbers is represented by $\mathbb{C}$.

Determine which of the following numbers are complex or real: $3-4i$, $-2+7i$, and 5.

Solution: For every number, we can clearly see:

$3-4i$: This number has a real part of 3 and an imaginary part of -4, making it a complex number.

$-2+7i$: This number also has both real and imaginary parts, -2 and 7 respectively, therefore, it is a complex number as well.

5: This number is a real number which can be seen as a special case of a complex number where the imaginary part is zero, i.e., $5 = 5+0i$.

Plotting point $z = 2+i$ on the complex coordinate plane.

Solution: To plot the complex number, $z = 2+i$, we move 2 units to the right on the real axis and 1 unit up on the imaginary axis.

Hence, the point $(2,1)$ on the complex plane corresponds to the complex number $z = 2+i$.

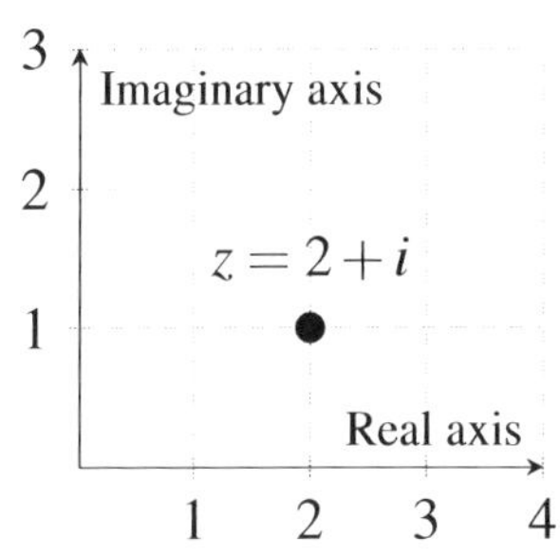

Compute the magnitude of the complex number $z = 3+4i$.

Solution: The magnitude of the complex number $z = 3+4i$ can be found using the formula mentioned above:

$$|z| = \sqrt{3^2+4^2} = \sqrt{9+16} = \sqrt{25} = 5.$$

Hence, the magnitude or absolute value of the complex number $z = 3+4i$ is 5.

12.4 Adding and Subtracting Complex Numbers

Complex numbers are represented as $a+bi$, where $a,b \in \mathbb{R}$ and i satisfies $i^2 = -1$. The operations of addition and subtraction are performed by separately combining the real (a and c) and imaginary (b and d) parts of complex numbers $z_1 = a+bi$ and $z_2 = c+di$.

Key Point

The sum and difference of $z_1 = a + bi$, and $z_2 = c + di$ are computed as:

$$z_1 + z_2 = (a+c) + (b+d)i, \quad \text{and} \quad z_1 - z_2 = (a-c) + (b-d)i.$$

Two complex numbers can be added by adding together their real components and imaginary components separately. The same rule applies for subtraction.

 Solve $(8+4i)+(6-2i)$.

Solution: First, remove parentheses:

$$(8+4i)+(6-2i) = 8+4i+6-2i.$$

Next, combine like terms:

$$8+4i+6-2i = 14+2i.$$

 Solve $(10+8i)-(8-3i)$.

Solution: Start by removing parentheses:

$$(10+8i)-(8-3i) = 10+8i-8+3i.$$

Finally, group-like terms:

$$10+8i-8+3i = 2+11i.$$

 Solve $(-5-3i)-(2+4i)$.

Solution: Initiate by removing parentheses by multiplying -1 by the second parentheses:

$$(-5-3i)-(2+4i) = -5-3i-2-4i.$$

Combine like terms to find the final answer:

$$-5-3i-2-4i = -7-7i.$$

12.5 Multiplying and Dividing Complex Numbers

When dealing with complex numbers, multiplication and division introduce more steps. In the multiplication of complex numbers, you can make use of the FOIL (First-Out-In-Last) method and always remember that $i^2 = -1$.

Key Point

Suppose that $z_1 = (a+bi)$, and $z_2 = (c+di)$. Then, for calculating z_1z_2 we use the following rule:

$$z_1z_2 = (a+bi)(c+di) = (ac-bd)+(ad+bc)i.$$

Division involves using the concept of conjugates, where the conjugate of $z = a+bi$ is $\bar{z} = a-bi$, aiding in the elimination of i in the denominator.

Key Point

The process of dividing a complex number $z_1 = a+bi$ by another complex number $z_2 = c+di$ as follows:

$$\frac{z_1}{z_2} = \frac{a+bi}{c+di} = \frac{a+bi}{c+di} \times \frac{c-di}{c-di} = (\frac{ac+bd}{c^2+d^2}) + (\frac{bc-ad}{c^2+d^2})i.$$

Solve $\frac{6-2i}{2+i}$.

Solution: To find the solution, we use the rule for dividing complex numbers and compute as follows:

$$\frac{6-2i}{2+i} \times \frac{2-i}{2-i} = \frac{(6(2)+(-2)1)}{4+1} + \frac{-2(2)-6(1)}{4+1}i = 2-2i.$$

Solve $(2-3i)(6-3i)$.

Solution: We use the multiplication rule for imaginary numbers to solve this:

$$(2-3i)(6-3i) = [(2(6)-(-3)(-3))+(2(-3)+(-3)6)i] = 3-24i.$$

Solve $\frac{3-2i}{4+i}$.

Solution: We apply the rule for dividing complex numbers and get the following:

$$\frac{3-2i}{4+i} \times \frac{4-i}{4-i} = \frac{3(4)+(-2)(1)+(-2(4)-3(1))i}{4^2+1} = \frac{10}{17} - \frac{11}{17}i.$$

12.6 Rationalizing Imaginary Denominators

When simplifying fractions with imaginary numbers in the denominator, we use rationalization to eliminate the complex number from the denominator, keeping the fraction's value unchanged. The process involves three steps:

1. *Find the conjugate* of the denominator. The conjugate of $a+bi$ is $a-bi$, and vice versa.
2. *Multiply both numerator and denominator* by the conjugate of the denominator. This step does not change the fraction's value but removes the imaginary part from the denominator.
3. *Simplify the fraction* to its simplest form after the imaginary part is eliminated.

Key Point

To rationalize a complex fraction $\frac{a+bi}{c+di}$, multiply both the numerator and the denominator by the conjugate of the denominator, $c-di$.

Rationalize $\dfrac{6i}{2-i}$.

Solution: Start by multiplying both numerator and denominator by the conjugate $\dfrac{2+i}{2+i}$:

$$\frac{6i}{2-i} = \frac{6i(2+i)}{(2-i)(2+i)}.$$

Apply the complex arithmetic rule $(a+bi)(a-bi) = a^2+b^2$ to get $2^2+(-1)^2 = 5$. Therefore,

$$\frac{6i(2+i)}{5} = -\frac{6}{5} + \frac{12}{5}i.$$

Rationalize $\dfrac{8-2i}{2i}$.

Solution: We start by factoring 2 from both sides:

$$\frac{8-2i}{2i} = \frac{2(4-i)}{2i},$$

then divide both sides by 2: $\frac{2(4-i)}{2i} = \frac{(4-i)}{i}$, next we multiply both numerator and denominator by $\frac{i}{i}$:

$$\frac{(4-i)}{i} \times \frac{i}{i} = \frac{(4i-i^2)}{i^2} = \frac{1+4i}{-1} = -1-4i.$$

Example Rationalize $\frac{4-3i}{6i}$.

Solution: Multiply both numerator and denominator by $\frac{i}{i}$:

$$\frac{4-3i}{6i} = \frac{(4-3i)(i)}{6i(i)}.$$

Simplify using the identity $i^2 = -1$ to get

$$\frac{4i-3i^2}{6(-1)} = \frac{4i-3(-1)}{-6} = -\frac{1}{2} - \frac{2}{3}i.$$

12.7 Polar Form of Complex Numbers

Exploring the polar or trigonometric form of complex numbers facilitates easier computation of operations such as multiplication, division, and exponentiation. The polar form of a complex number $z = a + bi$ is given by:

$$z = r(\cos(\theta) + i\sin(\theta)),$$

where r is the magnitude and θ is the angle in radians. The magnitude r is calculated as $\sqrt{a^2+b^2}$, and the angle θ, known as the argument, is measured from the positive real axis to the line segment that represents the complex number.

As shown in the Figure, the real part a and the imaginary part b correspond to the coordinates of the point representing the complex number in the complex plane, while r and θ represent the polar coordinates of the same point.

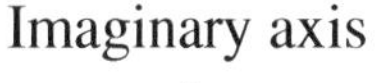

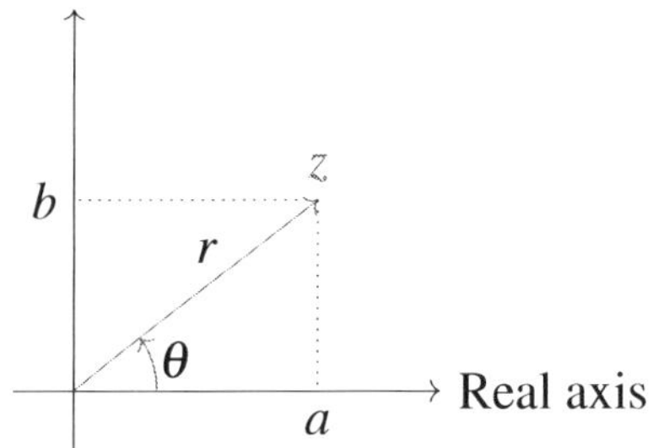

Key Point

To convert from Cartesian to polar form, use $r = \sqrt{a^2+b^2}$ and $\theta = \arctan\left(\frac{b}{a}\right)$. For conversion from polar to Cartesian, apply $a = r\cos(\theta)$ and $b = r\sin(\theta)$.

Adjust θ based on the point's quadrant:

1. Quadrant 1: $\theta = \arctan\left(\frac{y}{x}\right)$
2. Quadrants 2 and 3: $\theta = \pi + \arctan\left(\frac{y}{x}\right)$
3. Quadrant 4: $\theta = 2\pi + \arctan\left(\frac{y}{x}\right)$

 Find the trigonometric form of the complex number $8+6i$, where $0<\theta<\pi$.

Solution: First, we need to find its magnitude (r) and argument (θ). The magnitude (r) of a complex number $z=x+yi$ is given by: $r=\sqrt{x^2+y^2}$. In our case, $x=8$ and $y=6$, so we have:

$$r=\sqrt{6^2+8^2}=10.$$

The argument (θ) of a complex number is calculated using the arctan function: $\theta=\arctan\left(\frac{y}{x}\right)$. Plugging our values in, we get:

$$\theta=\arctan\left(\frac{6}{8}\right)=\arctan(0.75)\approx 0.6435 \text{ radians}.$$

Since our angle θ satisfies the given condition $0<\theta<\pi$, we write the number in trigonometric form as follows: $z=10(\cos(0.6435)+i\sin(0.6435))$.

12.8 Multiplying and Dividing in Polar Form

Multiplication and division of complex numbers are simplified in their trigonometric (polar) form, avoiding direct manipulation of the imaginary unit i. Consider two complex numbers in polar form:

$$z_1=r_1(\cos(\theta_1)+i\sin(\theta_1)), \text{ and } z_2=r_2(\cos(\theta_2)+i\sin(\theta_2)).$$

Here, r_1 and r_2 represent the magnitudes, and θ_1 and θ_2 are the angles of z_1 and z_2, respectively.

Key Point

To multiply z_1 and z_2, the magnitudes are multiplied and their angles are added:

$$z_1z_2=(r_1r_2)(\cos(\theta_1+\theta_2)+i\sin(\theta_1+\theta_2)).$$

Key Point

To divide z_1 by z_2, the magnitudes are divided, and their angles are subtracted:

$$\frac{z_1}{z_2}=\left(\frac{r_1}{r_2}\right)(\cos(\theta_1-\theta_2)+i\sin(\theta_1-\theta_2)).$$

These operations showcase the efficiency of polar form for complex number arithmetic, particularly for multiplication and division.

 Given the complex numbers:

$$z_1 = 4\left(\cos\left(30^\circ\right) + i\sin\left(30^\circ\right)\right), \text{ and } z_2 = 5\left(\cos\left(150^\circ\right) + i\sin\left(150^\circ\right)\right),$$

calculate the product and quotient of these two numbers.

Solution: We aim to calculate the product and quotient of two complex numbers when expressed in trigonometric form. The process involves the following steps: For the product, multiply their magnitudes and add their angles. For the quotient, divide their magnitudes and subtract their angles

The product is:

$$\begin{aligned} z_1 z_2 &= \left(r_1 r_2\right)\left(\cos\left(\theta_1 + \theta_2\right) + i\sin\left(\theta_1 + \theta_2\right)\right) \\ &= (4 \times 5)\left(\cos\left(30^\circ + 150^\circ\right) + i\sin\left(30^\circ + 150^\circ\right)\right) \\ &= 20\left(\cos\left(180^\circ\right) + i\sin\left(180^\circ\right)\right). \end{aligned}$$

The quotient is:

$$\begin{aligned} \frac{z_1}{z_2} &= \left(\frac{r_1}{r_2}\right)\left(\cos\left(\theta_1 - \theta_2\right) + i\sin\left(\theta_1 - \theta_2\right)\right) \\ &= \left(\frac{4}{5}\right)\left(\cos\left(30^\circ - 150^\circ\right) + i\sin\left(30^\circ - 150^\circ\right)\right) \\ &= 0.8\left(\cos\left(-120^\circ\right) + i\sin\left(-120^\circ\right)\right) \\ &= 0.8\left(\cos\left(240^\circ\right) + i\sin\left(240^\circ\right)\right). \end{aligned}$$

12.9 Powers and Roots in Polar Form

To find powers and roots of complex numbers we can use the polar form of complex numbers and De Moivre's Theorem.

Key Point

Using De Moivre's Theorem, a complex number $z = r(\cos\theta + i\sin\theta)$ raised to a positive integer n is:

$$z^n = r^n(\cos(n\theta) + i\sin(n\theta)).$$

Key Point

The n-th roots of a complex number are given by:

$$z_k = r^{\frac{1}{n}} \left[\cos\left(\frac{\theta + 2k\pi}{n} \right) + i \sin\left(\frac{\theta + 2k\pi}{n} \right) \right],$$

for $k = 0, 1, \ldots, n-1$. This results in n distinct n-th roots.

Example Find the fourth power of the complex number $z = 3\left(\cos\frac{\pi}{6} + i\sin\frac{\pi}{6}\right)$ in polar form.

Solution: We are given $z = 3\left(\cos\frac{\pi}{6} + i\sin\frac{\pi}{6}\right)$, so the magnitude r is 3, and the angle θ is $\frac{\pi}{6}$. We want to find z^4. Using De Moivre's theorem, we have: $z^4 = \left[3^4\left(\cos\frac{4\pi}{6} + i\sin\frac{4\pi}{6}\right)\right]$. First, let us raise the magnitude to the power of 4:

$$r^4 = 3^4 = 81.$$

Next, we multiply the angle θ by 4:

$$\theta \times 4 = \left(\frac{\pi}{6}\right) \times 4 = \frac{2\pi}{3}.$$

Now, we have the magnitude 81 and the angle $\frac{2\pi}{3}$ for z^4, and so we get:

$$z^4 = 81\left(\cos\frac{2\pi}{3} + i\sin\frac{2\pi}{3}\right).$$

Example Find the three cube roots of the complex number $z = 8\left(\cos\frac{\pi}{4} + i\sin\frac{\pi}{4}\right)$ in polar form.

Solution: We are given $z = 8\left(\cos\frac{\pi}{4} + i\sin\frac{\pi}{4}\right)$, so the magnitude r is 8, and the angle θ is $\frac{\pi}{4}$. We want to find $z^{\frac{1}{3}}$, which is the same as finding the three cube roots of z. Using the formula, we have:

$$z_k = 8^{\frac{1}{3}} \left[\cos\left(\frac{\frac{\pi}{4} + 2k\pi}{3} \right) + i \sin\left(\frac{\frac{\pi}{4} + 2k\pi}{3} \right) \right],$$

for $k = 0$, 1, 2. First, let us find the cube root of the magnitude:

$$r^{\frac{1}{3}} = 8^{\frac{1}{3}} = 2.$$

Next, we substitute the values of (k) and simplify the angles:

$$z_0 = 2\left[\cos\left(\frac{\pi}{12}\right) + i\sin\left(\frac{\pi}{12}\right)\right],$$
$$z_1 = 2\left[\cos\left(\frac{9\pi}{12}\right) + i\sin\left(\frac{9\pi}{12}\right)\right] = 2\left[\cos\left(\frac{3\pi}{4}\right) + i\sin\left(\frac{3\pi}{4}\right)\right],$$
$$z_2 = 2\left[\cos\left(\frac{17\pi}{12}\right) + i\sin\left(\frac{17\pi}{12}\right)\right].$$

12.10 Practices

1) Which of the following polar equations represents a circle centered at the origin with radius 3?

☐ A. $r = 6\cos(\theta)$
☐ B. $r = 3$
☐ C. $r = 2 + \cos(\theta)$
☐ D. $r = 3\sin(\theta)$

2) Identify the polar graph that corresponds to the polar equation $r = 1 + \sin(\theta)$.

☐ A. Circle with radius 1 centered at $(1, 0)$
☐ B. Cardioid centered at the origin touching the initial line
☐ C. Lemniscate centered at the origin
☐ D. Spiral increasing in distance from the origin

3) Convert the polar coordinates $\left(4, \frac{2\pi}{3}\right)$ to rectangular coordinates.

☐ A. $\left(2, 2\sqrt{3}\right)$
☐ B. $\left(-2, 2\sqrt{3}\right)$
☐ C. $\left(-2\sqrt{3}, 2\right)$
☐ D. $\left(-2, -2\sqrt{3}\right)$

4) What are the polar coordinates of the rectangular point $(-3, -3)$?

☐ A. $\left(3\sqrt{2}, \frac{5\pi}{4}\right)$
☐ B. $\left(3\sqrt{2}, \frac{7\pi}{4}\right)$
☐ C. $\left(3\sqrt{2}, \frac{3\pi}{4}\right)$
☐ D. $\left(3\sqrt{2}, \frac{\pi}{4}\right)$

5) Which of the following numbers is a complex number with a non-zero imaginary part?

☐ A. $5 - 0i$
☐ B. -3
☐ C. $\sqrt{2} + 5i$
☐ D. π

6) If $z = 3 - 2i$ and $w = -1 + 4i$, what is $z + w$?

☐ A. $2+2i$
☐ B. $4+6i$
☐ C. $2-6i$
☐ D. $4-2i$

7) Calculate the sum of the complex numbers $3+5i$ and $2-7i$.

☐ A. $6-2i$
☐ B. $1-12i$
☐ C. $5-2i$
☐ D. $5+12i$

8) If $z=-1+4i$ and $w=7-2i$, what is $z-w$?

☐ A. $-8+6i$
☐ B. $8-6i$
☐ C. $6+2i$
☐ D. $-6+2i$

9) Multiply the complex numbers $(3-2i)$ and $(1+4i)$.

☐ A. $11+10i$
☐ B. $11-10i$
☐ C. $7+14i$
☐ D. $5+2i$

10) What is the result of dividing the complex number $5+i$ by $2-3i$?

☐ A. $\frac{7}{13}+\frac{15}{13}i$
☐ B. $\frac{1}{13}+\frac{17}{13}i$
☐ C. $\frac{7}{13}-\frac{15}{13}i$
☐ D. $\frac{7}{13}+\frac{17}{13}i$

11) Rationalize the denominator of the complex fraction $\frac{3}{1+2i}$.

☐ A. $\frac{3-6i}{-5}$
☐ B. $\frac{3+6i}{5}$
☐ C. $\frac{1-2i}{5}$
☐ D. $\frac{3-6i}{5}$

12) What is the result of rationalizing the denominator of $\frac{2+i}{3i}$?

☐ A. $-i+\frac{2}{3}$
☐ B. $-\frac{2}{3}-i$
☐ C. $-\frac{1}{3}-\frac{2}{3}i$
☐ D. $\frac{1}{3}-\frac{2}{3}i$

13) Which of the following is the polar form of the complex number $-5+5\sqrt{3}i$?

☐ A. $10\left(\cos\left(\frac{\pi}{3}\right)+i\sin\left(\frac{\pi}{3}\right)\right)$
☐ B. $10\left(\cos\left(\frac{\pi}{6}\right)+i\sin\left(\frac{\pi}{6}\right)\right)$
☐ C. $10\left(\cos\left(\frac{2\pi}{3}\right)+i\sin\left(\frac{2\pi}{3}\right)\right)$
☐ D. $10\left(\cos\left(-\frac{\pi}{3}\right)+i\sin\left(-\frac{\pi}{3}\right)\right)$

14) Given the complex number in polar form $6\left(\cos\left(\frac{5\pi}{4}\right)+i\sin\left(\frac{5\pi}{4}\right)\right)$, what are the Cartesian coordinates (a,b)?

☐ A. $(3\sqrt{2},3\sqrt{2})$
☐ B. $(3\sqrt{2},-3\sqrt{2})$
☐ C. $(-3\sqrt{2},3\sqrt{2})$
☐ D. $(-3\sqrt{2},-3\sqrt{2})$

15) Given two complex numbers $z_1=2(\cos(45^\circ)+i\sin(45^\circ))$ and $z_2=3(\cos(60^\circ)+i\sin(60^\circ))$, what is the product $z_1\cdot z_2$?

☐ A. $6(\cos(105^\circ)+i\sin(105^\circ))$
☐ B. $6(\cos(135^\circ)+i\sin(135^\circ))$
☐ C. $6(\cos(105^\circ)-i\sin(105^\circ))$
☐ D. $5(\cos(105^\circ)+i\sin(105^\circ))$

16) If $z_1=8(\cos(270^\circ)+i\sin(270^\circ))$ and $z_2=2(\cos(90^\circ)+i\sin(90^\circ))$, what is the quotient $\frac{z_1}{z_2}$?

☐ A. $4(\cos(180^\circ)+i\sin(180^\circ))$
☐ B. $4(\cos(360^\circ)+i\sin(360^\circ))$
☐ C. $3(\cos(-180^\circ)+i\sin(-180^\circ))$
☐ D. $3(\cos(180^\circ)+i\sin(180^\circ))$

17) Calculate the sixth power of the complex number $z=2\left(\cos\frac{\pi}{3}+i\sin\frac{\pi}{3}\right)$ in polar form.

☐ A. $64\left(\cos\pi+i\sin\pi\right)$
☐ B. $64\left(\cos\frac{2\pi}{3}+i\sin\frac{2\pi}{3}\right)$
☐ C. $64\left(\cos 2\pi+i\sin 2\pi\right)$
☐ D. $64\left(\cos\pi+i\sin\frac{\pi}{3}\right)$

18) Which of the following is a square root of the complex number $z=4\left(\cos\frac{3\pi}{2}+i\sin\frac{3\pi}{2}\right)$ in polar form?

☐ A. $2\left(\cos\frac{3\pi}{4}+i\sin\frac{3\pi}{4}\right)$
☐ B. $2\left(\cos\frac{7\pi}{4}+i\sin\frac{7\pi}{4}\right)$
☐ C. $2\left(\cos\frac{5\pi}{2}+i\sin\frac{5\pi}{2}\right)$
☐ D. Both A and B are correct.

Answer Keys

1) B. $r=3$

2) B. Cardioid centered at the origin touching the initial line

3) B. $\left(-2,2\sqrt{3}\right)$

4) A. $\left(3\sqrt{2},\frac{5\pi}{4}\right)$

5) C. $\sqrt{2}+5i$

6) A. $2+2i$

7) C. $5-2i$

8) A. $-8+6i$

9) A. $11+10i$

10) D. $\frac{7}{13}+\frac{17}{13}i$

11) D. $\frac{3-6i}{5}$

12) D. $\frac{1}{3}-\frac{2}{3}i$

13) C. $10\left(\cos\left(\frac{2\pi}{3}\right)+i\sin\left(\frac{2\pi}{3}\right)\right)$

14) D. $(-3\sqrt{2},-3\sqrt{2})$

15) A. $6(\cos(105^\circ)+i\sin(105^\circ))$

16) A. $4(\cos(180^\circ)+i\sin(180^\circ))$

17) C. $64\left(\cos 2\pi+i\sin 2\pi\right)$

18) D. Both A and B are correct.

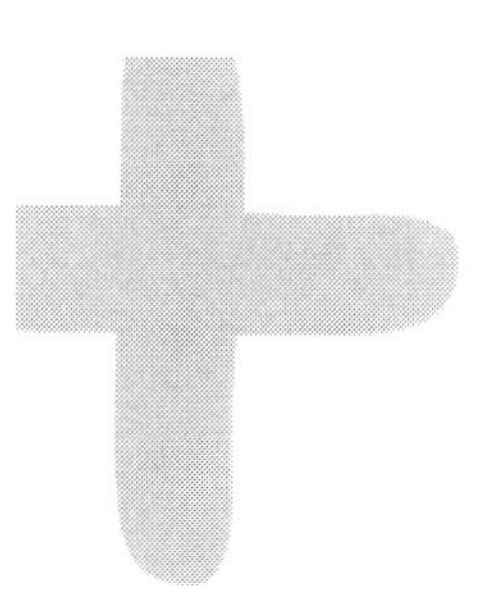

13. Real Numbers and Relation

13.1 Real Numbers

Real numbers include all the numbers that can be represented on a number line, covering both rational and irrational numbers.

Rational numbers are expressible as $\frac{p}{q}$, where p and q are integers and $q \neq 0$, and include:

- **Integers**: Numbers such as $0, 1, -1, 2, -2, \cdots$.
- **Fractions**: Quotients of integers, e.g., $\frac{1}{2}, \frac{3}{4}, \frac{-2}{5}, \cdots$.

Irrational numbers, which cannot be expressed as $\frac{p}{q}$ with p and q as integers, include:

- **Transcendental Numbers**: Non-algebraic numbers like e (base of natural logarithms) and π (circumference to diameter ratio of a circle).
- **Algebraic Irrationals**: Solutions to polynomial equations with rational coefficients but are irrational, such as $\sqrt{2}$.

Key Point

Real numbers encompass a broad range, including rational numbers (such as integers and fractions) and irrational numbers (like transcendental and algebraic irrationals).

Is 0 a real number?

Solution: Yes, 0 is a real number, specifically a rational number because it can be written as $\frac{0}{1}$, where $p = 0$ and $q = 1$, fitting the rational numbers' definition.

Is $2.17231234\cdots$ a real number?

Solution: Yes, $2.17231234\cdots$ is a real number, specifically an irrational number. This follows from the property that non-repeating and non-terminating decimals are classified as irrational numbers.

13.2 Real Number Line

The real number line is a line that shows all real numbers, including both rational and irrational numbers. Each point on the line matches a unique real number. The line is split into two equal parts by the origin, marked as 0, with positive numbers to the right and negative numbers to the left. This line goes on forever in both directions, meaning there is no highest or lowest real number.

Key Point

The distance between two points on the real number line is the absolute value of the difference between their real numbers.

Example Plot the number $\frac{4}{5}$ on the real number line.

Solution: To find $\frac{4}{5}$ on the real number line, start at 1, as $\frac{4}{5}$ is less than 1. Divide the section between 0 and 1 into five equal parts. $\frac{4}{5}$ is located four parts from 0, towards 1. Mark this point to show $\frac{4}{5}$, which is nearer to 1 than to 0.

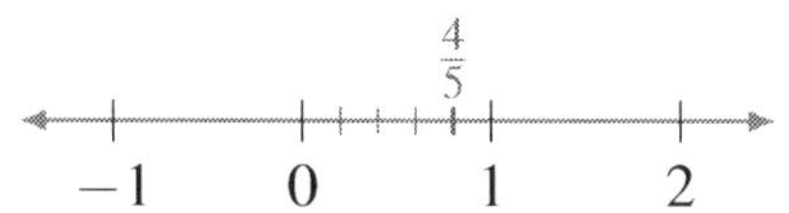

13.3 Coordinate Plane

The coordinate plane, or Cartesian plane, extends the concept of the real number line by introducing a perpendicular line through the real number line's zero point, creating two intersecting lines at right angles. This setup forms the Cartesian plane, composed of a horizontal line (the x-axis) and a vertical line (the y-axis), intersecting at a point called the origin, with coordinates $(0,0)$.

Key Point

The coordinates of a point (x,y) in the Cartesian plane represent the positions in the horizontal and vertical directions from the origin, respectively.

In a coordinate plane, quadrants are regions divided by axes.

First quadrant: $x > 0$, $y > 0$.

Second quadrant: $x < 0$, $y > 0$.

Third quadrant: $x < 0$, $y < 0$.

Fourth quadrant: $x > 0$, $y < 0$.

Plot the point $(2,3)$ on the coordinate plane.

Solution: Starting at the origin, move two units to the right along the x-axis.

Then, ascend three units along the y-axis to mark the point. Below is the graphical representation of the point $(2,3)$ on the Cartesian plane.

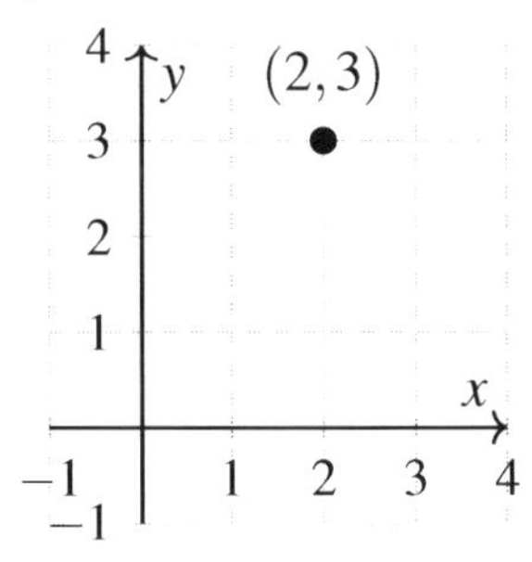

Find the coordinates of point 'A' in the following.

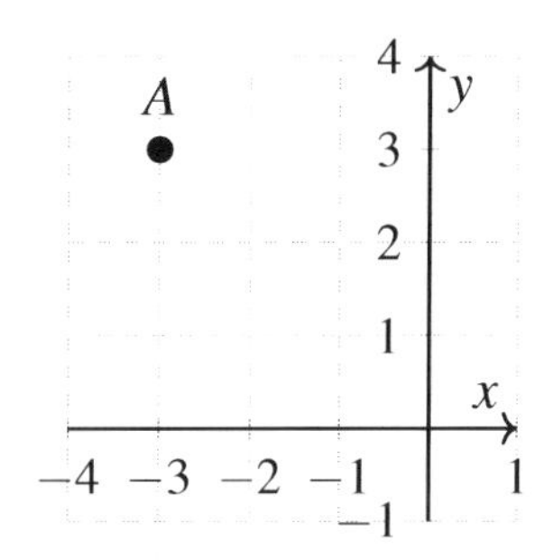

Solution: The point A, positioned three units left and three units up from the origin, is represented by the coordinates $(-3,3)$.

13.4 Relation

In mathematics, an ordered pair (a,b) represents two elements with a specific sequence, distinct from (b,a) unless $a = b$. The Cartesian product $A \times B$ forms all possible ordered pairs from sets A and B. For example, if $A = \{a,b\}$ and $B = \{1\}$, then $A \times B = \{(a,1),(b,1)\}$.

A relation from set A to set B is a subset of $A \times B$, i.e., a set of multiple pairs (a,b), where a is an element from set A, and b is an element from set B.

Key Point

A relation between sets A and B is defined by ordered pairs (a,b), where $a \in A$ and $b \in B$. Relations serve to establish a relationship between two sets.

Example Consider two sets of ordered pairs, $\{(4,a),(2,x),(6,b)\}$ and $\{(2,b),(6,b),(4,a)\}$ which are representations of the same relation R from $A = \{2,4,6\}$ and $B = \{a,b,c\}$. Find the value of x.

Solution: To find the value of x in this scenario, we look at the elements of the two sets that correspond to each other. The ordered pairs $(4,a)$ and $(6,b)$ can be found in both sets, indicating that a and b are independent of the changes in the set. The remaining pair $(2,x)$ in the first set corresponds to the pair $(2,b)$ in the second set. Therefore, for the two sets of ordered pairs to represent the same relation, x must be equal to b. So, $x = b$.

Example Is the set of ordered pairs of $R = \{(1,-2),(-3,8),(2,0),(4,4),(2,-2)\}$ a relation from the set $A = \{1,2,-3,4,-5\}$ to set $B = \{-2,4,-6,8\}$?

Solution: To determine whether the set of ordered pairs R constitutes a relation from set A to set B, every first element in the ordered pairs of R must belong to set A, and every second element in the ordered pairs of R must belong to set B. Upon examination, it is observed that all the first elements of the ordered pairs in R are elements of set A. However, among the second elements in R, there exists 0, which is not an element of set B. Therefore, R is not a relation from A to B.

13.5 Showing the Relation in the Coordinate Plane

Given a relation represented as a set of ordered pairs (a,b), we can visualize this relationship on a coordinate plane, where each ordered pair is treated as a point (x,y).

Key Point

Every unique ordered pair (a,b) in the relation between two sets corresponds to a unique point (x,y) on the coordinate plane.

Given the set of ordered pairs for relation

$$R = \{(-2,4),(-1,3),(1,0),(-1,-2),(4,2)\}.$$

How is this relation represented on the coordinate plane?

Solution: To visualize a relation on a coordinate plane, we plot the given ordered pairs as points, where the first value of the pair is the x-coordinate, and the second value is the y-coordinate.

For instance, the ordered pair $(-2,4)$ is plotted at $x=-2$ and $y=4$. Each plotted point corresponds to an ordered pair in the relation R, demonstrating the graphical representation of the relation on the coordinate plane.

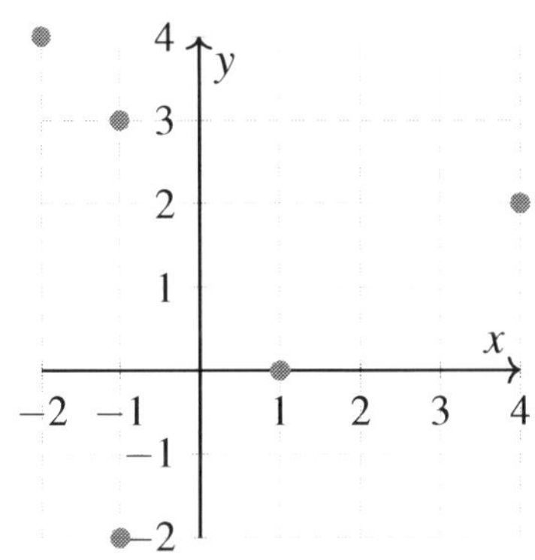

13.6 Domain and Range of Relation

The Domain of a relation R is the set of all x-coordinates from its ordered pairs, representing the inputs. Conversely, the Range of R includes all y-coordinates, indicating the outputs. Graphically, the Domain and Range correspond to the set of x and y values of points in R, respectively.

Key Point

The domain of a relation is the set of all the x-coordinates and the range of a relation is the set of all the y-coordinates.

Example Consider the relation $R=\{(-1,2),(3,1),(2,-5)\}$. Determine the domain and range of the relation.

Solution: Given the relation R, we identify the first and second elements in the ordered pairs. We find the domain by taking the set of all first elements in the ordered pairs of the relation. So, Domain $=\{-1,2,3\}$. The Range is determined by the set of all second elements in the ordered pairs of the relation. So, Range $=\{1,2,-5\}$.

 Consider the relation represented in the below graph.

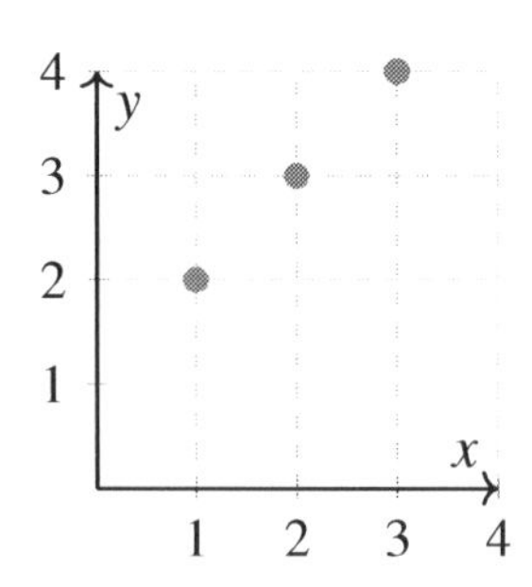

Determine the Domain and Range of the relation.

Solution: Consider the set of ordered pairs $R = \{(1,2),(2,3),(3,4)\}$. The domain is the set of all 'x' values: Domain $= \{1,2,3\}$. The range is the set of all 'y' values: Range $= \{2,3,4\}$.

13.7 Functions

A function is a special type of relation that uniquely maps elements from one set, called the domain, to another set, the range. An entity in the domain is linked to exactly one element in the range. For instance, you can think of a function as a machine with an input and an output. Every input you place into this machine produces precisely one output.

Key Point

A function assigns each element in the domain to exactly one element in the range.

Functions can be represented in various ways: through an equation, a table, a graph, or even a verbal description. This allows us to visualize and understand relations easily and practically.

 Given the set of ordered pairs $R = \{(-1,4),(2,-3),(3,4),(2,0),(1,6)\}$, does R represent a function?

Solution: Each input, or x value, in a function should map to exactly one unique output, or y value. As relation R demonstrates, the input 2 is associated with two different outputs: -3 and 0. This means R does not satisfy the criteria for a function because input 2 does not have a unique output.

13.8 Identifying the Function from the Graph

In mathematics, identifying a function from its graph goes beyond just looking at the graph. It involves examining the graphical representation in terms of its shape, points, and behaviors to infer the relationship between the input values, represented on the x-axis, and the output values, represented on the y-axis. This is an extension to our previous topic on functions, where we saw that each input corresponds precisely with one output.

Key Point

A graph represents a function if, when you draw a vertical line across the graph, the line touches the graph at no more than one point at any location. If the line intersects the graph more than once at any point, it is not a function.

Does the following graph represent a function?

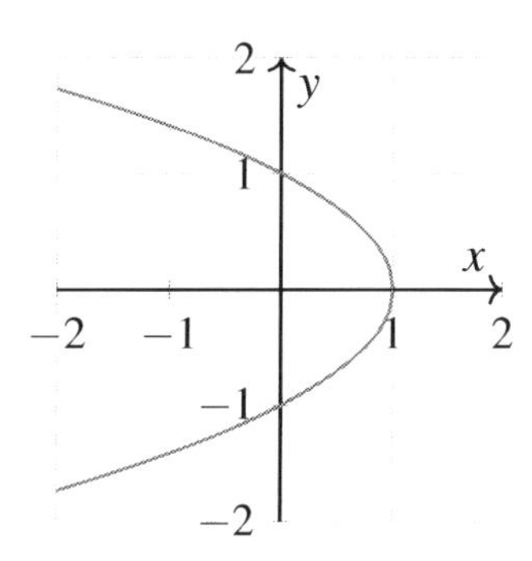

Solution: Consider the vertical line running through $x = 0$. It intersects the graph at two points, implying that a single input ($x = 0$) matches to two different outputs. On applying the vertical line test, we can conclude that the graph does not represent a function.

Given the graph below, can we determine whether it represents a function or not?

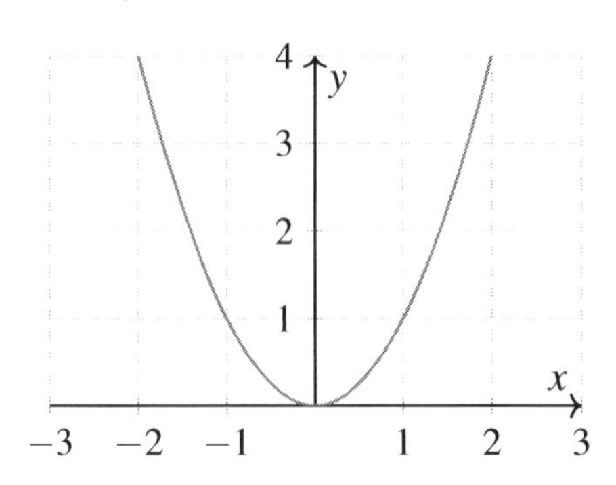

Solution: A vertical line drawn at any position would intersect the graph only once. Since there is just one intersection point, this graph satisfies the vertical line test, and hence, the graph does indeed represent a function.

13.9 Graphs of Basic Functions

Understanding the nature and behavior of basic functions is crucial. Our exploration will focus on linear and quadratic functions.

Linear functions are of the form $y = mx + c$ where m is the slope and c is the y-intercept. The graph of a linear function is a straight line.

For instance, the equation $y = \frac{1}{2}x - 1$ graphically represented would look as follows:

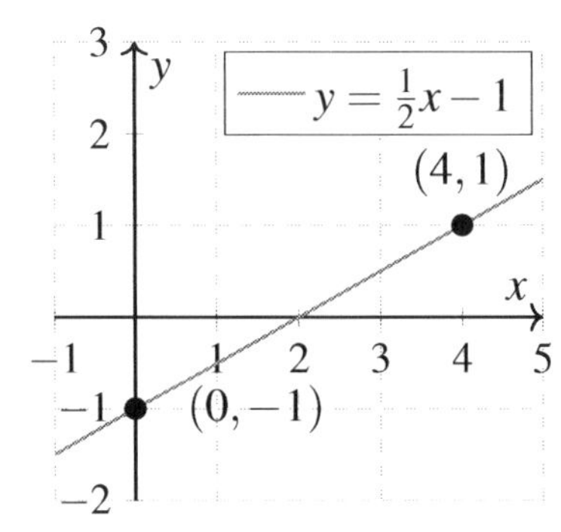

Key Point

Straight line represent the graph of linear functions. Changing the slope alters the angle of the line, and adjusting the y-intercept moves it vertically.

Quadratic functions take the form $y = ax^2 + bx + c$. Their graphs are parabolas. The peaks and troughs of parabolas signify the maximum and minimum points. The positions of these points are determined by the signs and magnitudes of the coefficients.

Sketch the graphs of the following functions and classify them:

$$a)\ y = x+3, \qquad b)\ y = x^2+2x-1.$$

Solution:

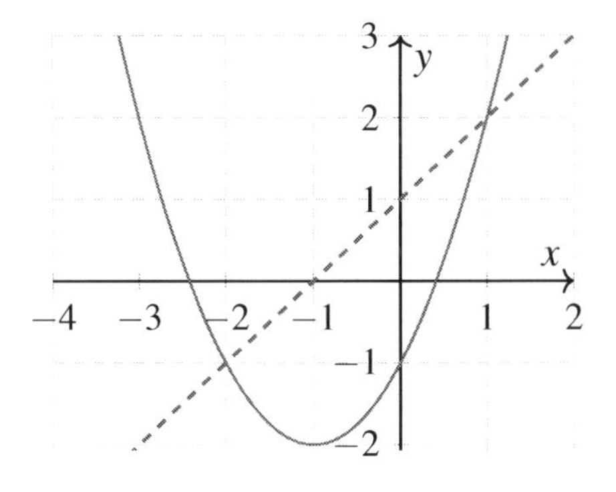

$a)$ This function is a linear function with a slope of 1 and a y-intercept at 3. Use this knowledge to sketch the graph.
$b)$ This equation is of a quadratic function. The curve will open upwards since the coefficient of x^2 is positive.

13.10 Practices

1) Which of the following numbers is a rational number?

☐ A. $\sqrt[3]{5}$
☐ B. π
☐ C. -7.125
☐ D. $\sqrt{2}$

2) Identify the irrational number from the list below.

☐ A. $\frac{9}{4}$
☐ B. 10
☐ C. 1.414213...
☐ D. 0

3) Identify the real number that the point P represents on the number line below, assuming each segment represents a unit length.

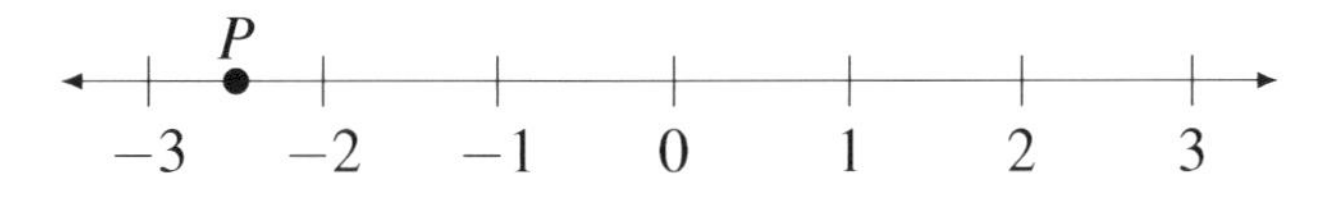

☐ A. -2.5
☐ B. 2.5
☐ C. -3
☐ D. 3

4) Which of the following inequalities best represents the location of the number x, if it is to the left of -1.5 on the real number line?

☐ A. $x \geq -1.5$
☐ B. $x \leq -1.5$
☐ C. $x > -1.5$
☐ D. $x < -1.5$

5) If a point in the Cartesian plane has an x-coordinate of 5 and lies on the x-axis, which of the following represents its coordinates?

☐ A. $(0,5)$
☐ B. $(5,0)$
☐ C. $(5,5)$
☐ D. $(-5,0)$

6) Which of the following points is located in the second quadrant of the Cartesian plane?

☐ A. $(2,-4)$
☐ B. $(-3,-2)$
☐ C. $(-5,1)$
☐ D. $(3,5)$

7) Given sets $A = \{1,2\}$ and $B = \{x,y,z\}$, which of the following represents the Cartesian product $A \times B$?

☐ A. $\{(1,x),(2,y),(1,z)\}$
☐ B. $\{(1,x),(1,y),(1,z),(2,x),(2,y),(2,z)\}$
☐ C. $\{(x,1),(y,2),(z,1)\}$
☐ D. $\{(x,y),(y,z),(z,x)\}$

8) If $A = \{2,4\}$ and $B = \{3,5,7\}$, which of the following is not a member of the Cartesian product $A \times B$?

☐ A. $(2,3)$
☐ B. $(4,7)$
☐ C. $(4,5)$
☐ D. $(5,4)$

9) Given the set of ordered pairs for relation $S = \{(0,0),(2,4),(3,9),(4,16),(5,25)\}$, which represents a functional relationship between 'x' and 'y'?

☐ A. $y = x^2$
☐ B. $y = \sqrt{x}$
☐ C. $y = 2x$
☐ D. $y = x+1$

10) Which point would be included in the graph of the relation $y = -2x+3$?

☐ A. $(0,3)$
☐ B. $(1,5)$
☐ C. $(-1,-5)$
☐ D. $(2,1)$

11) What is the domain of the relation R represented by the graph below?

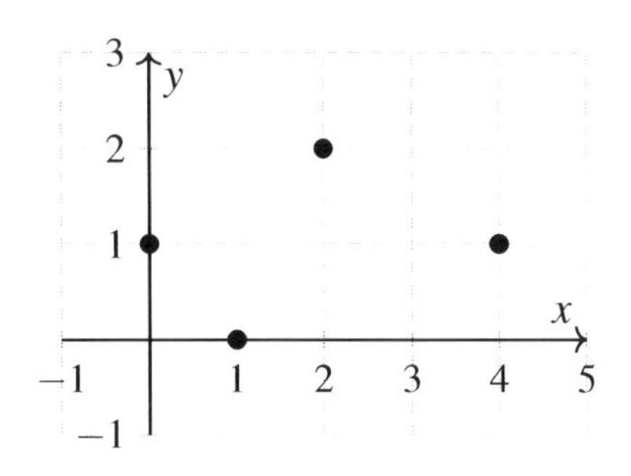

☐ A. {0, 1, 2, 4}

☐ B. {−1, 0, 1, 2, 4}

☐ C. {1}

☐ D. {0, 1, 2, 3, 4}

12) A relation R on a coordinate plane includes the point $(5,-3)$. Which of the following must be true about R?

☐ A. 5 is not in the domain, −3 is not in the range.

☐ B. −3 is in the domain, but not in the range.

☐ C. 5 is in the range, but not in the domain.

☐ D. −3 is in the range, and 5 is in the domain.

13) Which of the following sets of ordered pairs represents a function?

☐ A. $\{(3,4),(-1,6),(3,5),(-1,6)\}$

☐ B. $\{(0,1),(2,1),(3,1),(0,1)\}$

☐ C. $\{(-2,0),(2,0),(2,-2),(3,-1)\}$

☐ D. $\{(4,3),(4,-2),(5,0),(3,4)\}$

14) The relation given by the equation $y = x^2 + 2x - 8$ is a function. For which x value does the function have a 0 as the output?

☐ A. 1

☐ B. −4

☐ C. 0

☐ D. No such x value exists

15) Which of the following graphs does **not** represent a function?

☐ A. A straight line that passes through the origin.

☐ B. A circle centered at the origin.

☐ C. A parabola that opens upwards.

☐ D. An exponential curve that never touches the x-axis.

16) Examine the graph below. Does it represent a function?

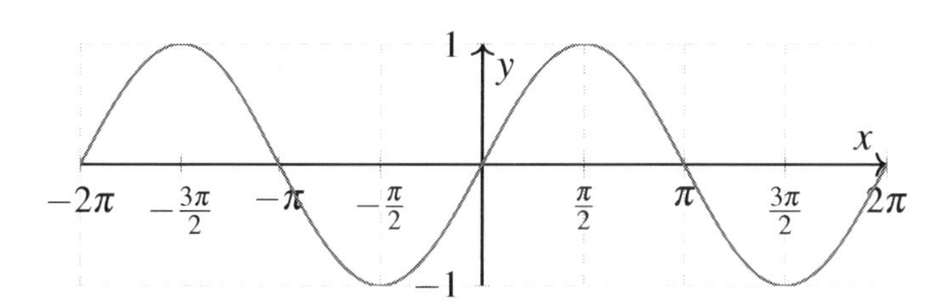

☐ A. Yes

☐ B. No

☐ C. Only on the interval $[-\frac{\pi}{2}, \frac{\pi}{2}]$

☐ D. Only on the interval $[0, \pi]$

17) Which of the following represents the equation of a line with a slope of -3 and a y-intercept of 2?

☐ A. $y = 3x + 2$

☐ B. $y = -3x + 2$

☐ C. $y = -\frac{1}{3}x + 2$

☐ D. $y = -3x - 2$

18) What is the slope of the line $y = 2x - 1$?

☐ A. 1

☐ B. 2

☐ C. -1

☐ D. -2

Answer Keys

1) C. -7.125

2) C. $1.414213\ldots$

3) A. -2.5

4) D. $x < -1.5$

5) B. $(5,0)$

6) C. $(-5,1)$

7) B. $\{(1,x),(1,y),(1,z),(2,x),(2,y),(2,z)\}$

8) D. $(5,4)$

9) A. $y = x^2$

10) A. $(0,3)$

11) A. $\{0, 1, 2, 4\}$

12) D. -3 is in the range, and 5 is in the domain.

13) B. $\{(0,1),(2,1),(3,1),(0,1)\}$

14) B. -4

15) B. A circle centered at the origin.

16) A. Yes

17) B. $y = -3x + 2$

18) B. 2

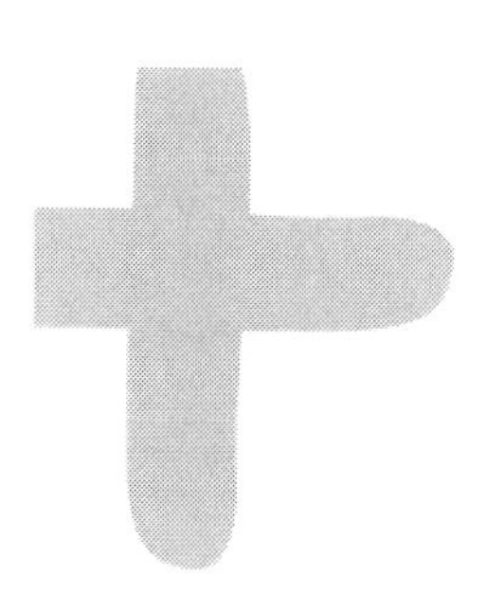

14. Vectors in Two Dimension

14.1 Introduction to Vectors: Vectors in Two Dimensions

Vectors are fundamental in various fields like physics, computer science, and engineering, representing quantities with both magnitude and direction. In two dimensions, a vector is depicted as an arrow from one point to another. The arrow's length indicates magnitude, and its direction signifies the vector's direction. In mathematical terms, a 2-dimensions vector is denoted as (x, y) or $\begin{pmatrix} x \\ y \end{pmatrix}$, with x and y as horizontal and vertical components, respectively.

Key Point

Vectors can be manipulated in a variety of ways such as addition, subtraction, and multiplication by scalars (real numbers) as well as other vectors via the dot or cross product.

Example Consider the vector $A = (2, 3)$. Display this vector in the coordinate plane.

Solution: To represent vector A on the graph, we simply plot the components of A as displacements along the x and y axes, starting from the origin. The horizontal component extends two units to the right, while the vertical component extends three units upwards. Here is how we visualize the vector A:

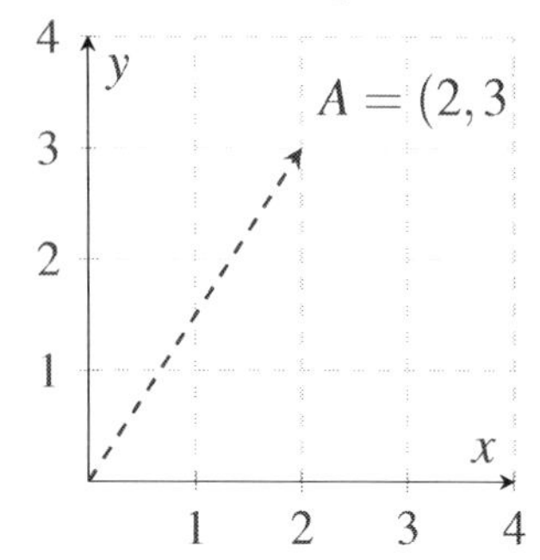

The dashed arrow represents the vector A. It starts from the origin and ends at the point $(2, 3)$, which corresponds to the components of vector A.

Show vector $B = (-3,5)$ in coordinate plane.

Solution: Vector B has components $(-3,5)$, represented as an arrow starting at the origin and ending at the point $(-3,5)$.

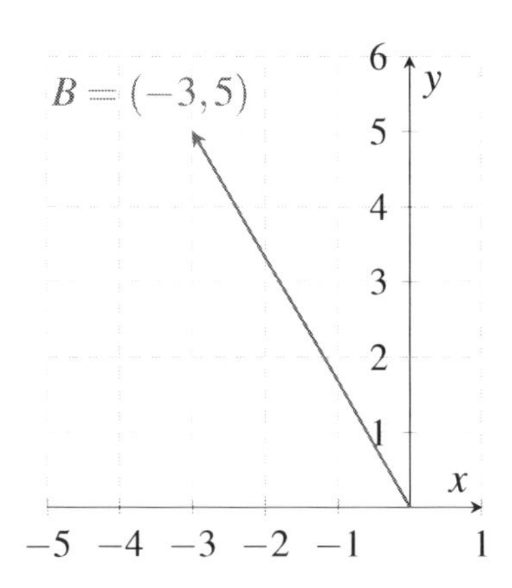

14.2 Equality of Vectors in Two Dimensions

In a 2-dimensional space (the plane), vectors have both magnitude (length) and direction. When discussing vector equality, we focus on these attributes: magnitude and direction. Two 2D vectors are equal only if they have the same magnitude and direction, regardless of their starting points. To make this clearer, when vectors are represented as ordered pairs of numbers (components), this concept becomes more tangible.

Key Point

Vectors $A = (a_1, a_2)$ and $B = (b_1, b_2)$ are equal if and only if their corresponding components are the same. This translates into the equations $a_1 = b_1$ and $a_2 = b_2$.

Note that these conditions must both be fulfilled. A failure in equality for either component would render the vectors unequal.

Are the vectors $A = (4,6)$ and $B = (2^2, \frac{12}{2})$ equal?

Solution: These vectors are indeed equal. This conclusion is derived from matching both the x-component and y-component of A and B. Because both pairs of corresponding components satisfy $a_1 = b_1$ and $a_2 = b_2$, we can say with certainty that $A = B$.

Given vectors $E = (k,5)$ and $F = (3,5)$, for what value of k are the vectors equal?

Solution: For E to be equal to F, k must be 3.

14.3 Scalar Multiplication

Scalar multiplication stretches or shrinks vectors by multiplying each component of the vector by a scalar, which can be positive, negative, or zero. This operation alters the magnitude of a vector while potentially preserving or reversing its direction. Two vectors are considered equal if they share the same magnitude and direction.

Key Point

The formula for scalar multiplication is straightforward. If $v = (x, y)$ is a two-dimensional vector and k is a scalar, then $kv = k(x, y) = (kx, ky)$.

Scalar multiplication can alter a vector's magnitude and potentially its direction, but it cannot displace the vector in space. The resulting vector's direction is determined by the scalar value: a positive scalar maintains the vector's direction, while a negative scalar reverses it. If the scalar is zero, the vector collapses to the origin, forming the zero vector.

Let us say we have the vector $u = (1, 3)$, calculate $2u$, $-3u$, and $0u$.

Solution: The scalar multiplication will modify each component of the vector as follows:

$2u = 2(1, 3) = (2 \times 1, 2 \times 3) = (2, 6)$,
$-3u = -3(1, 3) = (-3 \times 1, -3 \times 3) = (-3, -9)$,
$0u = 0(1, 3) = (0 \times 1, 0 \times 3) = (0, 0)$.

14.4 Vector Addition and Subtraction

Vector addition involves placing two vectors head to tail and drawing the resultant vector from the tail of the first to the head of the second. Remember that the tail of a vector is its initial point, and the head is its terminal point. For vectors $a = (x_1, y_1)$ and $b = (x_2, y_2)$, their sum is given by: $a + b = (x_1 + x_2, y_1 + y_2)$.

Key Point

The sum of two vectors a and b is found by adding their corresponding components.

Vector subtraction finds a vector such that adding it to the second vector results in the first vector. Using the same vectors as above, the difference between a and b is: $a - b = (x_1 - x_2, y_1 - y_2)$.

Key Point

The difference between two vectors a and b is found by subtracting the components of b from the corresponding components of a.

 Given vectors $u = (2,4)$ and $v = (3,-1)$, compute $u+v$.

Solution: We use the vector addition formula to get:

$$u+v = (2,4)+(3,-1) = (2+3,4+(-1)) = (5,3).$$

The figure displays the vector addition $u+v$ graphically.

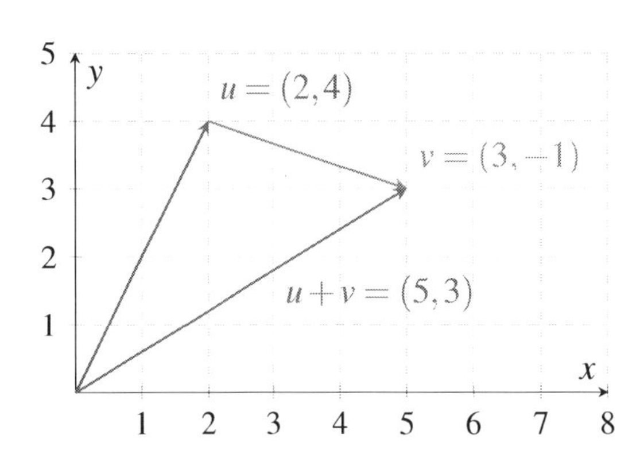

 Find $u-v$ using vectors $u = (2,4)$ and $v = (3,-1)$.

Solution: Using the vector subtraction formula, we get:

$$u-v = (2,4)-(3,-1) = (2-3,4-(-1)) = (-1,5).$$

14.5 Representation of Addition and Subtraction

In two-dimensional space, the parallelogram rule is a valuable tool for visualizing vector addition and subtraction, particularly when vectors lack parallelism or perpendicularity. This visualization aids in comprehending vectors and their operations more effectively.

To apply the parallelogram rule for vector addition, begin by drawing both vectors from the same starting point. Next, construct a parallelogram with these vectors as adjacent sides. The diagonal of the parallelogram, starting from the common initial point, represents the resultant vector or the sum of the two vectors.

The parallelogram rule of vector addition states that if two vectors are considered as two adjacent sides of a parallelogram, their resultant or sum is represented by the closing diagonal of the parallelogram from their common point.

Vector subtraction follows a similar rule. To subtract one vector from another, start by drawing both vectors with the same initial point. Then, create a parallelogram using these vectors as adjacent sides. The diagonal of the parallelogram, starting from the common initial point, represents the resultant vector or the difference between the two vectors.

Key Point

To subtract one vector from another using the parallelogram rule, first find the negative of the vector to be subtracted. Then, add this negative vector to the original vector using the parallelogram rule for vector addition.

Example Given vectors $A = (4,1)$ and $B = (1,3)$, find $A+B$ and $B-A$ geometrically.

Solution: For the addition, i.e., $C = A+B$, perform these steps:

1. Plot both vectors with a common starting point.
2. Draw a parallelogram using these vectors as adjacent sides.
3. The diagonal from the common start point gives the resultant vector C, which can be computed algebraically as $(5,4)$.

For subtraction, i.e., $D = B-A$, follow these steps:

1. Plot vector B and the negative of A, i.e., $-A$ from the same point.
2. Construct a parallelogram using B and $-A$ as adjacent sides.
3. The diagonal emerging from the common start point gives $B-A$, which can be computed algebraically as $(-3,2)$.

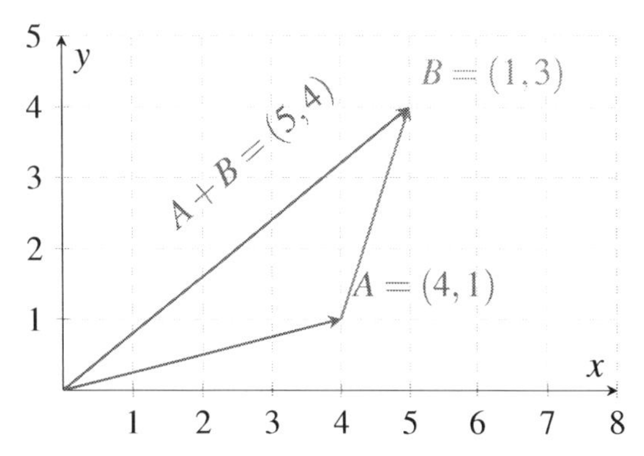

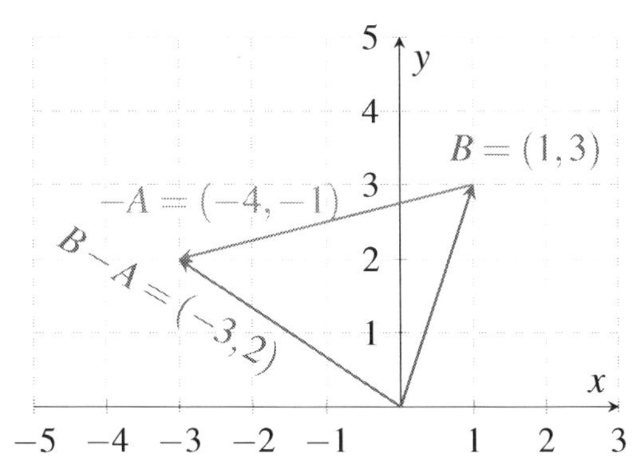

14.6 Length of a Vector

In this section, we explore the concept of vector length, which measures the size of a vector regardless of its direction. The length of a vector is crucial in understanding physical quantities like velocity or force, where it indicates speed or strength, respectively.

Visualize a vector as an arrow on a plane; its length is the arrow's size. This can be found using the Pythagorean theorem. Notably, vector length is a scalar, meaning it is only influenced by magnitude, not direction.

Key Point

The length (or magnitude) of a vector represents its size, independent of direction. Consider a vector $v = (x, y)$, its length, denoted as $|v|$, is given by the formula: $|v| = \sqrt{x^2 + y^2}$.

Find the length of the vector $a = (3, 4)$.

Solution: By substituting in length formula, we have:

$$|a| = \sqrt{3^2 + 4^2} = \sqrt{9 + 16} = \sqrt{25} = 5.$$

Thus, the length of the vector a is 5 units.

14.7 Dot Product and Cross Product

The dot product, also known as the scalar product, combines two vectors into a single scalar value by multiplying their magnitudes and the cosine of the angle between them. To compute it:

1. Multiply the x-components of the two vectors.
2. Multiply the y-components of the two vectors.
3. Add the results from steps 1 and 2.

Key Point

The dot product of vectors $A = (x_1, y_1)$ and $B = (x_2, y_2)$ is given by:

$$A \cdot B = (x_1)(x_2) + (y_1)(y_2).$$

The cross product in two dimensions gives a scalar representing the area of the parallelogram formed by two vectors. It contrasts with the dot product by focusing on the vectors' orientation rather than their alignment. To find the cross product:

1. Multiply the x-component of the first vector by the y-component of the second vector.
2. Multiply the y-component of the first vector by the x-component of the second vector.
3. Subtract the second product from the first.

Key Point

For vectors $A = (x_1, y_1)$ and $B = (x_2, y_2)$, the cross product formula is:

$$A \times B = (x_1)(y_2) - (y_1)(x_2).$$

 Assuming we have vectors $P = (2,3)$ and $Q = (4,-1)$, calculate the dot product.

Solution: $P \cdot Q = (2)(4) + (3)(-1) = 8 - 3 = 5$.

 Given vectors $R = (1,2)$ and $S = (3,4)$, find the cross product.

Solution: $R \times S = (1)(4) - (2)(3) = 4 - 6 = -2$.

14.8 Parallel Vectors

Parallel vectors move in the same or exactly opposite directions, but their lengths might differ. Vectors are parallel if one is simply a scaled version of the other.

Key Point

Two vectors $A = (x_1, y_1)$ and $B = (x_2, y_2)$ are parallel if there is a number k such that $x_1 = kx_2$ and $y_1 = ky_2$.

A vector parallel to $E = (e_1, e_2)$ is obtained by scaling E's components by any scalar $k \neq 0$, denoted as $kE = (ke_1, ke_2)$. Interestingly, the zero vector is unique because it is parallel to all vectors, derived from multiplying any vector by 0. This makes it the sole vector parallel to others regardless of their directions.

 Given that $A = (2,4)$ and $B = (1,2)$, are these vectors parallel?

Solution: The vectors A and B are parallel because $A = 2B$, indicating they have the same direction.

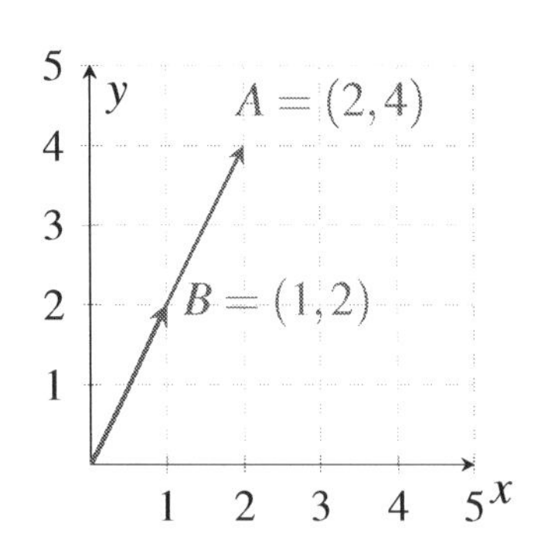

14.9 Orthogonal Vectors

Orthogonal vectors are two vectors that are perpendicular to each other, forming a 90° angle. Their main feature is that their dot product equals zero. To check if two vectors are orthogonal, follow these steps:

1. Calculate the dot product $A \cdot B$.
2. If the result is zero, the vectors are orthogonal.

Key Point

For vectors $A = (x_1, y_1)$ and $B = (x_2, y_2)$ to be orthogonal, the condition $A \cdot B = 0$ must be met.

Remember, a nonzero vector cannot be orthogonal to itself since its dot product is always positive. The only exception is the zero vector, which is orthogonal to every vector, including itself. To find a vector orthogonal to a given vector $E = (e_1, e_2)$, swap its components and change the sign of one. Therefore, vectors orthogonal to E can be $(-e_2, e_1)$ or $(e_2, -e_1)$.

Example Given vectors $P = (3,2)$ and $Q = (-4,6)$, determine if P and Q are orthogonal.

Solution: First, calculate the dot product for vectors P and Q:

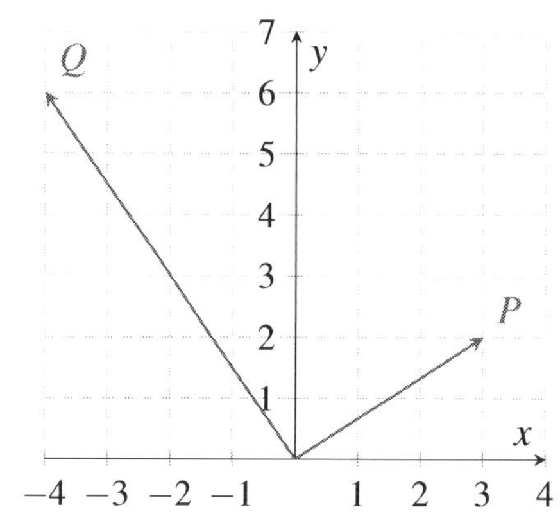

$$P \cdot Q = (3)(-4) + (2)(6) = -12 + 12 = 0.$$

Since the dot product is zero, vectors P and Q are indeed orthogonal.

Example Let $M = (1,-2)$ and $N = (-2,1)$. Are these vectors orthogonal?

Solution: Compute the dot product:

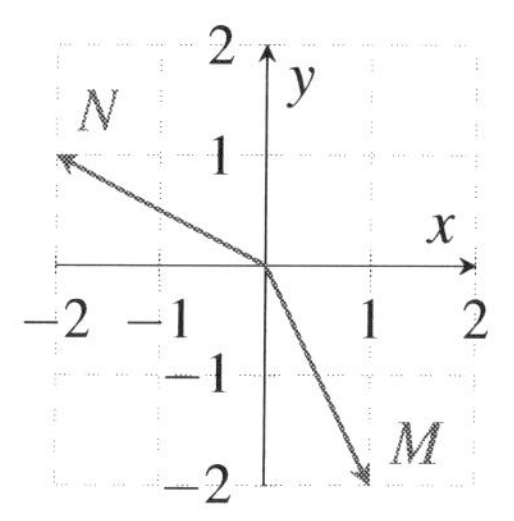

$$M \cdot N = (1)(-2) + (-2)(1) = -4 \neq 0.$$

Since the dot product is non-zero, the vectors are not orthogonal.

14.10 Parametric Equations and Graphs

Parametric equations provide a method to represent a point's coordinates using a single parameter, usually denoted as t. This approach is particularly useful for describing a vector's trajectory in a plane.

For a given point P in a two-dimensional setting, if the x-coordinate is given by a function $f(t)$ and the y-coordinate by a function $g(t)$, then the pair of equations $x = f(t)$ and $y = g(t)$ serves as the parametric equations for point P. The vector form $r(t) = (f(t), g(t))$ then represents the location of P in relation to t.

Key Point

Parametric equations efficiently describe the trajectory or movement of an object in space, providing an explicit relationship between the two dimensions coordinated with a single parameter.

To graph a curve defined by parametric equations:

1. Determine the range of t. The range of the parameter will often be given.
2. For several values of t within this range, calculate the coordinates x and y.
3. Plot these points on the Cartesian plane and trace the curve connecting the points.

Key Point

Graphing from parametric equations involves finding coordinates for specific values of the parameter, plotting these on a plane, and linking them to visualize the curve.

Another crucial aspect of working with parametric equations is converting them to a Cartesian equation. To achieve this, you can eliminate the parameter t by solving one equation for t and substituting it into the other equation.

Given the parametric equations: $x(t) = t$, $y(t) = t^2$. Sketch the curve for $-2 \leq t \leq 2$.

Solution: The range of t is from -2 to 2. Next, select a few values of t and calculate $x(t)$ and $y(t)$.

For instance, when $t = -2$, $x(-2) = -2$ and $y(-2) = 4$. Thus, one point on the curve is $(-2, 4)$. Continue calculating points for various t values and plot them. Joining these points will give a parabolic curve opening upwards.

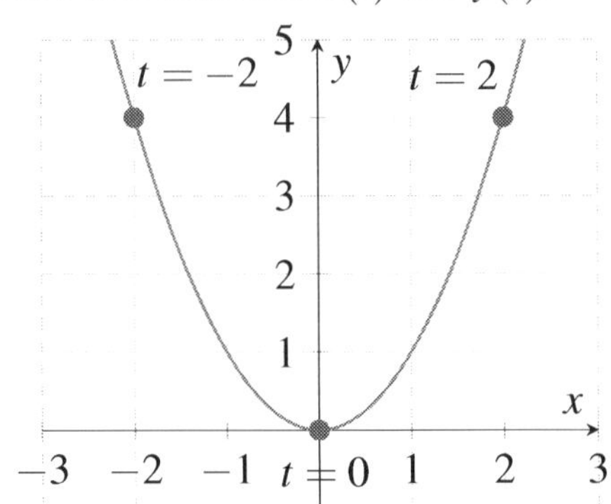

14.11 Applications of Vectors

Vectors are powerful tools used across various fields such as physics, engineering, and computer graphics due to their ability to represent direction and magnitude. In this section, we explore their applications in two-dimensional space, focusing on concepts that are fundamental in understanding physical movements and forces.

Vectors are crucial in depicting displacement, which is the change in position from one point to another. This displacement is represented by a straight line—known as the displacement vector—that shows the shortest route and direction between two points.

Key Point

Displacement vectors show the shortest path and direction from one point to another.

In physics, understanding forces is made easier with vectors. A force vector indicates how strong the force is (magnitude) and where it's applied (direction). This makes it possible to analyze forces acting at angles by breaking them down into horizontal and vertical components, a process known as resolving the force.

Key Point

Vectors help in breaking down forces into horizontal and vertical components, simplifying the study of forces in physics.

Moreover, vectors play a vital role in motion, particularly in describing velocity and acceleration. The velocity vector shows the direction of motion and how fast an object is moving, while the acceleration vector indicates how quickly the velocity is changing, both in speed and direction.

Key Point

Velocity and acceleration vectors provide insight into the direction and magnitude of motion and how quickly it changes.

These applications underscore the significance of vectors in understanding and analyzing real-world phenomena, especially in physics and engineering.

Example Suppose a person walks 4 *km* east and then 3 *km* north. Determine the resulting displacement vector.

Solution: Using the Cartesian plane, represent the eastward movement as vector $A = (4,0)$ and the northward movement as vector $B = (0,3)$. The resultant displacement R is $A + B$, giving $R = (4,3)$.

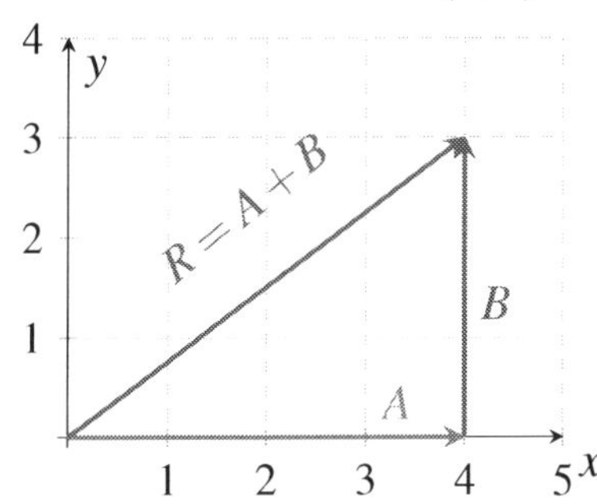

14.12 Practices

1) If vector $C = (4,-2)$, what is the magnitude of C?

☐ A. 2
☐ B. 4
☐ C. $2\sqrt{5}$
☐ D. $4\sqrt{2}$

2) Which of the following represents the vector obtained by adding vectors $U = (3,1)$ and $V = (2,-4)$?

☐ A. $(5,-3)$
☐ B. $(1,5)$
☐ C. $(5,3)$
☐ D. $(1,-3)$

3) Consider vectors $A = (3,-2)$ and $B = (x,y)$. For which values of x and y are vectors A and B equal?

☐ A. $x = 3,\ y = -2$
☐ B. $x = -2,\ y = 3$
☐ C. $x = 3,\ y = 2$
☐ D. $x = -3,\ y = 2$

4) Given vectors $G = (9,m)$ and $H = (n,4)$, if $G = H$, what is the value of $m+n$?

☐ A. 5
☐ B. 9
☐ C. 13
☐ D. 17

5) If the vector a is $(4,-3)$ and the scalar k is -2, what is the result of the scalar multiplication ka?

☐ A. $(8,-6)$
☐ B. $(-8,6)$
☐ C. $(4,-6)$
☐ D. $(-8,-6)$

6) If vector b scales by a factor of 0.5 resulting in the vector $(1.5,2)$, what was the original vector b?

☐ A. $(3,4)$
☐ B. $(0.75,1)$
☐ C. $(2,2.5)$
☐ D. $(1,1.333)$

7) If $p = (4,2)$ and $q = (1,3)$, what is $p+q$?

☐ A. $(3,5)$
☐ B. $(5,-1)$
☐ C. $(5,5)$
☐ D. $(3,-1)$

8) What is the result of subtracting $a = (-2,5)$ from $b = (7,-4)$?

☐ A. $(9,-9)$
☐ B. $(9,-1)$
☐ C. $(-9,1)$
☐ D. $(-9,9)$

9) Given vectors $P=(2,-1)$ and $Q=(-3,4)$, what is the result of $P+Q$?

☐ A. $(-1,3)$
☐ B. $(5,3)$
☐ C. $(1,-5)$
☐ D. $(-1,5)$

10) What is the vector subtraction result of $P=(6,-2)$ from $Q=(3,5)$?

☐ A. $(-3,7)$
☐ B. $(3,-7)$
☐ C. $(9,3)$
☐ D. $(-9,-3)$

11) What is the length of the vector $v=(-6,8)$?

☐ A. 2
☐ B. 10
☐ C. 14
☐ D. 48

12) Given the vector $w=(2,-7)$, select the correct expression for its length.

☐ A. $2-7$
☐ B. $2+(-7)$
☐ C. $\sqrt{2^2-7^2}$
☐ D. $\sqrt{2^2+(-7)^2}$

13) Calculate the dot product of vectors $A=(-3,1)$ and $B=(4,-2)$.

☐ A. -10
☐ B. -14
☐ C. 12
☐ D. -12

14) Given vectors $A=(5,0)$ and $B=(0,7)$, what is the angle θ between them?

☐ A. 0°
☐ B. 45°
☐ C. 90°
☐ D. 180°

15) Which of the following vectors is parallel to the vector $A=(3,9)$?

☐ A. $(-1,3)$
☐ B. $(3,3)$
☐ C. $(-2,6)$
☐ D. $(-1,-3)$

16) Are the vectors $P=(2,-4)$ and $Q=(1,-2)$ parallel?

☐ A. Yes
☐ B. No
☐ C. Cannot be determined with the given information
☐ D. The question is invalid because one of the vectors has negative components

17) Which of the following pairs of vectors is orthogonal?

☐ A. $(2,3)$ and $(3,2)$
☐ B. $(1,0)$ and $(0,-1)$
☐ C. $(6,8)$ and $(4,3)$
☐ D. $(7,-5)$ and $(-10,14)$

18) If vector $A=(a,b)$ is orthogonal to vector $B=(2,-3)$ and $a=6$, what is the value of b?

☐ A. -4
☐ B. 4
☐ C. $\frac{9}{2}$
☐ D. $-\frac{9}{2}$

19) Consider the parameterized curve defined by the parametric equations $x(t)=3t-2$ and $y(t)=2-t^2$ for $t\in\mathbb{R}$. Which of the following points lies on the curve?

☐ A. $(1,-1)$
☐ B. $(7,-5)$
☐ C. $(-2,2)$
☐ D. $(4,-6)$

20) Which of the following parametric equations describes a line passing through the origin?

☐ A. $x(t)=2t+1$, $y(t)=3t-1$
☐ B. $x(t)=t^2+1$, $y(t)=t^2$
☐ C. $x(t)=t$, $y(t)=2t$
☐ D. $x(t)=1-t$, $y(t)=1+t$

21) An airplane is flying with a velocity 500 km/h in the north direction and 100 km/h in the east direction. Represent the airplane's resultant velocity as a vector.

☐ A. $(500,100)$
☐ B. $(100,600)$
☐ C. $(600,500)$
☐ D. $(100,500)$

22) Suppose a person walks 4 km east and then 3 km north. Determine the resulting displacement vector.

☐ A. $(4,3)$

☐ B. $(3,4)$

☐ C. $(1,-2)$

☐ D. $(-3,4)$

Answer Keys

1) C. $2\sqrt{5}$

2) A. $(5,-3)$

3) A. $x=3, y=-2$

4) C. 13

5) B. $(-8,6)$

6) A. $(3,4)$

7) C. $(5,5)$

8) A. $(9,-9)$

9) A. $(-1,3)$

10) A. $(-3,7)$

11) B. 10

12) D. $\sqrt{2^2+(-7)^2}$

13) B. -14

14) C. $90°$

15) D. $(-1,-3)$

16) A. Yes

17) B. $(1,0)$ and $(0,-1)$

18) B. 4

19) C. $(-2,2)$

20) C. $x(t)=t,\ y(t)=2t$

21) D. $(100,500)$

22) A. $(4,3)$

15. Analytic Geometry

15.1 Distance and Midpoint Formulas

Analytic Geometry in two dimensions leverages the Distance and Midpoint Formulas for computations on the coordinate plane. The Distance Formula for points (x_1, y_1) and (x_2, y_2), rooted in the Pythagorean theorem, is expressed as:

$$D = \sqrt{(x_2 - x_1)^2 + (y_2 - y_1)^2},$$

where D denotes the Euclidean distance, attributed to Euclid. Meanwhile, the Midpoint Formula, for determining the midpoint (M_x, M_y) between two points (x_1, y_1) and (x_2, y_2), is given by:

$$M_x = \frac{x_1 + x_2}{2}, \quad M_y = \frac{y_1 + y_2}{2}.$$

Example Given two points $A(1,2)$ and $B(3,5)$, find the distance between A and B and the midpoint of the line segment joining A and B.

Solution: First, we will find the distance using the Distance Formula:

$$D = \sqrt{(3-1)^2 + (5-2)^2} = \sqrt{4+9} = \sqrt{13}\,\text{units}.$$

Next, we will find the midpoint using the Midpoint Formula:

$$M_x = \frac{1+3}{2} = 2, \quad M_y = \frac{2+5}{2} = \frac{7}{2}.$$

So, the midpoint is $M(2, 3.5)$.

15.2 Circles

Building upon the fundamentals of distance and midpoint, we delve into the geometry of circles, which are defined as the locus of all points equidistant from a central point. The discussion encompasses the circle's equation in standard and general forms.

Key Point

The standard equation of a circle is $(x-h)^2+(y-k)^2=r^2$, where (h,k) indicates the center's coordinates and r the radius.

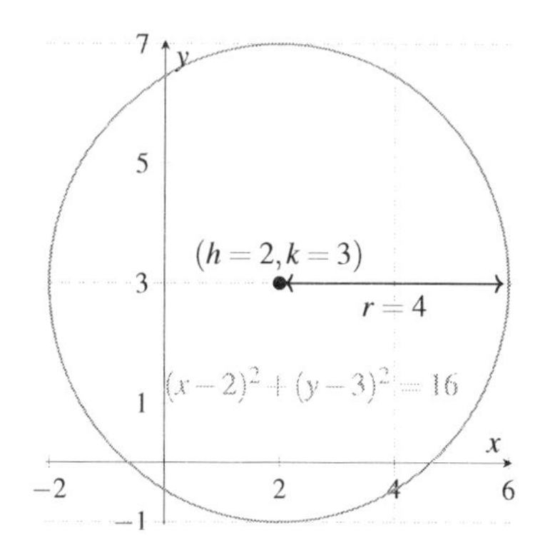

The general form of a circle's equation is $ax^2+by^2+cx+dy+e=0$. To convert the general form to the standard form, follow these steps:

1. Group the x and y terms on one side, and move the constant term e to the other side.
2. For x terms, complete the square by adding and subtracting $(\frac{c}{2})^2$, where c is the coefficient of x.
3. For y terms, apply the same process using $(\frac{d}{2})^2$, where d is the coefficient of y.
4. Factor the left side into perfect square trinomials and simplify the right side to achieve the standard form.

Key Point

With the circle's center and radius, we can formulate the standard equation. Conversely, the general form can be converted to the standard form by completing the square.

Example Write the standard form equation of the circle with center $(-9,-12)$ and radius 4.

Solution: Substituting $h=-9$, $k=-12$, and $r=4$ into the standard form equation, we have:

$$(x-(-9))^2+(y-(-12))^2=(4)^2 \Rightarrow (x+9)^2+(y+12)^2=16.$$

 Write the equation in standard form for this circle $x^2+y^2-8x-6y+21=0$.

Solution: Starting from the given equation, rearrange the equation to group x and y variables together, and completing the square:

$$\begin{aligned}&(x^2-8x)+(y^2-6y)=-21,\\ \Rightarrow&(x^2-8x+16)+(y^2-6y+9)=-21+16+9,\\ \Rightarrow&(x-4)^2+(y-3)^2=4.\end{aligned}$$

Therefore, the circle's equation is $(x-4)^2+(y-3)^2=(2)^2$.

15.3 Finding the Center and the Radius of Circles

This section focuses on deducing the center and radius of a circle from its equation, highlighting the transition from the general to the standard form of a circle's equation and the subsequent extraction of these vital parameters.

For a circle defined by the equation $(x-h)^2+(y-k)^2=r^2$, the center is at (h,k) and the radius is r.

To find the center and radius from a general circle equation of the form

$$ax^2+by^2+cx+dy+e=0,$$

convert it into the standard form.

 Find the center and the radius of the circle defined by the equation

$$(x-3)^2+(y+2)^2=25.$$

Solution: In this case, comparing the given equation $(x-3)^2+(y+2)^2=25$ with the standard equation of a circle $(x-h)^2+(y-k)^2=r^2$, we find the center of the circle is $(h,k)=(3,-2)$ and the radius of the circle is $r=\sqrt{25}=5$.

 Identify the center and radius of the circle $x^2+y^2-4y+3=0$.

Solution: Starting from the given equation and rearranging it to obtain the standard form, we have:

$$x^2+(y^2-4y)=-3,$$
$$\Rightarrow x^2+(y^2-4y+4)=-3+4,$$
$$\Rightarrow (x-0)^2+(y-2)^2=(1)^2.$$

Reading from this equation, the center of the circle is $(0,2)$ and the radius is 1.

Find the center and the radius of the circle defined by the equation

$$x^2+y^2-6x+4y+4=0.$$

Solution: To find the center and the radius, first, we rewrite the given equation in the standard form. Doing so, we have $(x-3)^2+(y+2)^2=9$. Now, comparing this with the standard form, we find that the center of the circle is $(h,k)=(3,-2)$, and the radius of the circle is $r=\sqrt{9}=3$.

Identify the center and radius of the circle $4x+x^2-6y=24-y^2$.

Solution: Starting from the given equation, rearranging, and simplifying into standard form for the circle's equation, we obtain:

$$(x-(-2))^2+(y-3)^2=(\sqrt{37})^2.$$

Reading from this equation, the center is $(-2,3)$ and the radius is $\sqrt{37}$.

15.4 Parabolas

A parabola is a U-shaped curve represented as the set of points equidistant from a focus and a directrix. Its equation in standard form is $(x-h)^2=4p(y-k)$ for vertical parabolas, indicating an upward or downward orientation, and $(y-k)^2=4p(x-h)$ for horizontal parabolas, indicating a rightward or leftward orientation. The direction is determined by the sign of p: $p>0$ signifies upward or rightward orientation, while $p<0$ signifies downward or leftward.

Key Point

For vertical parabolas, the directrix is $y=k-p$, and the focus is at $(h,k+p)$. For horizontal parabolas, the directrix is $x=h-p$, and the focus is at $(h+p,k)$. The vertex (h,k) is central to the parabola's geometry and graphing.

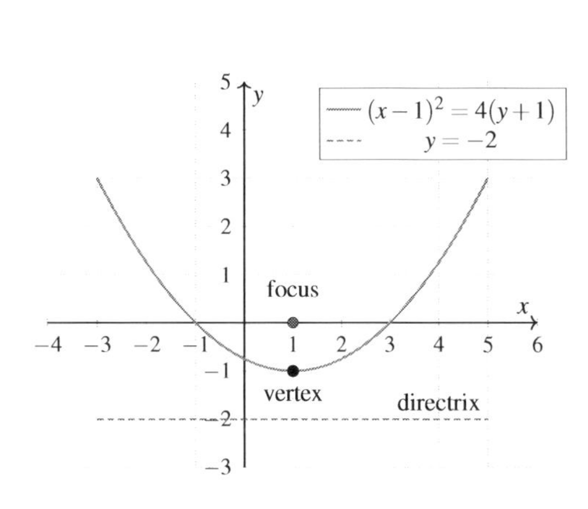

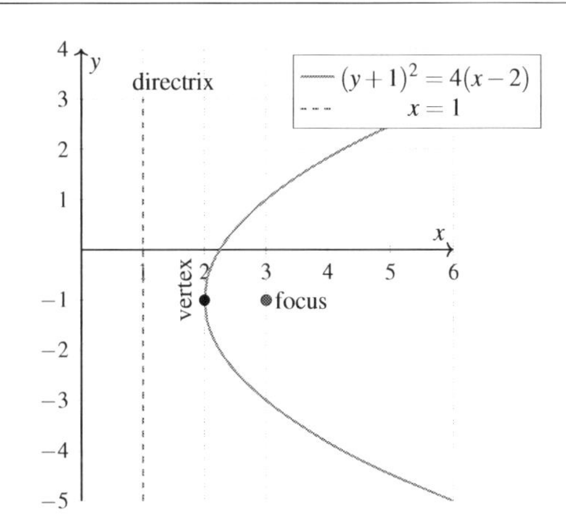

Example Write the equation of the parabola with vertex $(0,0)$ and focus $(0,4)$.

Solution: First, we identify that the standard form for this parabola is $(x-h)^2=4p(y-k)$, because it opens up. We have that $(h,k)=(0,0)$ and $(h,k+p)=(0,4)$. Using these values in the standard form equation, we obtain:

$$(x-0)^2=(4)(4)(y-0) \Rightarrow x^2=16y.$$

Hence, the equation of the parabola is $x^2=16y$.

Example Write the equation of the parabola with vertex $(2,3)$ and focus $(2,4)$.

Solution: We know that this parabola opens up, so its standard form is $(x-h)^2=4p(y-k)$. We have that $(h,k)=(2,3)$ and $(h,k+p)=(2,4)$. Replacing these values in the standard form equation we get:

$$(x-2)^2=(4)(1)(y-3) \Rightarrow (x-2)^2=4(y-3).$$

Thus, the equation of the parabola is $(x-2)^2=4(y-3)$.

Example Write the equation of the parabola with vertex $(-2,1)$ and focus $(-3,1)$.

Solution: We know that this parabola opens left, so its standard form is $(y-k)^2=4p(x-h)$. We have that $(h,k)=(-2,1)$ and $(h+p,k)=(-3,1)$. Replacing these values in the standard form equation, we get:

$$(y-1)^2=4(-1)(x+2) \Rightarrow (y-1)^2=-4(x+2).$$

Thus, the equation of the parabola is $(y-1)^2=-4(x+2)$.

15.5 Focus, Vertex, and Directrix of a Parabola

This section delves deeper into the essential characteristics of parabolas, specifically focusing on identifying the vertex, focus, and directrix.

Key Point

For parabolas opening vertically, the standard equation is $(x-h)^2 = 4p(y-k)$, where (h,k) marks the vertex, $y = k-p$ the directrix, and $(h, k+p)$ the focus.

Key Point

For parabolas opening horizontally, the equation is $(y-k)^2 = 4p(x-h)$, with the vertex at (h,k), the directrix at $x = h-p$, and the focus at $(h+p, k)$.

Note that:

- For vertical parabolas, the directrix is a horizontal line. The focus lies above the vertex if $p > 0$ and below if $p < 0$.
- For horizontal parabolas, the directrix is vertical. The focus is to the vertex's right if $p > 0$ and to the left if $p < 0$.

Additionally, for a parabola described by $y = ax^2 + bx + c$, converting to standard form requires completing the square technique. The vertex's x-coordinate is determined by $x_v = -\frac{b}{2a}$, providing a systematic approach to find the parabola's peak.

Example Consider the equation $(y-3)^2 = 8(x-1)$. Find the vertex, focus, and the equation of the directrix.

Solution: First, the standard form of the given equation is $(y-k)^2 = 4p(x-h)$.

The vertex (h,k) is $(1,3)$. For this parabola, $4p = 8$, resulting in $p = 2$. Thus, the directrix is $x = h-p = 1-2 = -1$ and the focus is at point $(h+p, k) = (1+2, 3) = (3,3)$. See the figure for the graphical representation of the given parabola along with its focus and directrix.

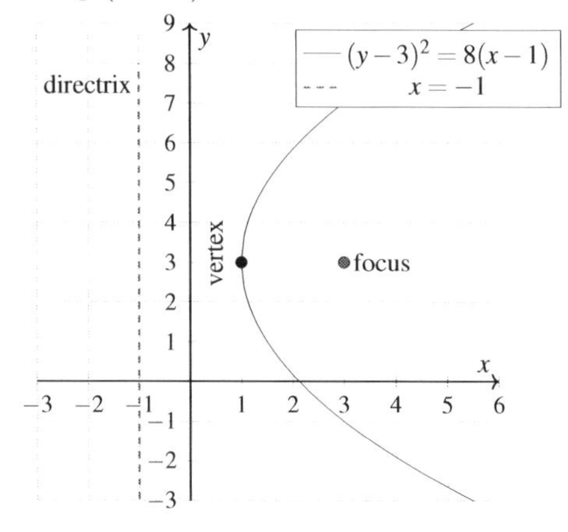

Find the x-value of the vertex of the parabola $y = x^2 + 4x$.

Solution: The parabola parameters are: $a = 1$, $b = 4$ and $c = 0$. So: $x_v = -\frac{b}{2a} = -2$.

15.6 Ellipses

Ellipses are curves on a plane where the sum of the distances from two fixed points (the foci) to any point on the curve is constant. This section introduces the fundamental aspects of ellipses, including the major and minor axes, the foci, and the standard equation of an ellipse.

An ellipse has two axes: the major axis, its longest diameter, and the minor axis, its shortest. The foci, $F1$ and $F2$, lie along the major axis, equidistant from the center.

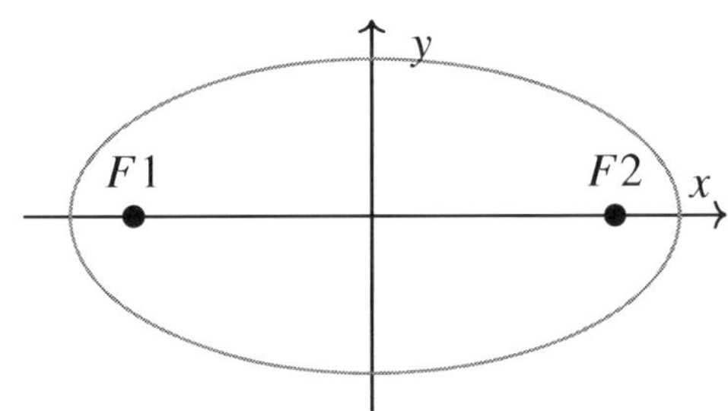

Key Point

The standard equation of an ellipse centered at (h,k) is: $\frac{(x-h)^2}{a^2} + \frac{(y-k)^2}{b^2} = 1$, where a and b are half the lengths of the major and minor axes, respectively. An ellipse is termed horizontal if $a > b$, and vertical if $a < b$.

Note that:

- The center of the ellipse at the origin simplifies the equation to $\frac{x^2}{a^2} + \frac{y^2}{b^2} = 1$.
- The vertices, located at the major axis's intersections with the ellipse, are $(h \pm a, k)$ for a horizontal ellipse and $(h, k \pm a)$ for a vertical ellipse.
- The foci are at a distance of $c = \sqrt{a^2 - b^2}$ from the center along the major axis, resulting in coordinates $(h \pm c, k)$ for a horizontal ellipse and $(h, k \pm c)$ for a vertical ellipse.

Remember, a always pertains to the major axis and b to the minor, regardless of the ellipse's orientation.

Find the vertices and foci of the ellipse given by $\frac{x^2}{169} + \frac{y^2}{64} = 1$.

Solution: To find the vertices and foci of the ellipse, we first rewrite the given equation in standard form:

$$\frac{(x-0)^2}{13^2} + \frac{(y-0)^2}{8^2} = 1.$$

From this equation, we have $h = 0 = k$, $a = 13$, and $b = 8$. Hence, $c = \sqrt{a^2 - b^2} = \sqrt{105}$. The vertices, therefore, are given by: $(h+a,k) = (13,0)$, and $(h-a,k) = (-13,0)$. Hence, the vertices of the ellipse are at $(13,0)$ and $(-13,0)$. Also, the foci are given by: $(h+c,k) = (\sqrt{105},0)$, and $(h-c,k) = (-\sqrt{105},0)$.

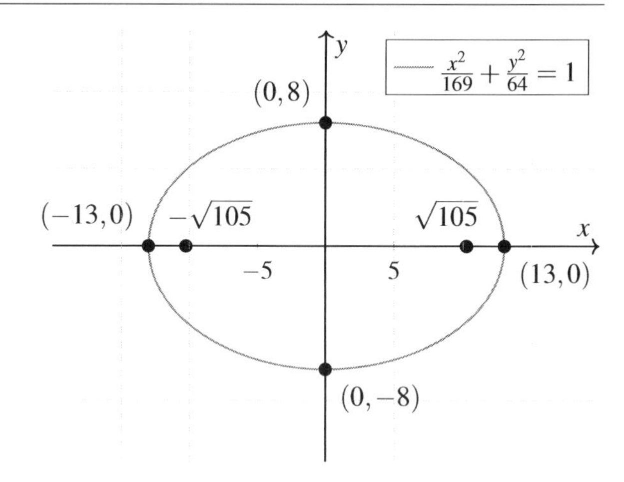

15.7 Hyperbolas

A hyperbola, a type of conic section, is defined as the locus of all points in a plane where the difference in distances from two fixed points (foci) is constant.

The standard equation of a hyperbola depends on its orientation:

- For hyperbolas oriented vertically, the equation is $\frac{(y-k)^2}{a^2} - \frac{(x-h)^2}{b^2} = 1$,
- For those oriented horizontally, the equation is $\frac{(x-h)^2}{a^2} - \frac{(y-k)^2}{b^2} = 1$.

The center of the hyperbola is at the point (h,k). The foci are located at $(h \pm c, k)$ for hyperbolas that open left and right, and $(h, k \pm c)$ for those that open up and down, where c is defined as $c = \sqrt{a^2 + b^2}$. The vertices, which are the closest points on the hyperbola to its center, are situated at $(h \pm a, k)$ for hyperbolas aligned horizontally, and $(h, k \pm a)$ for those aligned vertically.

The alignment of the transverse axis is parallel to the x-axis for horizontally-oriented hyperbolas and parallel to the y-axis for vertically-oriented hyperbolas. The asymptotes, which are straight lines that the hyperbola approaches but never touches, are defined by the equations:

- For vertically-oriented hyperbolas, $y - k = \pm\frac{a}{b}(x-h)$,
- For horizontally-oriented hyperbolas, $y - k = \pm\frac{b}{a}(x-h)$.

A distinctive feature of hyperbolas is their two separate, mirror-image branches.

Example Given the hyperbola equation $\frac{x^2}{25} - \frac{y^2}{9} = 1$, find the center, vertices and foci, and then sketch the graph.

Solution: Here, $a = 5$, $b = 3$, and $h = k = 0$, so the center is at the origin, $(0,0)$. The vertices are then at $(\pm a, 0) = (\pm 5, 0)$. The distance from the center to each focus is $c = \sqrt{a^2+b^2} = \sqrt{25+9} = \sqrt{34}$ (approximately 5.83). So the foci are at $(\pm c, 0) = (\pm\sqrt{34}, 0)$.

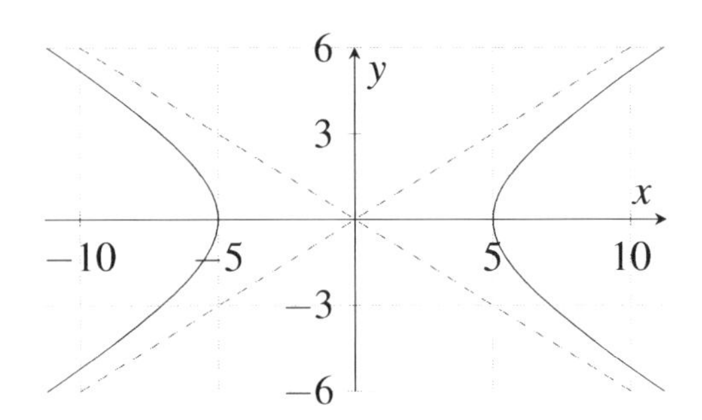

Example Find the center and foci of the hyperbola $-x^2+y^2-18x-14y-132=0$.

Solution: To find the center and foci, we need to rewrite this equation in standard form. Start by reorganizing and grouping like terms on both sides: $-x^2+y^2-18x-14y=132$. Next, factor out the coefficients of the square terms: $(y^2-14y)-(x^2+18x)=132$. Then, complete the square for x and y terms:

$$(y^2-14y+49)-(x^2+18x+81)=132+49-81 \Longrightarrow (y-7)^2-(x+9)^2=100.$$

So, the standard form of the given equation is:

$$\frac{(y-7)^2}{10^2}-\frac{(x+9)^2}{10^2}=1.$$

Hence, the center of the hyperbola is at the point $(-9,7)$. Since $c=\sqrt{a^2+b^2}$, we have $c=\sqrt{10^2+10^2}=10\sqrt{2}$. Therefore, the foci of the hyperbola are at $(-9,7+10\sqrt{2})$ and $(-9,7-10\sqrt{2})$.

15.8 Classifying a Conic Section (in Standard Form)

All conic sections have a general form. The general form of a conic section is represented as $Ax^2+Bxy+Cy^2+Dx+Ey+F=0$. In this section, we want to learn about classifying conic sections, which means that if we have the general form of a conic section, how we can distinguish whether it is an ellipse, a circle, a parabola, or a hyperbola.

Key Point

For the general form $Ax^2+Bxy+Cy^2+Dx+Ey+F=0$, we have three cases:

1. If $B^2-4AC<0$, it is an ellipse (or circle if $A=C$ and $B=0$).
2. If $B^2-4AC=0$, it is a parabola.
3. If $B^2-4AC>0$, it is a hyperbola.

To convert the general equation to standard form, the method of completing the square is applied.

 **Example** Classify the equation $6x^2 - 2\sqrt{3}xy + y^2 - 16 = 0$.

Solution: Here, $A = 6$, $B = -2\sqrt{3}$, and $C = 1$. Substitute and calculate:

$$B^2 - 4AC = (-2\sqrt{3})^2 - 4(6)(1) = 12 - 24 = -12.$$

Since $B^2 - 4AC < 0$, this is an ellipse.

 **Example** Classify the equation $x^2 - y^2 - 2x + 4y - 1 = 0$.

Solution: Here, $A = 1$, $B = 0$, and $C = -1$. Substitute and calculate:

$$B^2 - 4AC = 0 - 4(1)(-1) = 4.$$

Since $B^2 - 4AC > 0$, this is a hyperbola.

 **Example** Classify the equation $x^2 + 4x + y^2 - 6y + 9 = 0$

Solution: Here, $A = 1$, $B = 0$, and $C = 1$. Calculate the discriminant to be:

$$B^2 - 4AC = 0 - 4(1)(1) = -4.$$

Since $B^2 - 4AC < 0$ and $A = C$, this is a circle.

 **Example** Classify the equation $y^2 - 4x - 4y = 0$.

Solution: Here, $A = 0$, $B = 0$, and $C = 1$. Calculate the discriminant to be:

$$B^2 - 4AC = 0 - 4(0)(1) = 0.$$

Since $B^2 - 4AC = 0$, this is a parabola.

 **Example** Classify and standardize the equation $x^2 - 4y^2 + 6x - 8y + 1 = 0$.

Solution: We identify it as a hyperbola, since $B^2 - 4AC = 0 - 4(1)(-4) > 0$. We can rewrite this into standard form. The result is:

$$\frac{(x-(-3))^2}{2^2} - \frac{(y-(-1))^2}{1^2} = 1.$$

The direction and shape of the hyperbola confirm the analysis.

Example Write the equation $-x^2+8x+y-17=0$ in standard form.

Solution: We recognize it as a parabola since $B^2-4AC=0-4(-1)(0)=0$. Let us rewrite this equation in standard form by completing the square. We first rewrite the equation as: $x^2-8x=-y+17$, which simplifies to $y=(x-4)^2+1$. Putting into standard form:

$$(x-4)^2=4(\frac{1}{4})(y-1).$$

The direction of the parabolic opening confirms the positive sign associated with the x^2 term in standard form.

15.9 Rotation of Axes and General Form of Conic Sections

Rotation of axes simplifies the analysis of conic sections (ellipses, hyperbolas, and parabolas) by aligning coordinate axes with the conic's symmetry axes, thus eliminating the xy term in the equation for clearer properties.

Key Point

For a rotation by angle θ, the coordinates transform as follows:

$$x=x'\cos(\theta)-y'\sin(\theta),$$
$$y=x'\sin(\theta)+y'\cos(\theta).$$

In these equations, x and y represent the coordinates in the original coordinate system, while x' and y' denote the coordinates in the rotated system.

The angle θ is the angle of rotation, measured from the positive x-axis to the new x'-axis. This transformation shifts the perspective to a new coordinate system where the axes may be aligned with the symmetry axes of the conic section, making its equation simpler and removing the xy cross-term. This realignment allows for an easier analysis of the conic section's properties, such as its orientation, without changing the conic's inherent characteristics.

This process transforms the standard form of an ellipse $\left(\frac{x^2}{a^2}+\frac{y^2}{b^2}=1\right)$ into the General Form of Conic Sections $\left(Ax^2+Bxy+Cy^2+Dx+Ey+F=0\right)$, complicating the identification of the conic type. Axis rotation, essential for simplification, does not alter the conic's identity.

 Rotate the point $P(3,4)$ by an angle of 45° counterclockwise.

Solution: Using the rotation of axes formulas, we have:

$$x' = 3\cos(45^\circ) - 4\sin(45^\circ) = \tfrac{3\sqrt{2}}{2} - \tfrac{4\sqrt{2}}{2} = -\tfrac{\sqrt{2}}{2},$$
$$y' = 3\sin(45^\circ) + 4\cos(45^\circ) = \tfrac{3\sqrt{2}}{2} + \tfrac{4\sqrt{2}}{2} = \tfrac{7\sqrt{2}}{2}.$$

The rotated point is $(-\frac{\sqrt{2}}{2}, \frac{7\sqrt{2}}{2})$.

 If we have the equation of an ellipse $\frac{x^2}{9} + \frac{y^2}{4} = 1$ and rotate the system counter-clockwise by $\theta = \frac{\pi}{4}$, how will the new equation look like?

Solution: We employ the formulas of rotation of axes with $\theta = \frac{\pi}{4}$:

$$x = x'\cos(\frac{\pi}{4}) - y'\sin(\frac{\pi}{4}), \quad \text{and} \quad y = x'\sin(\frac{\pi}{4}) + y'\cos(\frac{\pi}{4}).$$

Substituting these into the given ellipse equation yields:

$$\begin{aligned}
&\frac{(x'\cos(\frac{\pi}{4}) - y'\sin(\frac{\pi}{4}))^2}{9} + \frac{(x'\sin(\frac{\pi}{4}) + y'\cos(\frac{\pi}{4}))^2}{4} \\
&= \frac{x'^2}{9}\cos^2(\frac{\pi}{4}) - \frac{2x'y'}{9}\sin(\frac{\pi}{4})\cos(\frac{\pi}{4}) + \frac{y'^2}{9}\sin^2(\frac{\pi}{4}) \\
&\quad + \frac{x'^2}{4}\sin^2(\frac{\pi}{4}) + \frac{2x'y'}{4}\sin(\frac{\pi}{4})\cos(\frac{\pi}{4}) + \frac{y'^2}{4}\cos^2(\frac{\pi}{4}) \\
&= \left(\frac{1}{9} + \frac{1}{4}\right)x'^2 + \left(-\frac{1}{9} + \frac{1}{4}\right)y'^2 \\
&= \frac{13}{36}x'^2 - \frac{5}{36}y'^2 = 1.
\end{aligned}$$

Thus, the transformed equation $\frac{13}{36}x'^2 - \frac{5}{36}y'^2 = 1$ represents the same ellipse from a rotated perspective.

15.10 Practices

1) What is the distance between the points $(3,-2)$ and $(7,4)$?

- ☐ A. $2\sqrt{10}$
- ☐ B. $2\sqrt{13}$
- ☐ C. $2\sqrt{5}$
- ☐ D. $2\sqrt{8}$

2) Calculate the midpoint of the segment connecting $(0,0)$ and $(8,6)$.

☐ A. $(4,6)$
☐ B. $(4,3)$
☐ C. $(8,3)$
☐ D. $(2,1.5)$

3) Find the center of the circle with the equation $(x+4)^2+(y-3)^2=25$.

☐ A. $(-4,-3)$
☐ B. $(4,-3)$
☐ C. $(-4,3)$
☐ D. $(4,3)$

4) Which of the following is the radius of the circle with equation $x^2+y^2-6x+8y+9=0$?

☐ A. 4
☐ B. $\sqrt{5}$
☐ C. 2
☐ D. $\sqrt{2}$

5) Given the equation of a circle $(x+1)^2+(y-4)^2=16$, what are the center and radius of the circle?

☐ A. Center at $(-1,4)$ with radius 4
☐ B. Center at $(1,-4)$ with radius 4
☐ C. Center at $(-1,4)$ with radius 16
☐ D. Center at $(1,-4)$ with radius 16

6) What is the center of the circle whose equation is $4x^2+4y^2-24x+16y+36=0$?

☐ A. $(3,2)$
☐ B. $(-3,2)$
☐ C. $(-3,-2)$
☐ D. $(3,-2)$

7) Which of the following is the equation of a parabola with vertex at $(2,-3)$ and focus at $(2,-2)$?

☐ A. $(x-2)^2=4(y-2)$
☐ B. $(y+3)^2=4(x-2)$
☐ C. $(y+3)^2=-4(x-2)$
☐ D. $(x-2)^2=4(y+3)$

8) Given the parabola with equation $(y-4)^2=12(x+1)$, which of the following is the coordinate of the focus?

☐ A. $(-1,4)$
☐ B. $(1,4)$
☐ C. $(4,-1)$
☐ D. $(2,4)$

9) Given the equation of the parabola $(x+2)^2 = -12(y-1)$, which of the following represents the focus of the parabola?

☐ A. $(-2,-1)$
☐ B. $(-2,1)$
☐ C. $(-2,-2)$
☐ D. $(-2,4)$

10) The parabola defined by the equation $(y-4)^2 = -16(x+1)$ opens in which direction?

☐ A. Upward
☐ B. Downward
☐ C. To the left
☐ D. To the right

11) An ellipse is centered at the origin and passes through the point $(3,2)$. If the length of the semi-major axis is 5, which of the following is the length of the semi-minor axis?

☐ A. 3
☐ B. 2.5
☐ C. $\sqrt{3}$
☐ D. 4

12) Which of the following is the standard form equation of an ellipse with foci at $(5,4)$ and $(5,-6)$ and a major axis of length 22?

☐ A. $\frac{(x-5)^2}{96} + \frac{(y+1)^2}{121} = 1$
☐ B. $\frac{(x-5)^2}{11} + \frac{(y+1)^2}{96} = 1$
☐ C. $\frac{(x-5)^2}{121} + \frac{(y+1)^2}{146} = 1$
☐ D. $\frac{(x-5)^2}{121} + \frac{(y+1)^2}{11} = 1$

13) For the hyperbola given by $\frac{(x-2)^2}{16} - \frac{(y+3)^2}{25} = 1$, which is the equation of its asymptote that has a positive slope?

☐ A. $y+3 = \frac{4}{5}(x-2)$
☐ B. $y+3 = -\frac{5}{4}(x-2)$
☐ C. $y+3 = -\frac{4}{5}(x-2)$
☐ D. $y+3 = \frac{5}{4}(x-2)$

14) Which of the following represents the vertices of the hyperbola defined by the equation $\frac{y^2}{36} - \frac{x^2}{49} = 1$?

☐ A. $(\pm 6,0)$
☐ B. $(0,\pm 7)$
☐ C. $(0,\pm 6)$
☐ D. $(\pm 7,0)$

15) Classify the following conic section: $-9x^2 + y^2 - 72x - 153 = 0$.

☐ A. Ellipse ☐ C. Parabola

☐ B. Circle ☐ D. Hyperbola

16) Determine the type of conic section given by the equation: $x^2 - 6x + 9y^2 + 36y + 36 = 0$.

☐ A. Ellipse ☐ C. Parabola

☐ B. Circle ☐ D. Hyperbola

17) Rotate the conic section given by the equation $4x^2 + 9y^2 = 36$ by an angle of $30°$ counterclockwise. Which of the following new equations represents the conic section?

☐ A. $3x'^2 + 8x'y' + 3y'^2 = 144$ ☐ C. $6x'^2 + y'^2 = 36$

☐ B. $4x'^2 + 5\sqrt{3}x'y' + 9y'^2 = 36$ ☐ D. $9x'^2 + 4y'^2 = 36$

18) The transformation of the standard equation $x^2 + y^2 = 16$ by rotation of the axes by $45°$ introduces which of the following terms into the equation?

☐ A. $x'y'$ ☐ C. $2x'^2 + 2y'^2$

☐ B. $x'^2 - y'^2$ ☐ D. No new terms are introduced

Answer Keys

1) B. $2\sqrt{13}$

2) B. $(4,3)$

3) C. $(-4,3)$

4) A. 4

5) A. Center at $(-1,4)$ with radius 4

6) D. $(3,-2)$

7) D. $(x-2)^2 = 4(y+3)$

8) D. $(2,4)$

9) C. $(-2,-2)$

10) C. To the left

11) B. 2.5

12) A. $\frac{(x-5)^2}{96} + \frac{(y+1)^2}{121} = 1$

13) D. $y+3 = \frac{5}{4}(x-2)$

14) C. $(0,\pm 6)$

15) D. Hyperbola

16) A. Ellipse

17) D. $9x'^2 + 4y'^2 = 36$

18) D. No new terms are introduced

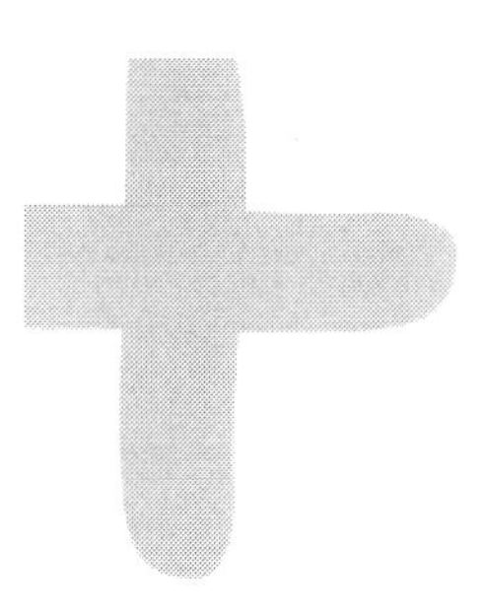

16. Matrices

16.1 Introduction to Matrices

Matrices, rectangular arrays of numbers, are fundamental in pre-calculus for solving various mathematical problems. These are denoted by capital letters (A, B, C, etc.). The structure of a matrix A is displayed as:

$$A = \begin{bmatrix} a_{11} & a_{12} & \cdots & a_{1n} \\ a_{21} & a_{22} & \cdots & a_{2n} \\ \vdots & \vdots & \ddots & \vdots \\ a_{m1} & a_{m2} & \cdots & a_{mn} \end{bmatrix},$$

with each a_{ij} indicating the entry in the ith row and j^{th} column. Key aspects include:

1. Rows and columns respectively denote horizontal and vertical entries.
2. The dimension is noted as $m \times n$ (rows $\times$ columns).
3. Matrices are equal if all corresponding entries match.
4. A square matrix has equal numbers of rows and columns, noted as $n \times n$.
5. The subscript (i, j) denotes the position of an entry within the matrix.

Key Point

Matrices are crucial for organizing data, representing linear equations, performing geometric transformations, and more, across various mathematical and interdisciplinary fields.

Solve the matrix equation for a and b:

$$\begin{bmatrix} -1 & a+2 \\ 4 & 2 \\ 1 & -1 \end{bmatrix} = \begin{bmatrix} -1 & 0 \\ 2b-a & 2 \\ 1 & -1 \end{bmatrix}.$$

Solution: According to the properties of matrices, two matrices are only equal if each of their corresponding entries are equal as well. By comparing the entries in the first row and the second column, we get: $a+2=0$. Solving for a yields $a=-2$. Similarly, by comparing the entries in the second row and first column, we get: $2b-a=4$. Substituting $a=-2$ into this equation, we get: $2b-(-2)=4$. Solving it gives $b=1$.

Example The following table shows the age and shoe size of each student in a class. Write a matrix corresponding to the table.

	Student 1	Student 2	Student 3
Shoe Size	37	38	32
Age	11	12	10

Solution: We can represent each student's data as a row in the matrix, with the first column as the shoe sizes and the second column as the ages. The matrix would then be a 3×2 dimension matrix, and it would look like this: $S = \begin{bmatrix} 37 & 11 \\ 38 & 12 \\ 32 & 10 \end{bmatrix}$.

16.2 Matrix Addition and Subtraction

Matrices are sets of numbers arranged in rows and columns. A fundamental operation with matrices is their addition or subtraction, which requires that both matrices have identical dimensions. To add or subtract matrices, simply add or subtract corresponding entries, placing the result in the corresponding position of the resultant matrix. This operation is straightforward and applied element-wise across the matrices.

Key Point

Matrix Addition and Subtraction require matrices to have the same dimensions, meaning the same number of rows and columns.

Calculate $A+B$, for $A = \begin{bmatrix} 1 & -4 & 6 \end{bmatrix}$ and $B = \begin{bmatrix} 2 & -3 & -9 \end{bmatrix}$.

Solution: We add the elements in the matching positions:

$$A+B = \begin{bmatrix} 1+2 & -4+(-3) & 6+(-9) \end{bmatrix} = \begin{bmatrix} 3 & -7 & -3 \end{bmatrix}.$$

So the sum of the two given matrices is $\begin{bmatrix} 3 & -7 & -3 \end{bmatrix}$.

Calculate $A-B$, for $A = \begin{bmatrix} 1 & -1 \\ 2 & 0 \end{bmatrix}$ and $B = \begin{bmatrix} 4 & 0 \\ 2 & -1 \end{bmatrix}$.

Solution: We subtract the elements in the matching positions:

$$A-B = \begin{bmatrix} 1-4 & -1-0 \\ 2-2 & 0-(-1) \end{bmatrix} = \begin{bmatrix} -3 & -1 \\ 0 & 1 \end{bmatrix}.$$

So, the difference of the two given matrices is $\begin{bmatrix} -3 & -1 \\ 0 & 1 \end{bmatrix}$.

16.3 Scalar Multiplication

Scalar multiplication entails multiplying each element a_{ij} of a matrix A by a scalar (a real number) c, producing a new matrix of identical size. This operation, represented as $(cA)_{ij} = ca_{ij}$, applies to every element within A, effectively scaling or transforming the matrix. A special instance is when $c = 0$, which converts A into a zero matrix of the same dimensions, with all entries set to 0.

Consider the matrix $A = \begin{bmatrix} 3 & 5 \\ 2 & 4 \end{bmatrix}$, and scalar $d = 3$. Get the corresponding scalar multiplication.

Solution: We obtain the scalar multiplication dA by multiplying each entry in A by 3:

$$3A = \begin{bmatrix} 9 & 15 \\ 6 & 12 \end{bmatrix}.$$

Consider the matrix $A = \begin{bmatrix} 2 & 4 \\ 1 & 3 \end{bmatrix}$, and scalar $d = -1$. Get the corresponding scalar multiplication.

Solution: We obtain the scalar multiplication dA by multiplying each entry in A by -1:

$$-1A = \begin{bmatrix} -2 & -4 \\ -1 & -3 \end{bmatrix}.$$

Example Consider the matrix $A = \begin{bmatrix} 1 & 3 & 5 \\ 2 & 4 & 6 \end{bmatrix}$, and scalar $d = 0$. Get the corresponding scalar multiplication.

Solution: We obtain the scalar multiplication dA by multiplying each entry in A by 0:

$$0A = \begin{bmatrix} 0 & 0 & 0 \\ 0 & 0 & 0 \end{bmatrix}.$$

16.4 Matrix Multiplication

Matrix multiplication, essential in fields like data processing, systems of equations, and computer graphics, requires the first matrix's columns to match the second's rows. To multiply:

1. Ensure the matrices meet the multiplication condition: the number of columns in the first matrix equals the number of rows in the second.
2. Multiply each row of the first matrix by each column of the second, element-wise.
3. Sum these products to form the resultant matrix's elements.

Key Point

When a matrix $A_{m\times n}$ is multiplied by $B_{n\times p}$, the resulting matrix C maintains the row count of A and the column count of B. The ij-th element of the product of matrices A and B is given by: $(AB)_{ij} = a_{i1}b_{1j} + a_{i2}b_{2j} + \cdots + a_{in}b_{nj} = \sum_{k=1}^{n} a_{ik}b_{kj}$.

Example Find the product of the matrices: $A = \begin{bmatrix} -1 & -3 \\ -4 & 0 \end{bmatrix}$, and $B = \begin{bmatrix} -3 & -2 \\ 4 & 4 \end{bmatrix}$.

Solution: The number of columns in A is 2, and the number of rows in B is also 2. Thus, the matrices are compatible for multiplication. Calculating the elements in the first row of the product:

$$\begin{bmatrix} (-1)(-3) + (-3)(4) & (-1)(-2) + (-3)(4) \end{bmatrix} = \begin{bmatrix} -9 & -10 \end{bmatrix}.$$

Calculating the elements in the second row of the product:

$$\begin{bmatrix}(-4)(-3)+(0)(4) & (-4)(-2)+(0)(4)\end{bmatrix} = \begin{bmatrix}12 & 8\end{bmatrix}.$$

Finally, $AB = \begin{bmatrix}-9 & -10 \\ 12 & 8\end{bmatrix}$.

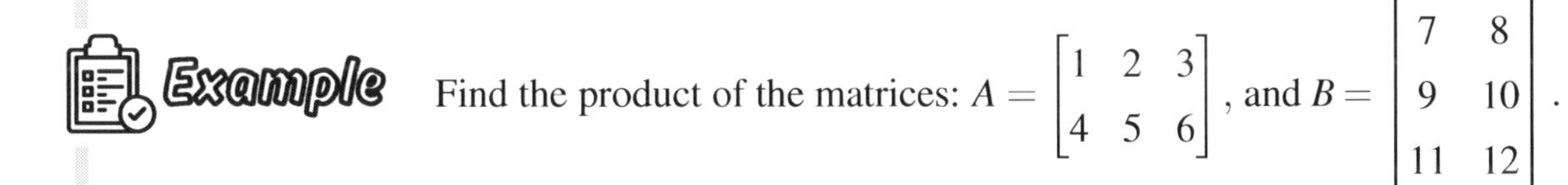

Example Find the product of the matrices: $A = \begin{bmatrix}1 & 2 & 3 \\ 4 & 5 & 6\end{bmatrix}$, and $B = \begin{bmatrix}7 & 8 \\ 9 & 10 \\ 11 & 12\end{bmatrix}$.

Solution: The number of columns in A is 3, and the number of rows in B is also 3. Thus, the matrices are compatible for multiplication. Calculating the elements in the first row of the product:

$$\begin{bmatrix}(1)(7)+(2)(9)+(3)(11) & (1)(8)+(2)(10)+(3)(12)\end{bmatrix} = \begin{bmatrix}58 & 64\end{bmatrix}.$$

Calculating the elements in the second row of the product:

$$\begin{bmatrix}(4)(7)+(5)(9)+(6)(11) & (4)(8)+(5)(10)+(6)(12)\end{bmatrix} = \begin{bmatrix}139 & 154\end{bmatrix}.$$

Finally, $AB = \begin{bmatrix}58 & 64 \\ 139 & 154\end{bmatrix}$.

16.5 Determinants of Matrices

Determinants, denoted as $|A|$ or $\det(A)$, are unique scalar values of square matrices, calculated by specific methods dependent on the matrix's dimensions.

Key Point

The determinant of a 2×2 matrix, $A = \begin{bmatrix}a & b \\ c & d\end{bmatrix}$, is calculated by $|A| = ad - bc$.

Key Point

The determinant of a 3×3 matrix, $A = \begin{bmatrix} a & b & c \\ d & e & f \\ g & h & i \end{bmatrix}$, is computed by

$$|A| = a(ei - fh) - b(di - fg) + c(dh - eg).$$

 Calculate the determinant of the matrix $A = \begin{bmatrix} 3 & 4 \\ 2 & 5 \end{bmatrix}$.

Solution: Using the formula for determinants $ad - bc$, we have

$$\det(A) = (3 \times 5) - (4 \times 2) = 15 - 8 = 7.$$

Thus, the determinant of the matrix A is 7.

 Evaluate the determinant of the matrix $\begin{bmatrix} 2 & 6 & 3 \\ 0 & 5 & 1 \\ 4 & 7 & 4 \end{bmatrix}$.

Solution: For a 3×3 matrix, we use the formula:

$$|A| = a(ei - fh) - b(di - fg) + c(dh - eg).$$

Plugging in the values from the matrix, we have

$$|A| = 2((5 \times 4) - (7 \times 1)) - 6((0 \times 4) - (4 \times 1)) + 3((0 \times 7) - (5 \times 4)) = -10.$$

16.6 Inverse of a Matrix

To understand the inverse of a matrix, one must first be familiar with determinants and the unit (identity) matrix I, characterized by ones on its diagonal and zeros elsewhere. For instance, a 3×3 identity matrix is:

$$I = \begin{bmatrix} 1 & 0 & 0 \\ 0 & 1 & 0 \\ 0 & 0 & 1 \end{bmatrix}.$$

Key Point

The conditions for a matrix A to have an inverse are:

1. A must be a square matrix, implying equal numbers of rows and columns.
2. A must be non-singular, indicated by a non-zero determinant ($|A| \neq 0$).

A matrix B, sharing dimensions with A and satisfying $AB = BA = I$, is termed the inverse of A, making A invertible.

Key Point

The inverse of a matrix is not the reciprocal of the matrix itself, instead it is a distinct matrix that when multiplied with the original matrix results in an identity matrix.

For a 2×2 matrix:

$$A = \begin{bmatrix} a & b \\ c & d \end{bmatrix} \Rightarrow A^{-1} = \frac{1}{|A|} \begin{bmatrix} d & -b \\ -c & a \end{bmatrix},$$

where $|A|$ is the determinant of matrix A. Computing the inverse becomes more complex with larger matrices.

Example Show that $A = \begin{bmatrix} 2 & 1 \\ -1 & 0 \end{bmatrix}$ and $B = \begin{bmatrix} 0 & -1 \\ 1 & 2 \end{bmatrix}$ are inverses of one another.

Solution: For two matrices to be inverses, they must satisfy $BA = AB = I$. Therefore:

$$A \times B = \begin{bmatrix} 2 & 1 \\ -1 & 0 \end{bmatrix} \times \begin{bmatrix} 0 & -1 \\ 1 & 2 \end{bmatrix} = \begin{bmatrix} 1 & 0 \\ 0 & 1 \end{bmatrix} = I,$$

and

$$B \times A = \begin{bmatrix} 0 & -1 \\ 1 & 2 \end{bmatrix} \times \begin{bmatrix} 2 & 1 \\ -1 & 0 \end{bmatrix} = \begin{bmatrix} 1 & 0 \\ 0 & 1 \end{bmatrix} = I.$$

From this, we can conclude that A and B are inverses of each other.

Example Find the inverse of the matrix $C = \begin{bmatrix} -1 & 3 \\ 0 & 2 \end{bmatrix}$.

Solution: The first step is to check if the matrix is a square and if the determinant is non-zero. Matrix C is indeed a 2×2, or square, matrix. Now we calculate the determinant:

$$|C| = (-1 \times 2) - (3 \times 0) = -2.$$

As we can see, the determinant is non-zero. The next step is to apply the formula to find the inverse of the matrix C:

$$C^{-1} = \frac{1}{|C|}\begin{bmatrix} 2 & -3 \\ 0 & -1 \end{bmatrix} = -\frac{1}{2}\begin{bmatrix} 2 & -3 \\ 0 & -1 \end{bmatrix} = \begin{bmatrix} -1 & \frac{3}{2} \\ 0 & \frac{1}{2} \end{bmatrix}.$$

So, the inverse of C is $\begin{bmatrix} -1 & \frac{3}{2} \\ 0 & \frac{1}{2} \end{bmatrix}$.

16.7 Solving Linear Systems of Equations with Matrices

This section explores solving linear systems by transforming them into the matrix equation $AX = B$, where A represents the coefficient matrix, X the variable matrix, and B the constant matrix.

Key Point

The conversion to $AX = B$ simplifies the representation and handling of linear systems.

The transformation involves three main steps:

1. Rewrite the linear system in standard form.
2. Express the system as a matrix equation.
3. Solve for X using $X = A^{-1}B$, ensuring the determinant of A is non-zero, which is essential for computing A^{-1}.

Key Point

A non-zero determinant of A is crucial for the existence and computation of the matrix inverse A^{-1}.

Example Find the value of y in the given system of equations: $\begin{cases} 2x - 1 = 3y \\ y = 4 + x \end{cases}$

Solution: To solve this system of equations, firstly rewrite it in standard form: $\begin{cases} 2x - 3y = 1 \\ x - y = -4 \end{cases}$ Then convert into a matrix equation:

$$\begin{bmatrix} 2 & -3 \\ 1 & -1 \end{bmatrix}\begin{bmatrix} x \\ y \end{bmatrix} = \begin{bmatrix} 1 \\ -4 \end{bmatrix}.$$

Next, evaluate the determinant of the coefficient matrix A: $|A| = (2 \times -1) - (-3 \times 1) = 1$.

The inverse of A hence can be calculated:

$$A^{-1} = \frac{1}{|A|}\begin{bmatrix} -1 & 3 \\ -1 & 2 \end{bmatrix} = \begin{bmatrix} -1 & 3 \\ -1 & 2 \end{bmatrix}.$$

Finally, solve for X:

$$X = A^{-1}B = \begin{bmatrix} -1 & 3 \\ -1 & 2 \end{bmatrix}\begin{bmatrix} 1 \\ -4 \end{bmatrix} = \begin{bmatrix} -13 \\ -9 \end{bmatrix}.$$

Thus, $x = -13$ and $y = -9$.

16.8 Practices

1) Determine the values of x and y that satisfy the following matrix equation:

$$\begin{bmatrix} x-3 & 5 \\ 2 & y+1 \end{bmatrix} = \begin{bmatrix} 1 & 5 \\ 2 & 4 \end{bmatrix}$$

☐ A. $x = 4,\ y = 3$

☐ B. $x = -1,\ y = 2$

☐ C. $x = 4,\ y = -3$

☐ D. $x = -1,\ y = 3$

2) What is the dimension of the matrix $Q = \begin{bmatrix} 3 & -1 & 2 \\ 0 & 4 & 5 \end{bmatrix}$?

☐ A. 2×2

☐ B. 2×3

☐ C. 3×2

☐ D. 3×3

3) Given the matrices $P = \begin{bmatrix} 5 & 0 \\ 3 & -7 \end{bmatrix}$ and $Q = \begin{bmatrix} 2 & 4 \\ -1 & 3 \end{bmatrix}$, what is $P + Q$?

☐ A. $\begin{bmatrix} 3 & 4 \\ 2 & 4 \end{bmatrix}$

☐ B. $\begin{bmatrix} 3 & -4 \\ 4 & -10 \end{bmatrix}$

☐ C. $\begin{bmatrix} 5 & 0 \\ 1 & 3 \end{bmatrix}$

☐ D. $\begin{bmatrix} 7 & 4 \\ 2 & -4 \end{bmatrix}$

4) If $R = \begin{bmatrix} -3 & 7 \end{bmatrix}$ and $S = \begin{bmatrix} 4 & -8 \end{bmatrix}$, which of the following represents the subtraction $R - S$?

☐ A. $\begin{bmatrix} 1 & -1 \end{bmatrix}$

☐ B. $\begin{bmatrix} -1 & -15 \end{bmatrix}$

☐ C. $\begin{bmatrix} -7 & 15 \end{bmatrix}$

☐ D. $\begin{bmatrix} -7 & 11 \end{bmatrix}$

5) If we have the matrix $B = \begin{bmatrix} -6 & 8 \\ 5 & -3 \end{bmatrix}$ and scalar $k = 2$, what is kB?

☐ A. $\begin{bmatrix} -12 & 16 \\ 10 & -6 \end{bmatrix}$

☐ B. $\begin{bmatrix} -3 & 4 \\ 2.5 & -1.5 \end{bmatrix}$

☐ C. $\begin{bmatrix} 12 & -16 \\ -10 & 6 \end{bmatrix}$

☐ D. $\begin{bmatrix} 3 & -4 \\ -2.5 & 1.5 \end{bmatrix}$

6) Calculate the scalar product when the matrix C is multiplied by the scalar $m = -1$: $C = \begin{bmatrix} 0 & 7 \\ -2 & 5 \\ 6 & -3 \end{bmatrix}$.

☐ A. $\begin{bmatrix} 0 & -7 \\ 2 & -5 \\ -6 & 3 \end{bmatrix}$

☐ B. $\begin{bmatrix} 0 & 7 \\ 2 & 5 \\ 6 & 3 \end{bmatrix}$

☐ C. $\begin{bmatrix} -1 & -7 \\ -2 & -5 \\ -6 & -3 \end{bmatrix}$

☐ D. $\begin{bmatrix} 0 & 7 \\ -2 & 5 \\ 6 & -3 \end{bmatrix}$

7) Given matrices $A = \begin{bmatrix} 3 & 1 \\ 2 & 4 \end{bmatrix}$ and $B = \begin{bmatrix} 5 \\ 6 \end{bmatrix}$, what is the product AB?

☐ A. $\begin{bmatrix} 21 \\ 14 \end{bmatrix}$

☐ B. $\begin{bmatrix} 15 & 18 \\ 20 & 24 \end{bmatrix}$

☐ C. $\begin{bmatrix} 21 \\ 34 \end{bmatrix}$

☐ D. $\begin{bmatrix} 15 & 6 \\ 10 & 24 \end{bmatrix}$

8) What is the product of the matrix $A = \begin{bmatrix} 2 & -1 \\ 0 & 3 \end{bmatrix}$ with the identity matrix $I = \begin{bmatrix} 1 & 0 \\ 0 & 1 \end{bmatrix}$?

☐ A. $\begin{bmatrix} 2 & -1 \\ 0 & 3 \end{bmatrix}$

☐ B. $\begin{bmatrix} 2 & 0 \\ 0 & -3 \end{bmatrix}$

☐ C. $\begin{bmatrix} 0 & 0 \\ 0 & 0 \end{bmatrix}$

☐ D. $\begin{bmatrix} 1 & 0 \\ 0 & 1 \end{bmatrix}$

9) What is the determinant of the matrix $\begin{bmatrix} 1 & 2 \\ 3 & 4 \end{bmatrix}$?

☐ A. -2

☐ B. 0

☐ C. 2

☐ D. 10

10) Find the determinant of the matrix $\begin{bmatrix} 1 & 4 & 2 \\ 3 & -1 & 5 \\ 2 & 0 & 3 \end{bmatrix}$.

☐ A. 5

☐ B. 11

☐ C. -11

☐ D. -5

11) Given the matrix $A = \begin{bmatrix} 4 & 7 \\ 2 & 6 \end{bmatrix}$, which one is the inverse of A?

☐ A. $\begin{bmatrix} 6 & -7 \\ -2 & 4 \end{bmatrix}$

☐ B. $\begin{bmatrix} 0.6 & -0.7 \\ -0.2 & 0.4 \end{bmatrix}$

☐ C. $\begin{bmatrix} 3 & -3.5 \\ -1 & 2 \end{bmatrix}$

☐ D. $\begin{bmatrix} 0.6 & 0.7 \\ 0.2 & -0.4 \end{bmatrix}$

12) If a 2×2 matrix $B = \begin{bmatrix} a & b \\ c & d \end{bmatrix}$ is singular, what can we conclude?

☐ A. The inverse of matrix B does not exist.

☐ B. The inverse of matrix B is the zero matrix.

☐ C. The inverse of matrix B is B itself.

☐ D. The determinant of matrix B is 1.

13) Given the system of equations below, which of the following is the matrix representation $AX = B$?

$$\begin{cases} 3x + 2y = 5 \\ -x + 4y = -2 \end{cases}$$

☐ A. $\begin{bmatrix} 3 & 2 \\ -1 & 4 \end{bmatrix} \begin{bmatrix} x \\ y \end{bmatrix} = \begin{bmatrix} 5 \\ -2 \end{bmatrix}$

☐ B. $\begin{bmatrix} 5 & -2 \\ 3 & 2 \end{bmatrix} \begin{bmatrix} x \\ y \end{bmatrix} = \begin{bmatrix} 3 \\ 4 \end{bmatrix}$

☐ C. $\begin{bmatrix} 3 & 2 \\ -1 & 4 \end{bmatrix} \begin{bmatrix} y \\ x \end{bmatrix} = \begin{bmatrix} 5 \\ -2 \end{bmatrix}$

☐ D. $\begin{bmatrix} 2 & 3 \\ 4 & -1 \end{bmatrix} \begin{bmatrix} x \\ y \end{bmatrix} = \begin{bmatrix} -2 \\ 5 \end{bmatrix}$

14) What is the determinant of the coefficient matrix for the following system of equations?

$$\begin{cases} x+2y=7 \\ 3x-y=4 \end{cases}$$

- ☐ A. −7
- ☐ B. −5
- ☐ C. 5
- ☐ D. 7

Answer Keys

1) A. $x = 4,\ y = 3$

2) B. 2×3

3) D. $\begin{bmatrix} 7 & 4 \\ 2 & -4 \end{bmatrix}$

4) C. $\begin{bmatrix} -7 & 15 \end{bmatrix}$

5) A. $\begin{bmatrix} -12 & 16 \\ 10 & -6 \end{bmatrix}$

6) A. $\begin{bmatrix} 0 & -7 \\ 2 & -5 \\ -6 & 3 \end{bmatrix}$

7) C. $\begin{bmatrix} 21 \\ 34 \end{bmatrix}$

8) A. $\begin{bmatrix} 2 & -1 \\ 0 & 3 \end{bmatrix}$

9) A. -2

10) A. 5

11) B. $\begin{bmatrix} 0.6 & -0.7 \\ -0.2 & 0.4 \end{bmatrix}$

12) A. The inverse of matrix B does not exist.

13) A. $\begin{bmatrix} 3 & 2 \\ -1 & 4 \end{bmatrix} \begin{bmatrix} x \\ y \end{bmatrix} = \begin{bmatrix} 5 \\ -2 \end{bmatrix}$

14) A. -7

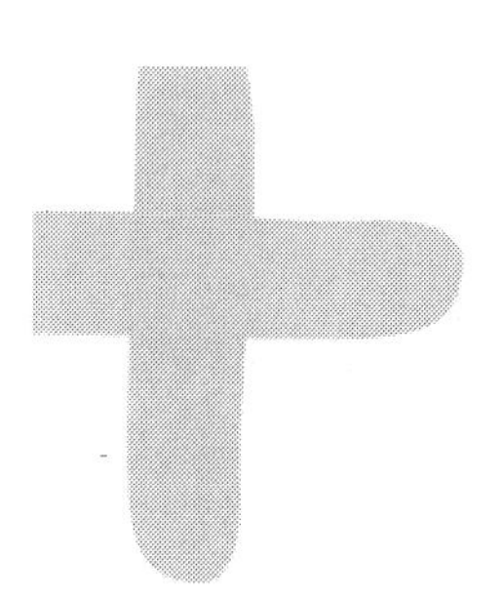

17. AP Precalculus Test Review and Strategies

17.1 The AP Precalculus Test Review

The *AP Precalculus Exam* is a standardized test designed for high school students who have completed a precalculus course and wish to earn college credit or advanced placement in college mathematics courses. This exam, introduced by the College Board, focuses on critical mathematical practices and concepts that are foundational for careers in mathematics, science, and related fields.

The exam structure is divided into two sections: multiple-choice and free-response, which together assess a student's understanding and ability to apply various mathematical skills.

- **Section I: Multiple Choice**
- **Section II: Free Response**

Section I of the AP Precalculus exam consists of 40 multiple-choice questions, which are split into two parts:

- Part A: 28 questions, 80 minutes, no calculator allowed, making up approximately 44% of the exam score.
- Part B: 12 questions, 40 minutes, where a graphing calculator is required, contributing about 19% to the total score.

Section II includes 4 free-response questions:

- Part A: 2 questions, 30 minutes, requiring a graphing calculator, contributing around 19% to the score.
- Part B: 2 questions, 30 minutes, where calculators are not allowed, also contributing about 19% to the total score.

The exam covers three major units from the AP Precalculus curriculum:

- **Polynomial and Rational Functions** (30%-40% of multiple-choice questions)
- **Exponential and Logarithmic Functions** (27%-40%)
- **Trigonometric and Polar Functions** (30%-35%)

The total duration of the exam is 3 hours, and the exam is scored on a scale from 1 to 5, with scores of 3 and above typically qualifying for college credit, depending on the institution's policy.

The AP Precalculus exam not only tests procedural and symbolic fluency but also evaluates a student's ability to translate mathematical information between different representations and communicate mathematical reasoning effectively.

Given the comprehensive nature of the exam, thorough preparation covering all units and practicing with both calculator and non-calculator sections is crucial for success. Students are encouraged to review all concepts in the course framework and take advantage of available resources like practice exams and AP Classroom materials.

17.2 AP Precalculus Test-Taking Strategies

Successfully navigating the AP Precalculus test requires not only a solid understanding of mathematical concepts but also effective problem-solving strategies. In this section, we explore a range of strategies to optimize your performance and outcomes on the AP Precalculus test. From comprehending the question and using informed guessing to finding ballpark answers and employing backsolving and numeric substitution, these strategies will empower you to tackle various types of math problems with confidence and efficiency.

#1 Understand the Questions and Review Answers

Below are a set of effective strategies to optimize your performance and outcomes on the AP Precalculus test.

- **Comprehend the Question:** Begin by carefully reviewing the question to identify keywords and essential information.
- **Mathematical Translation:** Translate the identified keywords into mathematical operations that will enable you to solve the problem effectively.
- **Analyze Answer Choices:** Examine the answer choices provided and identify any distinctions or patterns among them.
- **Visual Aids:** If necessary, consider drawing diagrams or labeling figures to aid in problem-solving.
- **Pattern Recognition:** Look for recurring patterns or relationships within the problem that can guide your solution.

- **Select the Right Method:** Determine the most suitable strategies for answering the question, whether it involves straightforward mathematical calculations, numerical substitution (plugging in numbers), or testing the answer choices (backsolving); see below for a comprehensive explanation of these methods.
- **Verification:** Before finalizing your answer, double-check your work to ensure accuracy and completeness.

Let's review some of the important strategies in detail.

#2 Use Educated Guessing

This strategy is particularly useful for tackling problems that you have some understanding of but cannot solve through straightforward mathematics. In such situations, aim to eliminate as many answer choices as possible before making a selection. When faced with a problem that seems entirely unfamiliar, there's no need to spend excessive time attempting to eliminate answer choices. Instead, opt for a random choice before proceeding to the next question.

As you can see, employing direct solutions is the most effective approach. Carefully read the question, apply the math concepts you've learned, and align your answer with one of the available choices. Feeling stuck? Make your best-educated guess and move forward.

Never leave questions unanswered! Even if a problem appears insurmountable, make an effort to provide a response. If necessary, make an educated guess. Remember, you won't lose points for an incorrect answer, but you may earn points for a correct one!

#3 Ballpark Estimates

A *"ballpark estimate"* is a *rough approximation.* When dealing with complex calculations and numbers, it's easy to make errors. Sometimes, a small decimal shift can turn a correct answer into an incorrect one, no matter how many steps you've taken to arrive at it. This is where ballparking can be incredibly useful.

If you have an idea of what the correct answer might be, even if it's just a rough estimate, you can often eliminate a few answer choices. While answer choices typically account for common student errors and closely related values, you can still rule out choices that are significantly off the mark. When facing a multiple-choice question, deliberately look for answers that don't even come close to the ballpark. This strategy effectively helps eliminate incorrect choices during problem-solving.

#4 Backsolving

A significant portion of questions on the AP Precalculus test are presented in multiple-choice format. Many test-takers find multiple-choice questions preferable since the correct answer is among the choices provided. Typically, you'll have four options to choose from, and your task is to determine the correct one. One effective approach for this is known as *"backsolving."*

As mentioned previously, solving questions directly is the most optimal method. Begin by thoroughly examining the problem, calculating a solution, and then matching the answer with one of the available choices. However, if you find yourself unable to calculate a solution, the next best approach involves employing *"backsolving."*

When employing backsolving, compare one of the answer choices to the problem at hand and determine which choice aligns most closely. Frequently, answer choices are arranged in either ascending or descending order. In such cases, consider testing options B or C first. If neither is correct, you can proceed either up or down from there.

#5 Plugging In Numbers

Using numeric substitution or *'plugging in numbers'* is a valuable strategy applicable to a wide array of math problems encountered on the AP Precalculus test. This approach is particularly helpful in simplifying complex questions, making them more manageable and comprehensible. By employing this strategy thoughtfully, you can arrive at the solution with ease.

The concept is relatively straightforward. Simply replace unknown variables in a problem with specific values. When selecting a number for substitution, consider the following guidelines:

- Opt for a basic number (though not overly basic). It's generally advisable to avoid choosing 1 (or even 0). A reasonable choice often includes selecting the number 2.
- Avoid picking a number already present in the problem statement.
- Ensure that the chosen numbers are distinct when substituting at least two of them.
- Frequently, the use of numeric substitution helps you eliminate some of the answer choices, so it's essential not to hastily select the first option that appears to be correct.
- When faced with multiple seemingly correct answers, you may need to opt for a different set of values and reevaluate the choices that haven't been ruled out yet.

It is Time to Test Yourself

It's time to refine your skills with a practice examination designed to simulate the AP Precalculus Test. Engaging with the practice tests will help you to familiarize yourself with the test format and timing, allowing for a more effective test day experience. After completing a test, use the provided answer key to score your work and identify areas for improvement.

Before You Start

To make the most of your practice test experience, please ensure you have:

- A pencil for marking answers on the answer sheet.
- A timer to manage pacing, replicating potential time constraints in other testing scenarios.

Please note the following important points as you prepare to take your practice test:

- It's okay to guess! There is no penalty for incorrect answers, so make sure to answer every question.
- After completing the test, review the answer key to understand any mistakes. This review is crucial for your learning and preparation.
- An answer sheet is provided for you to record your answers. Make sure to use it.
- For each multiple-choice question, you will be presented with possible choices. Your task is to choose the best one.

Good Luck! Your preparation and practice are the keys to success.

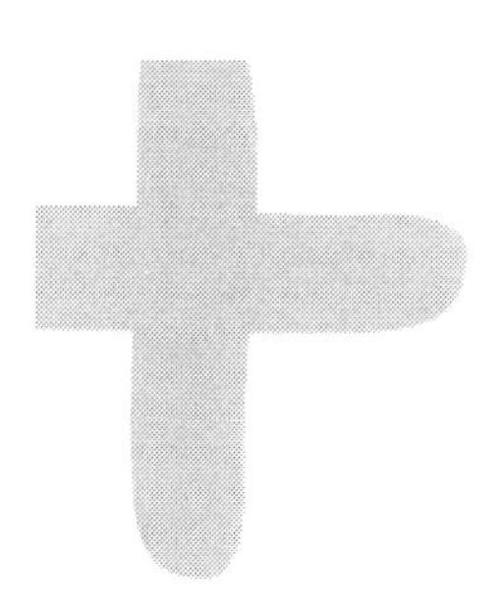

18. Practice Test 1

18.1 Practices

Section I. Part A: 28 questions (80 minutes). No calculator is allowed.

1) What is the domain and range of the function $f(x) = 3 + 4\cos\left(x + \frac{\pi}{4}\right)$?

☐ A. Domain: $\left[\frac{\pi}{4}, +\infty\right)$, Range: $[-7, 1]$

☐ B. Domain: $\mathbb{R}$, Range: $[-1, 7]$

☐ C. Domain: $\mathbb{R} - \{\frac{\pi}{4}\}$, Range: $[-1, 7]$

☐ D. Domain: $\left[-\frac{\pi}{4}, +\frac{\pi}{4}\right)$, Range: $[-7, 1]$

2) Which function is best represented by this graph?

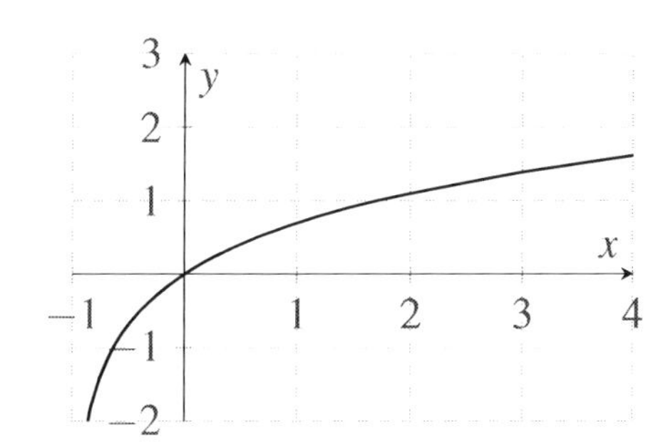

☐ A. $\ln(x)$

☐ B. $\ln(x+1)$

☐ C. $\ln(x-1)$

☐ D. $\ln(x) - 1$

3) Which function is best represented by this graph?

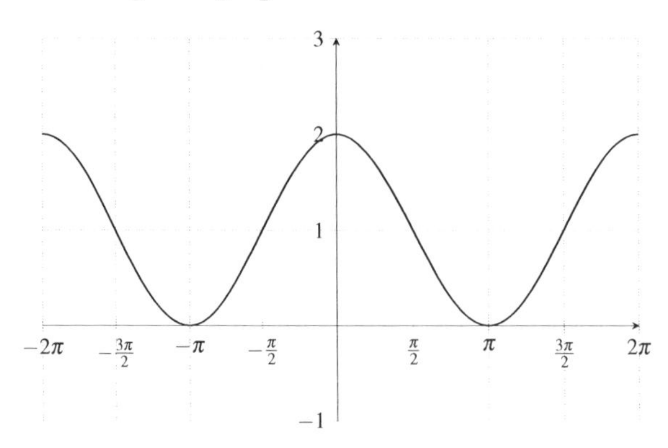

☐ A. $y = 2\sec(x)$

☐ B. $y = \cos(x) + 1$

☐ C. $y = \tan(2x) - 1$

☐ D. $y = 2\cos(x) - 1$

4) Find the equation of the given parabola:

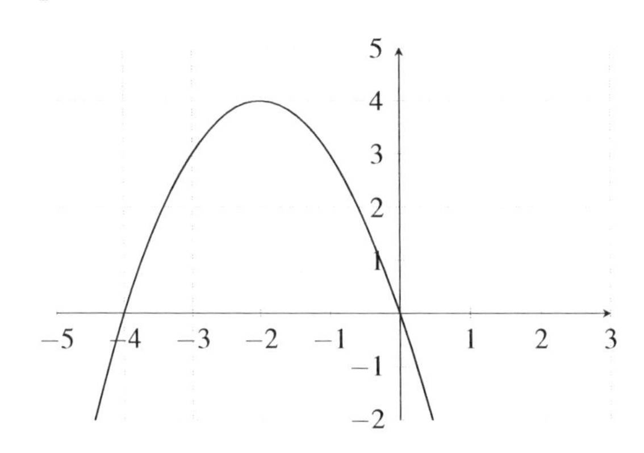

☐ A. $y = x^2 + 4x - 1$

☐ B. $y = -x^2 - 4x$

☐ C. $y = -2x^2 - 4x + 2$

☐ D. $y = -x^2 + 4$

5) What is the inverse of $g(x) = 2 + \cos(x)$?

☐ A. $g^{-1}(x) = \cos(2 + x)$

☐ B. $g^{-1}(x) = \arccos(x - 2)$

☐ C. $g^{-1}(x) = 2 + \arccos(x)$

☐ D. $g^{-1}(x) = \arccos(2 - x)$

6) The diagram shows two right-angled triangles clinging together on a common side. Find the length of the base of Triangle B, denoted as x.

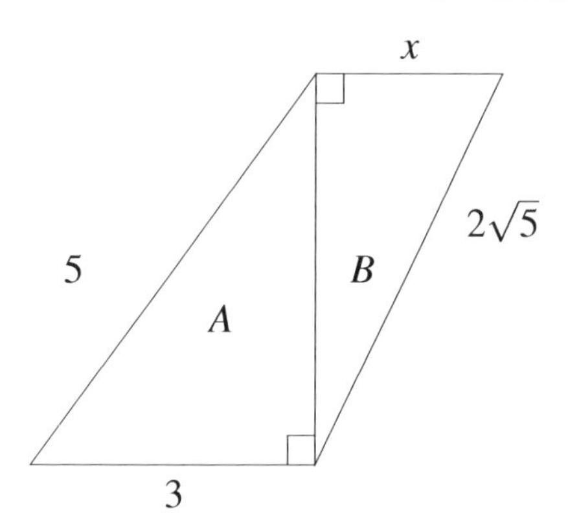

☐ A. $\frac{9}{\sqrt{2}}$

☐ B. 2

☐ C. $4\sqrt{2}$

☐ D. $3\sqrt{3}$

7) Which statement best describes these two functions?

$$h(x) = \sin x + 3$$
$$j(x) = 5 - \cos^2 x$$

☐ A. They have no common points.

☐ B. They have the same y-intercepts.

☐ C. The maximum of $h(x)$ is the same as the maximum of $j(x)$.

☐ D. They have the same minimum.

8) Which graph corresponds to $y = (x+1)^2 - 2$?

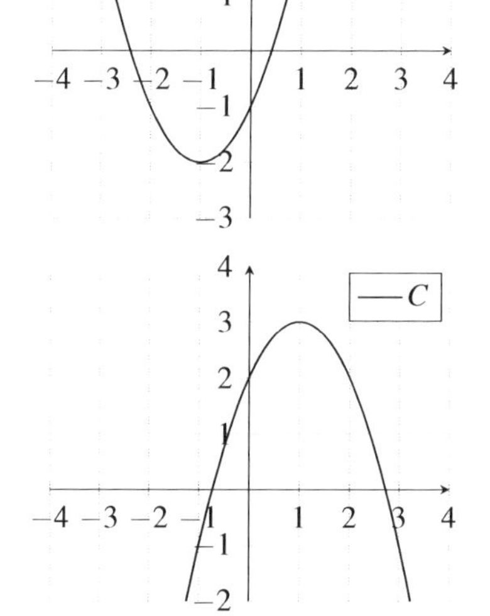

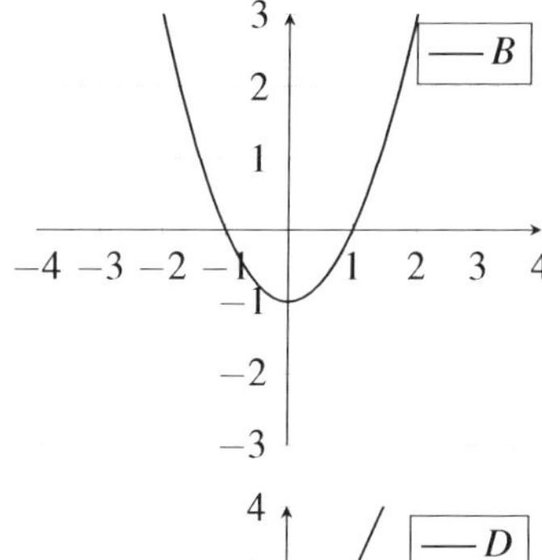

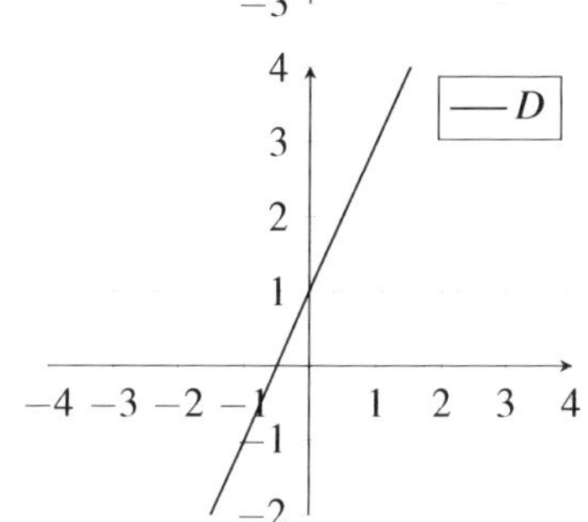

☐ A. Graph A

☐ B. Graph B

☐ C. Graph C

☐ D. Graph D

9) If $\sin\left(\frac{\pi}{2}\right) = x^2 - x + 1$, then the value of x is

☐ A. $x = -1$

☐ B. $x = 0$ and $x = 1$

☐ C. $x = 1$ and $x = -1$

☐ D. None of these

10) Which graph represents the inverse of $y = \cos\left(x + \frac{\pi}{4}\right)$ over the interval $\left[-\frac{\pi}{6}, \frac{5\pi}{6}\right]$?

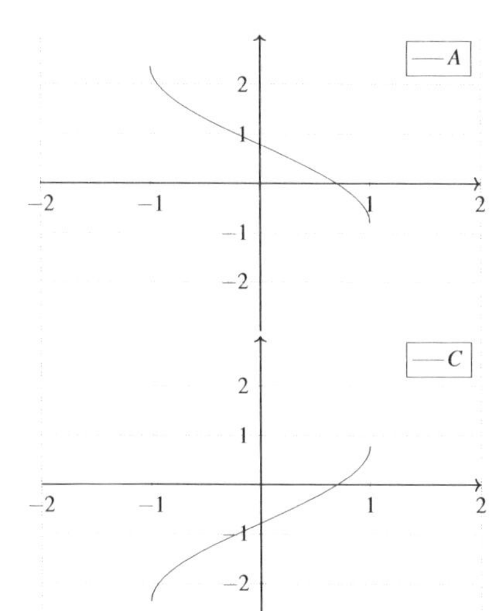

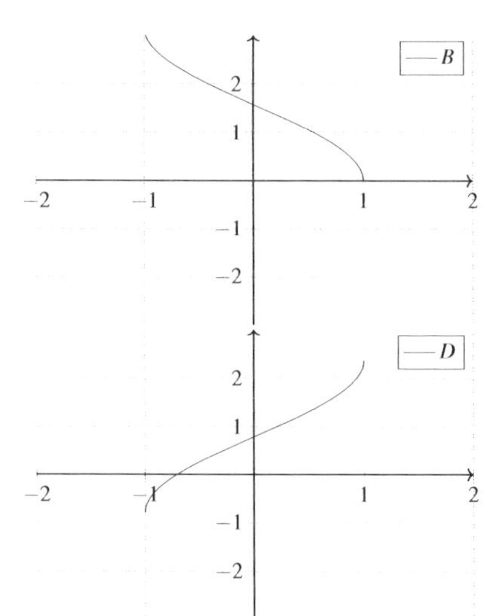

☐ A. Graph A

☐ B. Graph B

☐ C. Graph C

☐ D. Graph D

11) Find the slope of the line passing through the points $(2, -1)$ and $(7, -5)$.

☐ A. $-\frac{4}{5}$

☐ B. $\frac{4}{5}$

☐ C. $\frac{5}{4}$

☐ D. $-\frac{5}{4}$

12) The graph of a function $f(x)$ is a transformed trigonometric function. Which of the following represents the function $f(x)$?

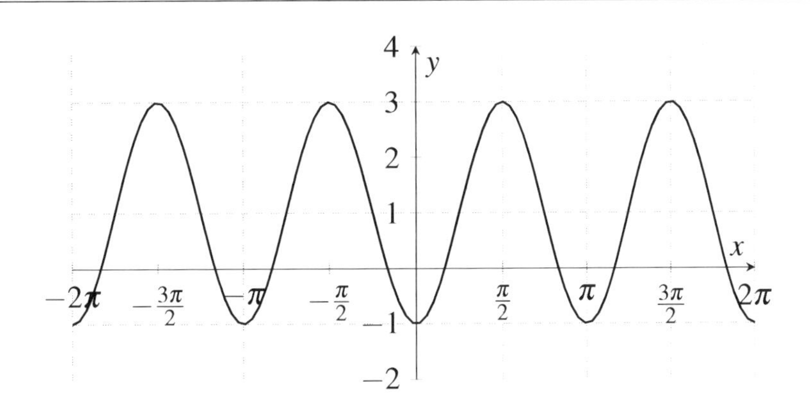

☐ A. $f(x)=2\cos\left(2\left(x+\frac{\pi}{2}\right)\right)+1$

☐ B. $f(x)=-2\cos\left(2\left(x-\frac{\pi}{2}\right)\right)+1$

☐ C. $f(x)=2\sin\left(2\left(x-\frac{\pi}{2}\right)\right)+1$

☐ D. $f(x)=-2\sin\left(2\left(x+\frac{\pi}{2}\right)\right)+1$

13) Which best describes the graph of $\frac{x^2}{25}+\frac{y^2}{36}=1$?

☐ A. Hyperbola

☐ B. Circle

☐ C. Parabola

☐ D. Ellipse

14) Find a positive and a negative coterminal angle to angle 60°.

☐ A. 60°, 300°

☐ B. 420°, -300°

☐ C. -300°, 770°

☐ D. 300°, -60°

15) If $\begin{bmatrix} 2 & 3 & 1 \\ -4 & 1 & x \end{bmatrix}\begin{bmatrix} y \\ 0 \\ -3 \end{bmatrix}=\begin{bmatrix} 5 \\ -11 \end{bmatrix}$, what are x and y?

☐ A. $x=2$ and $y=1$

☐ B. $x=-\frac{5}{3}$ and $y=4$

☐ C. $x=1$ and $y=-3$

☐ D. $x=-3$ and $y=2$

16) The value of $2(\sin 30^\circ \cos 60^\circ)$ is:

☐ A. $\frac{1}{2}$

☐ B. $\sqrt{2}$

☐ C. $\sqrt{2}+1$

☐ D. Not defined

17) What is the sum of the first 5 terms of the series $2, 6, 18, \cdots$?

☐ A. 242

☐ B. 726

☐ C. 362

☐ D. 90

18) What is the parent graph of the following function and what transformations have taken place on it?

$$y = -3 + \left(1 - 4x - x^2\right)$$

☐ A. The parent graph is $y = x^2$, reflected over the x-axis, shifted 2 units left, and shifted 2 units down.

☐ B. The parent graph is $y = x^2$, reflected over the x-axis, shifted 2 units right, and shifted 2 units down.

☐ C. The parent graph is $y = -x^2$, reflected over the x-axis, shifted 2 units right, and shifted 2 units up.

☐ D. The parent graph is $y = x^2$, reflected over the x-axis, shifted 2 units left, and shifted 2 units up

19) If $\sin\alpha \sec 30^\circ = 1$, the value of α is

☐ A. 30°

☐ B. 45°

☐ C. 60°

☐ D. 90°

20) If $|B| = 3$, where B is a matrix and represented by $\begin{bmatrix} 2 & 0 & 0 \\ 1 & p & 0 \\ -1 & 1 & 1 \end{bmatrix}$. What is p?

☐ A. -1

☐ B. $-\frac{1}{2}$

☐ C. $\frac{3}{2}$

☐ D. 2

21) Determine the equation of the graphed Hyperbola:

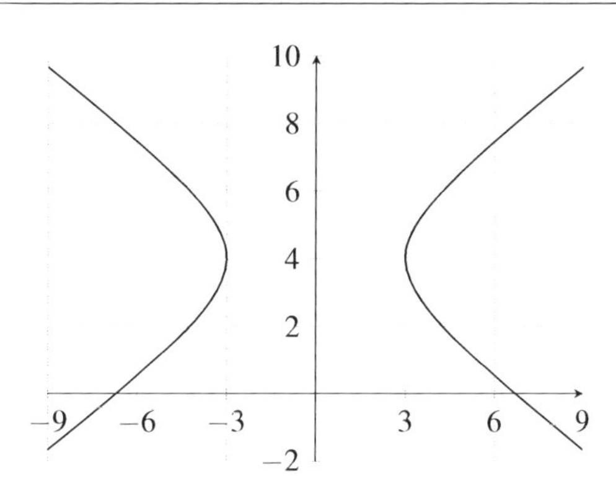

☐ A. $\frac{x^2}{9} - \frac{(y-4)^2}{4} = 1$
☐ B. $\frac{(x-3)^2}{4} - \frac{(y+4)^2}{9} = 1$
☐ C. $\frac{(y+4)^2}{4} - \frac{x^2}{9} = 1$
☐ D. $\frac{(y-4)^2}{4} - \frac{(x-3)^2}{9} = 1$

22) The measures of the angles of a triangle are in the ratio $4:5:7$. The measures of the angles are

☐ A. $24°$, $30°$, $126°$
☐ B. $45°$, $56°$, $79°$
☐ C. $28°$, $35°$, $117°$
☐ D. $36°$, $45°$, $99°$

23) What are the zeroes of the function $f(x) = x^3 + 4x^2 + 4x$?

☐ A. $\{0, -2\}$
☐ B. $\{0, -4\}$
☐ C. $\{2, -2\}$
☐ D. No roots

24) If $\frac{1}{2}\log_{2x} 81 = 2$, what is the value of x?

☐ A. -3
☐ B. $\frac{3}{2}$
☐ C. 2
☐ D. 3

25) The angles of depression of two ships from the top of a cliff are $60°$ and $30°$. If the ships are 300 meters apart, the height of the cliff is:

☐ A. 150 m
☐ B. $150\sqrt{3}$ m
☐ C. 300 m

☐ D. $150\left(\sqrt{3}+1\right)\ m$

26) If $2\left(\sin^2\theta-\cos^2\theta\right)=-1$, where θ is a positive acute angle, then the value of θ is

☐ A. 60°

☐ B. 45°

☐ C. 30°

☐ D. 22.5°

27) If $f(x)=3x^3-2x+1$ and $g(x)=x+3$, what is the value of $(f-g)(2)$?

☐ A. 17

☐ B. 21

☐ C. 25

☐ D. 16

28) How can the equation $y=x^2-3x+5$ be categorized?

☐ A. It represents a vertical parabola.

☐ B. It represents a vertical hyperbola.

☐ C. It represents a horizontal parabola.

☐ D. It represents a horizontal hyperbola.

Section I. Part B: 12 questions (40 minutes). A graphing calculator is required

29) If $h(x)=2x-4$ and $j(x)=x^2+4x+3$, what is $(j-h)(x)$?

☐ A. x^2+2x+7

☐ B. x^2+6x-1

☐ C. x^2-6x+7

☐ D. $-x^2+2x-1$

30) In circular measure, the value of 45° is

☐ A. $\frac{\pi}{6}$

☐ B. $\frac{\pi}{4}$

☐ C. $\frac{\pi}{3}$

☐ D. $\frac{\pi}{2}$

31) A ladder leans against a wall forming a 45° angle between the ground and the ladder. If the bottom of the ladder is 20 feet away from the wall, how long is the ladder?

☐ A. $20ft$

☐ B. $40ft$

☐ C. $20\sqrt{3}ft$

☐ D. $20\sqrt{2}ft$

32) If $|\vec{c}\cdot\vec{d}|=0$, $|\vec{c}|=3$, and $|\vec{d}|=4$, then what is $|\vec{c}-\vec{d}|$?

☐ A. 1

☐ B. 5

☐ C. 7

☐ D. 12

33) If $\begin{bmatrix} 3 & 5 \\ -2 & -6 \\ -1 & -3 \end{bmatrix} + \begin{bmatrix} c & 0 \\ 0 & 8 \\ 4 & 6 \end{bmatrix} = \begin{bmatrix} 2 & 5 \\ -2 & 2 \\ 3 & d \end{bmatrix}$, then $c, d = ?$

☐ A. $c=-1$ and $d=-1$

☐ B. $c=-1$ and $d=3$

☐ C. $c=1$ and $d=-3$

☐ D. $c=-1$ and $d=-3$

34) Convert the radian measure $\frac{5\pi}{6}$ to degree measure.

☐ A. 150°

☐ B. 75°

☐ C. 120°

☐ D. 90°

35) Find the value of x in this equation: $\log(4x-1)=\log(2x+3)$.

☐ A. 2

☐ B. -2

☐ C. $-2, 2$

☐ D. No solution

36) What are the coordinates at the minimum point of $f(x)=x^2-4x+3$?

☐ A. $(2,-1)$

☐ B. $(-2,1)$

☐ C. $(2,1)$

☐ D. $(-2,-1)$

37) Determine the asymptotic equation of the function given by $f(x)=\frac{3x^2-9x+1}{3x^2+6x-18}$.

☐ A. $y=-\frac{1}{3}$
☐ B. $y=0$
☐ C. $y=1$
☐ D. $y=3$

38) If $f(x)=3x^2-5$ and $g(x)=\frac{2}{x}$, what is the value of $f(g(x))$?

☐ A. $\frac{3}{2x^2}-5$
☐ B. $\frac{12}{x^2}-5$
☐ C. $3x^2-5+\frac{2}{x}$
☐ D. $3x-\frac{10}{x}$

39) Solve this equation for x: $e^{3x}=7$.

☐ A. $\frac{1}{3}\ln(7)$
☐ B. $\ln(7)$
☐ C. $\ln(3)$
☐ D. $\frac{1}{3}\ln(3)$

40) The diagram shows a right-angled triangle ABC. What is the length of AC?

☐ A. $4\ cm$
☐ B. $5\ cm$
☐ C. $4\sqrt{2}\ cm$
☐ D. $5\sqrt{2}\ cm$

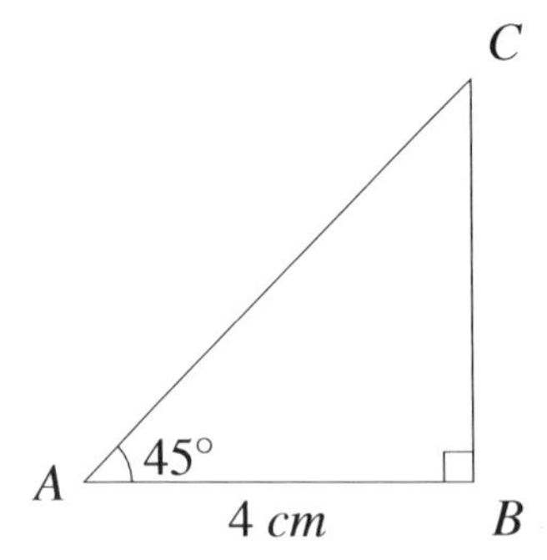

Section II. Part A: 2 questions (30 minutes). A graphing calculator is required.

41) Given the graph of function f defined on the domain $[-3.5, 3.5]$ and function $g(x)=2.916\cdot(0.7)^x$, answer the following questions:

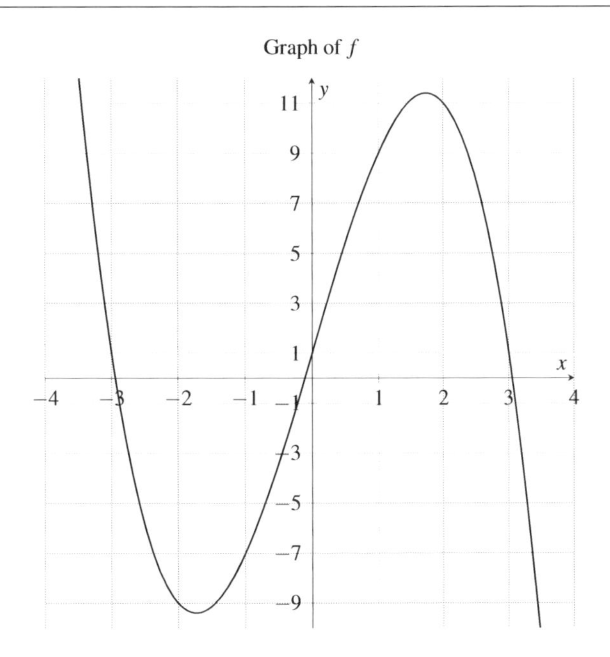

A.

(i) The function h is defined by $h(x) = (g \circ f)(x) = g(f(x))$. Find the value of $h(3)$ as a decimal approximation, or indicate that it is not defined.

(ii) Find all values of x for which $f(x) = 1$, or indicate that there are no such values.

B.

(i) Find all values of x, as decimal approximations, for which $g(x) = 2$, or indicate that there are no such values.

(ii) Determine the end behavior of g as x increases without bound. Express your answer using the mathematical notation of a limit.

C.

(i) Determine if f has an inverse function.

(ii) Give a reason for your answer based on the definition of a function and the graph of $y = f(x)$.

42) On the initial day of sales ($t = 0$) for a new video game, there were 40 thousand units of the game sold. Ninety-one days later ($t = 91$), there were 76 thousand units of the game sold that day. The sales model is given by $G(t) = a + b\ln(t+1)$, where $G(t)$ is the number of units sold, in thousands, on day t.

A.

(i) Use the given data to write two equations that can be used to find the values for constants a and b in the expression for $G(t)$.

(ii) Find the values for a and b as decimal approximations.

B.

(i) Use the given data to find the average rate of change of the number of units of the video game sold, in thousands per day, from $t = 0$ to $t = 91$. Express your answer as a decimal approximation. Show the computations that lead to your answer.

(ii) Use the average rate of change found in (i) to estimate the number of units of the video game sold, in thousands, on day $t = 50$. Show the work that leads to your answer.

(iii) Let A_t represent the estimate of the number of units of the video game sold, in thousands, using the average rate of change found in (i). For A_{50}, explain why, in general, $A_t < G(t)$ for all $0 < t < 91$.

C.

The makers of the video game reported that daily sales of the video game decreased each day after $t = 91$. Explain why the error in the model G increases after $t = 91$.

Section II. Part B: 2 questions (30 minutes). No calculator is permitted.

43) The tire of a car has a radius of 9 inches, and a person rolls the tire forward at a constant rate on level ground. The point W on the edge of the tire is at its highest point above the ground at $t = \frac{1}{2}$ second and again at $t = \frac{5}{2}$ seconds. As the tire rolls, the height of W periodically increases and decreases.

The sinusoidal function h models the height of point W above the ground, in inches, as a function of time, in seconds.

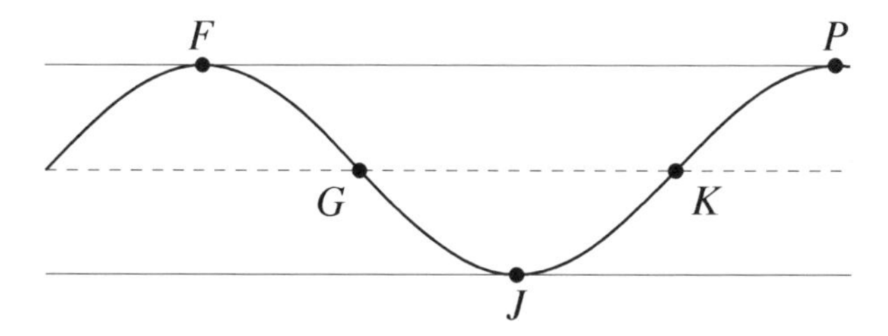

A. Given the graph, determine possible coordinates $(t, h(t))$ for the five points: F, G, J, K, and P.

B. The function h can be written in the form $h(t) = a\sin(b(t+c)) + d$. Find values of constants a, b, c, and d.

C.

(i) On the interval (t_1, t_2) (where t_1 is the t-coordinate of K and t_2 is the t-coordinate of P), which of the following is true about h?

a. h is positive and increasing.

b. h is positive and decreasing.

c. h is negative and increasing.

d. h is negative and decreasing.

(ii) Describe how the rate of change of h is changing on the interval (t_1, t_2).

44) The following problems involve various mathematical functions. Ensure to solve these using algebraic methods, and follow rules for exponents and logarithms where applicable.

A. Given the functions g and h defined as follows:

$$g(x) = e^{(x+3)}, \quad h(x) = \arcsin\left(\frac{x}{2}\right)$$

(i) Solve $g(x) = 10$ for values of x in the domain of g.

(ii) Solve $h(x) = \frac{\pi}{4}$ for values of x in the domain of h.

B. The functions j and k are given by:

$$j(x) = \log_{10}\left(8x^5\right) + \log_{10}\left(2x^2\right) - 9\log_{10}(x), \quad k(x) = \frac{1-\sin^2 x}{\sin x} \cdot \sec x.$$

(i) Rewrite $j(x)$ as a single logarithm base 10 without negative exponents in any part of the expression.

(ii) Rewrite $k(x)$ as a single term involving $\tan x$.

C. Consider the function m defined by:

$$m(x) = \cos^{-1}(\tan(2x)).$$

Find all values in the domain of m that yield an output value of 0.

18.2 Answer Keys

1) B. Domain: $\mathbb{R}$, Range: $[-1,7]$
2) A. x^2+2x+7
3) B. $\ln(x+1)$
4) B. $y=\cos(x)+1$
5) B. $y=-x^2-4x$
6) B. $g^{-1}(x)=\arccos(x-2)$
7) B. $\frac{\pi}{4}$
8) D. $20\sqrt{2}ft$
9) B. 5
10) B. 2
11) B. $c=-1$ and $d=3$
12) A. No common points.
13) A. Graph A
14) B. $x=0$ and $x=1$
15) A. Graph A
16) A. $-\frac{4}{5}$
17) A. 150°
18) A. 2
19) A. $(2,-1)$
20) C. $y=1$
21) B. $\frac{12}{x^2}-5$
22) A. $f(x)=2\cos\left(2\left(x+\frac{\pi}{2}\right)\right)+1$
23) A. $\frac{1}{3}\ln(7)$
24) C. $4\sqrt{2}\,cm$
25) D. Ellipse
26) B. 420°, -300°
27) B. $x=-\frac{5}{3}$ and $y=4$
28) A. $\frac{1}{2}$
29) A. 242
30) D. $y=x^2$, reflected over the x-axis, shifted 2 units left and 2 units up
31) C. 60°
32) c. $\frac{3}{2}$
33) A. $\frac{x^2}{9}-\frac{(y-4)^2}{4}=1$
34) B. 45°, 56°, 78°
35) A. $\{0,-2\}$
36) B. $\frac{3}{2}$
37) B. $150\sqrt{3}\,m$
38) C. 30°
39) D. 16
40) A. It represents a vertical parabola.
41) See answer details.
42) See answer details.
43) See answer details.
44) See answer details.

18.3 Answers with Explanation

1) The function $f(x) = 3 + 4\cos\left(x + \frac{\pi}{4}\right)$ is a cosine function, which is defined for all real numbers. Therefore, its domain is $\mathbb{R}$. The range of the cosine function is $[-1, 1]$, and after applying the transformation $3 + 4\cos(x)$, the range becomes $[3-4, 3+4]$ or $[-1, 7]$.

2) To find $(j-h)(x)$, subtract $h(x)$ from $j(x)$:

$$\begin{aligned} j(x) - h(x) &= (x^2 + 4x + 3) - (2x - 4) \\ &= x^2 + 4x + 3 - 2x + 4 \\ &= x^2 + 2x + 7. \end{aligned}$$

Therefore, the answer is A. $x^2 + 2x + 7$.

3) The graph shown represents the function $y = \ln(x+1)$. This function has a vertical asymptote at $x = -1$ and passes through the point (0,0), which is consistent with the graph presented.

4) The graph shows the cosine function shifted up by 1 unit. The function $y = \cos(x) + 1$ represents a cosine wave that oscillates between 0 and 2, centered at $y = 1$. This matches the graph presented.

5) To find the equation of the given parabola, we will use the vertex form of a parabola and analyze the graph.

The vertex form of a parabola is $y = a(x-h)^2 + k$, where (h, k) is the vertex of the parabola. In the given graph, it appears that the vertex of the parabola is $(-2, 4)$. This means that $h = -2$ and $k = 4$.

Substituting the vertex into the vertex form, we get: $y = a(x+2)^2 + 4$.

To find a, we need another point on the parabola. Since the parabola passes through the origin $(0,0)$, we can substitute these values into the equation:

$$0 = a(0+2)^2 + 4 \Rightarrow 0 = 4a + 4 \Rightarrow a = -1.$$

So we heve:

$$y = -1(x+2)^2 + 4 \Rightarrow y = -x^2 - 4x - 4 + 4 \Rightarrow y = -x^2 - 4x.$$

6) To find the inverse of $g(x) = 2 + \cos(x)$, we first replace $g(x)$ with y: $y = 2 + \cos(x)$. Now, we isolate the cosine function: $\cos(x) = y - 2$. Next, we need to reverse the cos operation by applying the arccos to both sides. Thus, $x = \arccos(y-2)$. Then we swap x and y to find the inverse: $y = g^{-1}(x) = \arccos(x-2)$.

7) To convert degrees to radians, use the conversion factor $\frac{\pi \text{ radians}}{180 \text{ degrees}}$. Therefore, 45° in radians is $45 \times \frac{\pi}{180} = \frac{\pi}{4}$.

8) The ladder forms a right-angled triangle with the ground and the wall. Using trigonometry, the length of the ladder l can be found using the cosine of the angle:

$$\cos(45^\circ) = \frac{\text{adjacent}}{\text{hypotenuse}} = \frac{20}{l}.$$

Solving for l gives $l = \frac{20}{\cos(45^\circ)} = 20\sqrt{2}$ ft.

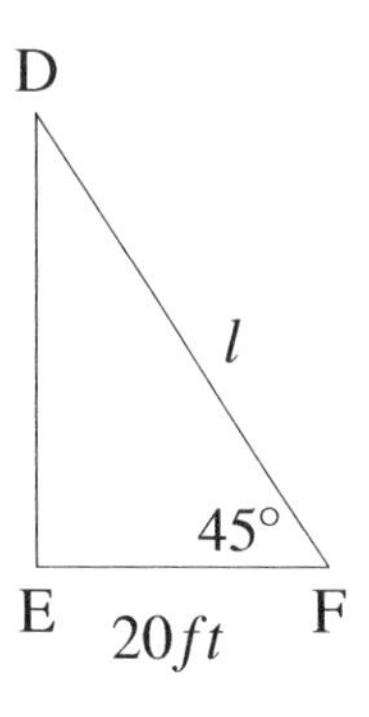

9) Since $\left|\vec{c}\cdot\vec{d}\right| = 0$, vectors $\vec{c}$ and $\vec{d}$ are perpendicular. The magnitude of $\vec{c}-\vec{d}$ can be found using the Pythagorean theorem: $\left|\vec{c}-\vec{d}\right| = \sqrt{|\vec{c}|^2 + |\vec{d}|^2} = \sqrt{3^2+4^2} = 5$.

10) In Triangle A, the sides are 3, 4, and 5 units long, which are the lengths of a Pythagorean triple. Triangle B shares a side with Triangle A, which is 4 units long, and has another side of length $2\sqrt{5}$. To find the base x of Triangle B, we use the Pythagorean theorem: $x^2+4^2 = (2\sqrt{5})^2$. Solving for x gives $x^2 = (2\sqrt{5})^2 - 4^2 = 20-16 = 4$. Therefore, $x = \sqrt{4} = 2$.

11) By equating the corresponding elements in the matrices we get:

$$3+c = 2 \Rightarrow c = -1,$$
$$-3+6 = d \Rightarrow d = 3.$$

Therefore, $c = -1$ and $d = 3$.

12) Analyze the properties of both functions. For $h(x) = \sin x + 3$, the range is from 2 to 4. For $j(x) = 5-\cos^2 x$, the range is from 4 to 5. According to the following graph, we see that they have no common points.

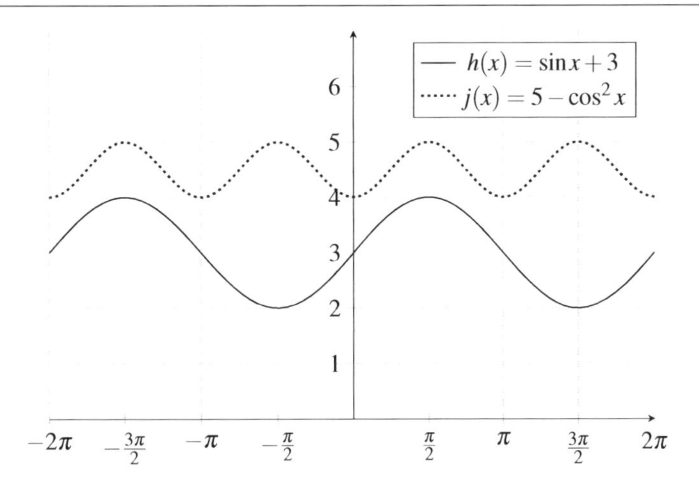

13) The graph of the function $y = (x+1)^2 - 2$ is a parabola opening upwards with its vertex shifted left by 1 unit and down by 2 units. Graph A shows this characteristic, as it depicts a parabola opening upwards with the vertex at $(-1, -2)$.

14) To find the value(s) of x, we need to solve the equation $\sin\left(\frac{\pi}{2}\right) = x^2 - x + 1$. Since $\sin\left(\frac{\pi}{2}\right) = 1$, the equation becomes:

$$1 = x^2 - x + 1.$$

Simplifying, we get:

$$0 = x^2 - x.$$

Factoring out x, we have:

$$x(x-1) = 0.$$

So, the value(s) of x are 0 and 1.

15) To identify the correct graph, first consider $y = \cos\left(x + \frac{\pi}{4}\right)$. Since the function is one-to-one in the interval $\left[-\frac{\pi}{6}, \frac{5\pi}{6}\right]$, its inverse can be determined.

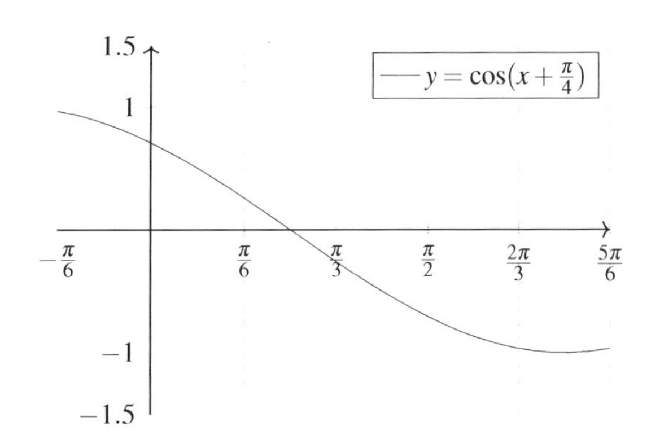

Reflecting this graph across the line $y = x$ yields the graph of its inverse. Among the choices, option A,

$y = \arccos(x) - \frac{\pi}{4}$, correctly represents this inverse.

16) The slope of a line through points (x_1, y_1) and (x_2, y_2) is given by $m = \frac{y_2 - y_1}{x_2 - x_1}$. Applying this to the given points $(2, -1)$ and $(7, -5)$, we have:

$$m = \frac{-5-(-1)}{7-2} = \frac{-5+1}{7-2} = \frac{-4}{5} = -\frac{4}{5}.$$

Therefore, the correct answer is A.

17) To convert from radians to degrees, we use the relationship 1 radian $= \frac{180^\circ}{\pi}$. Therefore,

$$\frac{5\pi}{6} \text{ radians} = \frac{5\pi}{6} \times \frac{180^\circ}{\pi} = \frac{5 \times 180^\circ}{6} = 150^\circ.$$

Hence, the correct answer is A.

18) To find the value of x, we set the expressions inside the logarithms equal to each other, because if $\log a = \log b$, then $a = b$. So,

$$4x - 1 = 2x + 3 \Rightarrow 4x - 2x = 3 + 1 \Rightarrow 2x = 4 \Rightarrow x = \frac{4}{2} = 2.$$

Therefore, the correct answer is A.

19) The minimum point of a quadratic function $f(x) = ax^2 + bx + c$ is found at $x = -\frac{b}{2a}$. For the function $f(x) = x^2 - 4x + 3$, we have $a = 1$ and $b = -4$. Therefore,

$$x = -\frac{-4}{2 \times 1} = \frac{4}{2} = 2.$$

Substituting $x = 2$ into the function, we get $f(2) = 2^2 - (4 \times 2) + 3 = 4 - 8 + 3 = -1$. Hence, the coordinates of the minimum point are $(2, -1)$, which corresponds to answer A.

20) The function $f(x) = \frac{3x^2 - 9x + 1}{3x^2 + 6x - 18}$ is a rational function. For large values of x, the behavior of the function is dominated by the leading terms in the numerator and the denominator. In this case, both the numerator and the denominator are quadratic polynomials with the same leading coefficient, 3. Therefore, the horizontal asymptote is given by the ratio of these coefficients, which is $\frac{3}{3} = 1$. Thus, the horizontal asymptote is $y = 1$.

21) To find $f(g(x))$, we substitute $g(x)$ into $f(x)$. Given $f(x) = 3x^2 - 5$ and $g(x) = \frac{2}{x}$, we have

$$f(g(x)) = f\left(\frac{2}{x}\right) = 3\left(\frac{2}{x}\right)^2 - 5 = 3\left(\frac{4}{x^2}\right) - 5 = \frac{12}{x^2} - 5.$$

Therefore, the correct answer is B.

22) The general form of a sinusoidal function is given by $f(x) = a\sin(b(x-c)) + d$ or $f(x) = a\cos(b(x-c)) + d$, where: a is the amplitude of the function. b determines the period of the function (period $= \frac{2\pi}{b}$). c is the phase shift (how much the function is shifted horizontally). d is the vertical shift (how much the function is shifted vertically).

Given the graph: The amplitude is 2 (so $a = 2$). The period is π (so $b = 2$, since $\frac{2\pi}{b} = \pi$). The phase shift is $\frac{\pi}{2}$ units to the left (so $c = -\frac{\pi}{2}$, note the negative sign because the shift is to the left). The vertical shift is 1 unit up (so $d = 1$).

The cosine function starts at its maximum at $x = 0$. Since the graph's first peak is at $x = -\frac{\pi}{2}$, this indicates a leftward shift of the cosine graph. Substituting all these values into the cosine equation, we get: $f(x) = 2\cos\left(2\left(x + \frac{\pi}{2}\right)\right) + 1$. Therefore, the correct option that matches this function is option A.

23) To solve the equation $e^{3x} = 7$ for x, we take the natural logarithm on both sides:

$$\ln(e^{3x}) = \ln(7) \Rightarrow 3x = \ln(7) \Rightarrow x = \frac{1}{3}\ln(7).$$

Therefore, the correct answer is A.

24) In right-angled triangle ABC, with angle $A = 45^\circ$ and side $AB = 4\ cm$, side AC is the hypotenuse. Since A is 45°, triangle ABC is an isosceles right-angled triangle. This means that sides AB and AC are equal. Hence, $AC = 4\ cm$. Applying the Pythagorean theorem, BC is calculated as:

$$AC = \sqrt{AB^2 + BC^2} = \sqrt{(4\ cm)^2 + (4\ cm)^2} = \sqrt{32\ cm^2} = 4\sqrt{2}\ cm.$$

Therefore, the length of AC is $4\sqrt{2}\ cm$.

25) The equation of the form $\frac{x^2}{a^2} + \frac{y^2}{b^2} = 1$ represents an ellipse, where a and b are the semi-major and semi-minor axes, respectively. In the given equation $\frac{x^2}{25} + \frac{y^2}{36} = 1$, both x^2 and y^2 have positive coefficients and are added together, which is characteristic of an ellipse. Therefore, the correct answer is D.

26) Coterminal angles are angles that differ by a multiple of 360°. For 60°, adding 360° gives a positive coterminal angle, $60^\circ + 360^\circ = 420^\circ$, and subtracting 360° gives a negative coterminal angle, $60^\circ - 360^\circ = -300^\circ$. Therefore, the correct answer is B.

27) We multiply the matrices and compare with the resulting matrix. For the first row, $2y + 1(-3) = 5$, which gives $y = 4$. For the second row, $-4y + x(-3) = -11$, substituting $y = 4$ we get $-4(4) + x(-3) = -11$, which gives $x = -\frac{5}{3}$. Therefore, the correct answer is B.

28) Using the values of the sine and cosine functions for common angles, we have

$$2(\sin 30^\circ \cos 60^\circ) = 2\left(\frac{1}{2}\right)\left(\frac{1}{2}\right) = \frac{1}{2}.$$

Therefore, the correct answer is A.

29) The given series is geometric with the first term $a = 2$ and common ratio $r = 3$. The sum of the first n terms of a geometric series is given by $S_n = a\frac{1-r^n}{1-r}$. For the first 5 terms, we have:

$$S_5 = 2\frac{1-3^5}{1-3} = 2\frac{1-243}{-2} = 2 \times 121 = 242.$$

Therefore, the correct answer is A.

30) The given function $y = -3 + \left(1 - 4x - x^2\right)$ simplifies to $y = -x^2 - 4x - 2$. completing the square gives: $y = -(x+2)^2 + 2$. This means that the parent graph is $y = x^2$, which indicates a reflection over the x-axis, a horizontal shift to the left by 2 units and a vertical shift of 2 units upwards. Therefore, the correct answer is D.

31) The equation $\sin\alpha \sec 30^\circ = 1$ simplifies to $\sin\alpha(\frac{2}{\sqrt{3}}) = 1$. Solving for α, we get $\sin\alpha = \frac{\sqrt{3}}{2}$. The angle α whose sine is $\frac{\sqrt{3}}{2}$ is 60°. Therefore, the correct answer is C.

32) The determinant of a matrix is calculated by the formula

$$|B| = b_{11}(b_{22}b_{33} - b_{23}b_{32}) - b_{12}(b_{21}b_{33} - b_{23}b_{31}) + b_{13}(b_{21}b_{32} - b_{22}b_{31}).$$

For the given matrix B,

$$3 = |B| = 2(p(1) - (1)(0)).$$

Simplifying, which gives $p = \frac{3}{2}$. Therefore, the correct answer is C.

33) The given graph represents a hyperbola. The standard form of a hyperbola with a horizontal transverse axis is $\frac{(x-h)^2}{a^2} - \frac{(y-k)^2}{b^2} = 1$, where (h,k) is the center of the hyperbola. For the graphed hyperbola, the center is at $(0,4)$, and the hyperbola opens horizontally. Therefore, the correct equation is $\frac{x^2}{9} - \frac{(y-4)^2}{4} = 1$, which is option A.

34) Since the angles of a triangle add up to 180°, we divide 180° by the sum of the ratios, which is $4+5+7 = 16$. This gives us $\frac{180^\circ}{16} = 11.25^\circ$ per unit ratio. Multiplying each ratio by 11.25° gives us the angle measures: $4 \times 11.25^\circ = 45^\circ$, $5 \times 11.25^\circ = 56.25^\circ$, and $7 \times 11.25^\circ = 78.75^\circ$. However, none of the options match these values exactly. The closest is option B, which is a rounded version of our calculated values.

35) To find the zeroes of the function, set $f(x) = 0$. This gives us $x^3 + 4x^2 + 4x = 0$. Factor out x to get $x(x^2 + 4x + 4) = 0$. The quadratic factor can be factored further as $x(x+2)^2 = 0$. This gives the roots $x = 0$ and $x = -2$. However, since $x = -2$ is a double root, the zeroes of the function are 0 and -2, corresponding to option A.

36) To find the value of x, we start with the equation $\frac{1}{2}\log_{2x} 81 = 2$. Simplifying, we get $\log_{2x} 81 = 4$. By the definition of logarithms, this means $(2x)^4 = 81$. Since $81 = 3^4$, we can write this as $(2x)^4 = 3^4$. Taking the fourth root of both sides, we find $2x = 3$. Solving for x gives $x = \frac{3}{2}$.

37)

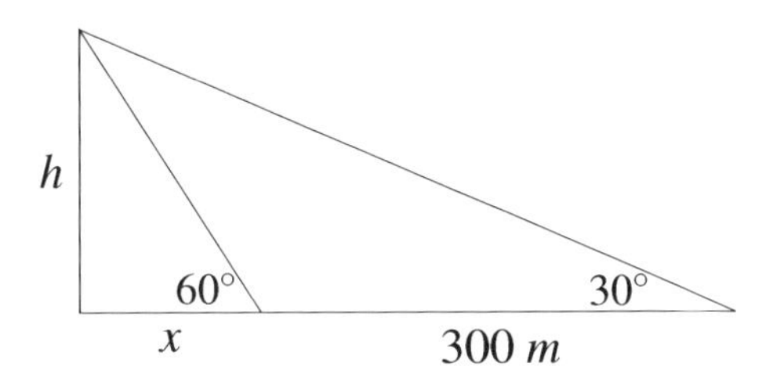

Let the height of the cliff be h. The distances of the ships from the cliff are denoted by x and $300 + x$. Using the tangent of the angles, we have $\tan 60° = \frac{h}{x}$ and $\tan 30° = \frac{h}{300+x}$. Solving for h gives $h = x\tan 60°$ and $h = (300+x)\tan 30°$. Substituting $\tan 60° = \sqrt{3}$ and $\tan 30° = \frac{1}{\sqrt{3}}$, we get $h = x\sqrt{3}$ and $h = \frac{300+x}{\sqrt{3}}$. Equating these two expressions for h and solving for x, we get $x\sqrt{3} = \frac{300+x}{\sqrt{3}}$, leading to $3x = 300 + x$, and $x = 150$ meters. Substituting x back into $h = x\sqrt{3}$ yields $h = 150\sqrt{3}$ meters.

38) Given the equation $2\left(\sin^2\theta - \cos^2\theta\right) = -1$, we simplify it (using $\cos^2\theta = 1 - \sin^2\theta$) as follows:

$$\sin^2\theta - \cos^2\theta = -\frac{1}{2} \Rightarrow \sin^2\theta - (1 - \sin^2\theta) = -\frac{1}{2}$$
$$\Rightarrow 2\sin^2\theta - 1 = -\frac{1}{2} \Rightarrow \sin^2\theta = \frac{1}{4} \Rightarrow \sin\theta = \pm\frac{1}{2}.$$

Since θ is a positive acute angle, the value of $\sin\theta$ is $\frac{1}{2}$ and the value of θ is $30°$, corresponding to Option C.

39) To find the value of $(f-g)(2)$, first evaluate $f(2)$ and $g(2)$. For $f(2)$, we have

$$3(2)^3 - 2(2) + 1 = (3 \times 8) - 4 + 1 = 24 - 4 + 1 = 21.$$

For $g(2)$, we have $2 + 3 = 5$. Therefore,

$$(f-g)(2) = f(2) - g(2) = 21 - 5 = 16.$$

40) The equation $y = x^2 - 3x + 5$ is a quadratic equation in the standard form $y = ax^2 + bx + c$. Quadratic equations of this form represent vertical parabolas, as they open upwards or downwards depending on the coefficient of x^2. Therefore, the correct answer is A. It represents a vertical parabola.

41) A.

(i) From the graph, $f(3) = 1$. Therefore, $h(3) = g(1) = 2.916 \cdot (0.7)^1 = 2.0412$.

(ii) From the graph, $f(x) = 1$ at $x = 3$ and $x = -3$.

B.

(i) Solving $2.916 \cdot (0.7)^x = 2$ yields

$$(0.7)^x = \frac{2}{2.916} \Rightarrow x \approx \log_{0.7}\left(\frac{2}{2.916}\right) \approx 1.057.$$

.

(ii) The end behavior of g as $x \to \infty$ is

$$\lim_{x\to\infty} g(x) = \lim_{x\to\infty} 2.916 \cdot (0.7)^x = 0,$$

because 0.7^x approaches 0 as x increases.

C.

(i) f does not have an inverse function.

(ii) The graph of f fails the horizontal line test, indicating that f is not one-to-one and thus does not have an inverse function.

42) A.

(i) Formulating the equations:

$$\begin{cases} a + b\ln(1) = 40, \\ a + b\ln(92) = 76. \end{cases}$$

(ii) Solving these equations gives (Note that $\ln(1) = 0$):

$$a = 40, \quad b = \frac{76 - 40}{\ln(92)} \approx 7.96.$$

B.

(i) Average rate of change from $t = 0$ to $t = 91$:

$$\frac{76 - 40}{91 - 0} \approx 0.3956 \text{ thousands per day.}$$

(ii) Estimating sales on $t = 50$ using average rate of change:

$$40 + 0.3956 \times 50 = 59.78 \text{ thousands.}$$

(iii) A_t generally underestimates $G(t)$ because the logarithmic model reflects increasing sales at a decreasing rate, which is not linear over time.

C.

The error in the model G increases after $t = 91$ because the model, based on logarithmic growth, does not account for the actual decline in sales reported after this time. As the model continues to predict a slowing growth rather than a decrease, its predictions become increasingly inaccurate, leading to larger errors as time progresses.

43) A.

1. F: $\left(\frac{1}{2}, 18\right)$ - Maximum height of W at peak.
2. G: $(1, 9)$ - Midway down the falling slope.
3. J: $\left(\frac{3}{2}, 0\right)$ - Point W touches the ground.
4. K: $(2, 9)$ - Midway up the rising slope.
5. P: $\left(\frac{5}{2}, 18\right)$ - Another maximum height of W at peak.

B.

1. $a = 9$ inches - Amplitude, equal to the radius of the tire.
2. $b = \pi$ - Frequency, corresponding to a period of 2 seconds per cycle.
3. $c = 0$ - No phase shift, as the sine wave starts at the top at $t = \frac{1}{2}$.
4. $d = 9$ inches - Vertical shift, the midline of the sinusoid.

C.

(i) h is positive and increasing on the interval $\left(2, \frac{5}{2}\right)$, as it ascends from midline to peak.

(ii) The rate of change of h is positive but decreasing in this interval, meaning h is increasing but at a slower rate as it approaches the peak.

44) A.

(i) Solve $e^{(x+3)} = 10$ gives $x + 3 = \ln(10)$, then $x = \ln(10) - 3$.

(ii) $\arcsin\left(\frac{x}{2}\right) = \frac{\pi}{4}$ leads to $x = 2\sin\left(\frac{\pi}{4}\right) = \sqrt{2}$.

B.

(i) $j(x) = \log_{10}\left(8x^5 \cdot 2x^2 \cdot x^{-9}\right) = \log_{10}\left(\frac{16}{x^2}\right)$.

(ii) $k(x) = \frac{\cos^2 x}{\sin x} \cdot \frac{1}{\cos x} = \cot x = \frac{1}{\tan x}$.

C. $\cos^{-1}(\tan(2x)) = 0$ when $\tan(2x) = 1$, thus $2x = \frac{\pi}{4} + k\pi$, giving $x = \frac{\pi}{8} + \frac{k\pi}{2}$ for integers k.

19. Practice Test 2

19.1 Practices

Section I. Part A: 28 questions (80 minutes). No calculator is allowed.

1) In $\triangle DEF$, $\angle F = 90^\circ$ and $DE = a$, $EF = b$, $DF = c$. What is the value of $\cos(E) \times \cot(D)$?

- ☐ A. $\frac{c}{a}$
- ☐ B. $\frac{a^2}{bc}$
- ☐ C. $\frac{b}{a}$
- ☐ D. $\frac{bc}{a^2}$

2) Determine the domain and range of the function $g(x) = \frac{3}{\cos(x - \frac{\pi}{2})}$.

- ☐ A. Domain: $\mathbb{R}$, Range: $[-3, 3]$
- ☐ B. Domain: $\mathbb{R} - \{0, \pi\}$, Range: $\mathbb{R}$
- ☐ C. Domain: $\{x \in \mathbb{R} : x \neq k\pi \quad \forall k \in \mathbb{Z}\}$, Range: $[-3, 3]$
- ☐ D. Domain: $(-\infty, +\infty)$, Range: $[-1.5, 1.5]$

3) Evaluate $\cos(45^\circ + \theta) - \sin(45^\circ - \theta)$.

- ☐ A. 1
- ☐ B. $\sqrt{2}\sin\theta$
- ☐ C. $\sqrt{2}\cos\theta$
- ☐ D. 0

4) Determine the equation (in standard form) of the graph depicted below:

☐ A. $\sin\left(x-\frac{\pi}{4}\right)$
☐ B. $\sin\left(x+\frac{\pi}{4}\right)-2$
☐ C. $\sin\left(x-\frac{\pi}{4}\right)+1$
☐ D. $\sin\left(x+\frac{\pi}{4}\right)-1$

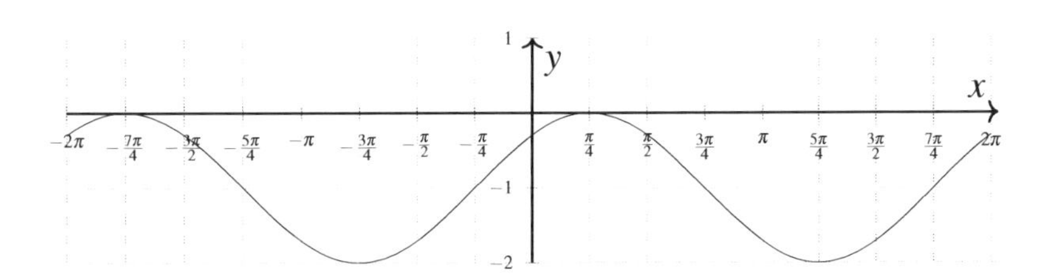

5) For which of the following values of x does the equation $\cos^2(x) = 1 - x^4$ hold true?

☐ A. $\frac{\pi}{6}$
☐ B. $\frac{\pi}{2}$
☐ C. 1
☐ D. 0

6) Determine the domain and range for the function $g(x) = x^2 - 5$.

A. $\mathbb{R}$ and $\mathbb{R}$

B. $\mathbb{R}$ and $(-5, +\infty)$

C. $[5, +\infty)$ and $(5, +\infty)$

D. $\mathbb{R}$ and $[-5, +\infty)$

7) If $h(x) = -2x+3$ and $k(x) = 3$, which graph corresponds to the function of $(h \circ k)(x)$?

☐ A.
☐ B.
☐ C.
☐ D.

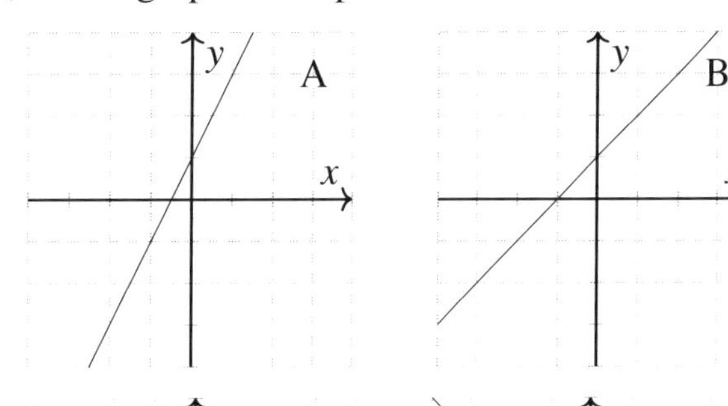

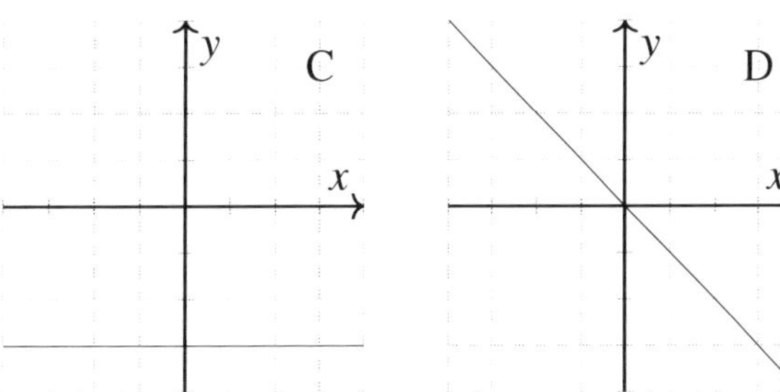

8) Examine the graph below, which represents a curve $y = a\sec(bx) + c$ for the interval $0 \le x \le 2\pi$. Determine the values of a, b, and c.

☐ A. $a = 3, b = 1, c = -2$

☐ B. $a = 2, b = 2, c = 3$

☐ C. $a = -3, b = 1, c = 2$

☐ D. $a = 1, b = 2, c = -3$

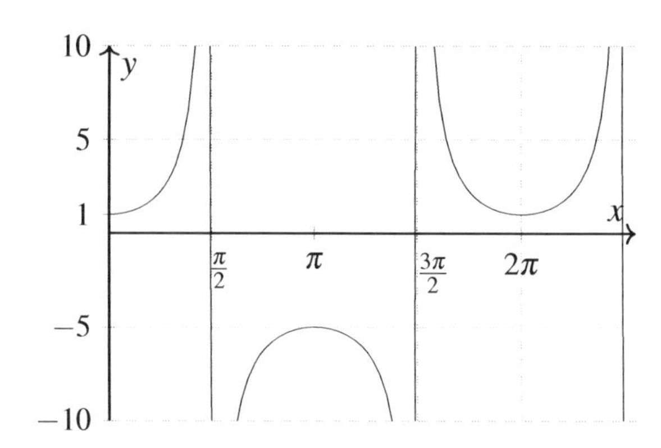

9) A child of height 3 $ft.$ wants to reach an apple hanging from a 15 $ft.$ high branch. If the child is standing 4 ft. away from the base of the tree, then at what angle should she look up to see the apple?

☐ A. 20°

☐ B. 72°

☐ C. 65°

☐ D. 75°

10) If $\cos 30° = y$, then what is the value of $(\sec 30° - \sin 60°)$?

☐ A. $\frac{1-y^2}{y}$

☐ B. $\frac{y}{1-y^2}$

☐ C. $1 - y^2$

☐ D. $\frac{1}{1-y^2}$

11) Which of the following is true for $90° < \theta < 180°$?

☐ A. $\cos\theta > \cos^2\theta$

☐ B. $\cos\theta < \cos^2\theta$

☐ C. $\cos\theta \leq \cos^2\theta$

☐ D. $\cos\theta \geq \cos^2\theta$

12) Which function can represent the following graph?

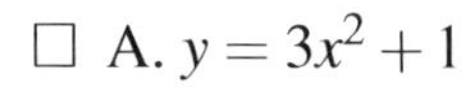

☐ A. $y = 3x^2 + 1$

☐ B. $y = -x^2 - 1$

☐ C. $y = x^2 - 3$

☐ D. $y = -2x^2 + 3$

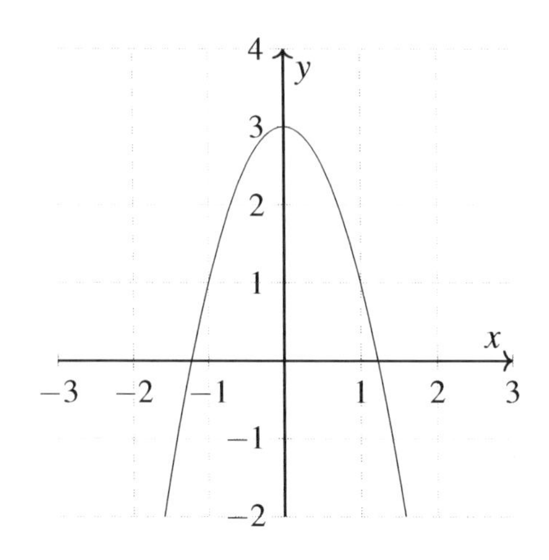

13) In right triangle DEF, angle E is a right angle. Angle EDF is 30°. $DE = 12\ cm$. What is the length of DF? (Nearest tenth)

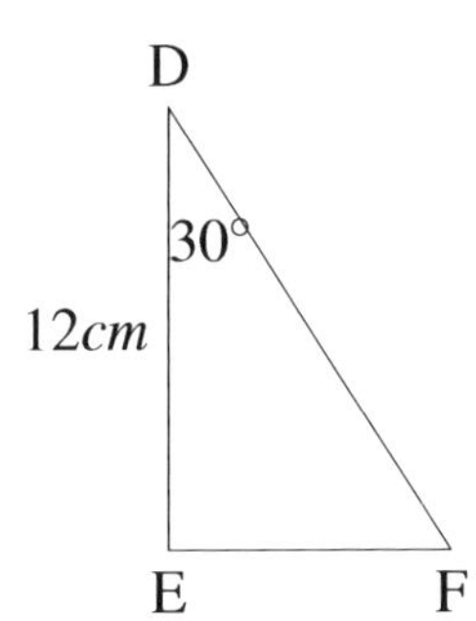

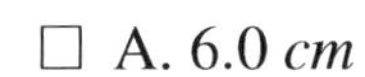

☐ A. 6.0 cm

☐ B. 13.9 cm

☐ C. 20.8 cm

☐ D. 24.0 cm

14) Find the function represented by the following graph.

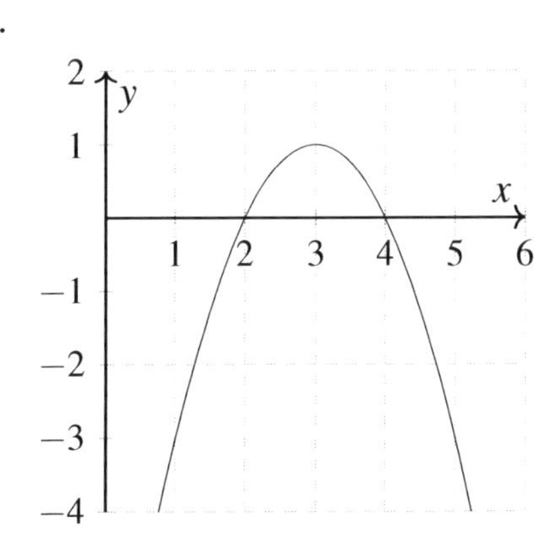

☐ A. $g(x) = -x^2 + 6x - 8$

☐ B. $g(x) = x^2 + 6x - 8$

☐ C. $g(x) = -x^2 - 6x - 8$

☐ D. $g(x) = x^2 - 6x - 8$

15) If $\tan\theta = \frac{5}{12}$, the value of $\frac{\sqrt{1-\sin\theta}}{\sqrt{1+\sin\theta}}$ is:

☐ A. $\frac{1}{12}$

☐ B. $\frac{4}{5}$

☐ C. $\frac{2}{3}$

☐ D. $\frac{5}{13}$

16) What is the graph of the linear inequality, $y < -3x - 3$?

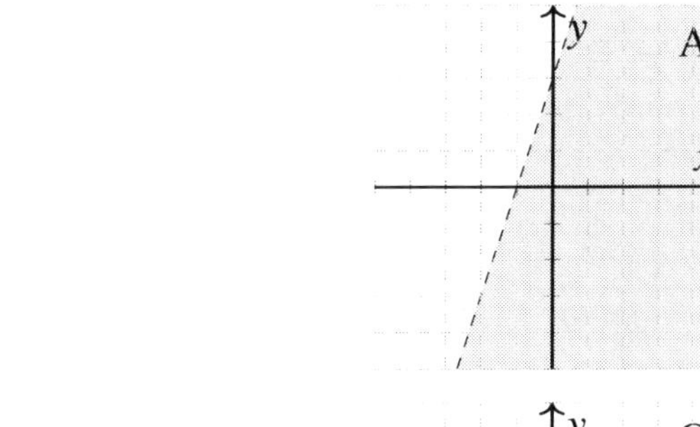

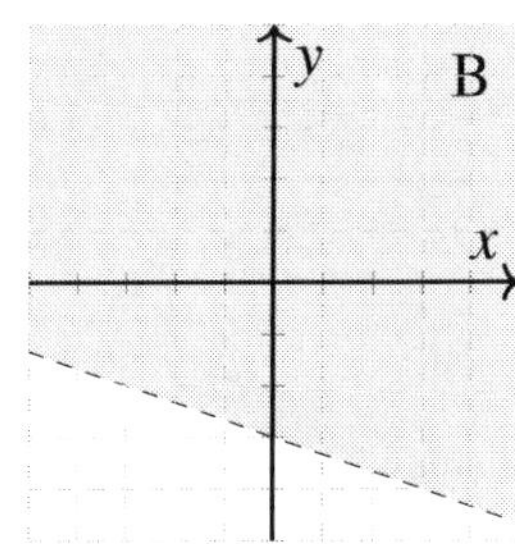

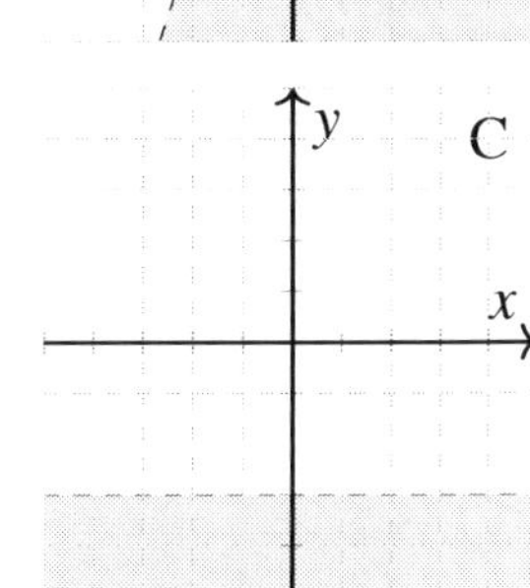

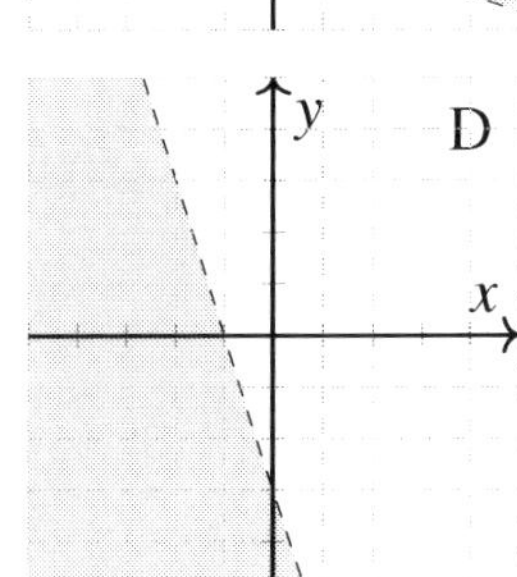

☐ A.

☐ B.

☐ C.

☐ D.

17) What is the vertex of the parabola $y = (x+1)^2 - 4$?

☐ A. $(-1,-4)$

☐ B. $(-1,4)$

☐ C. $(1,-4)$

☐ D. $(1,4)$

18) If $GHIJKL$ is a regular hexagon, find the value of $\overrightarrow{GH} + \overrightarrow{GI} + \overrightarrow{GK} + \overrightarrow{GL}$.

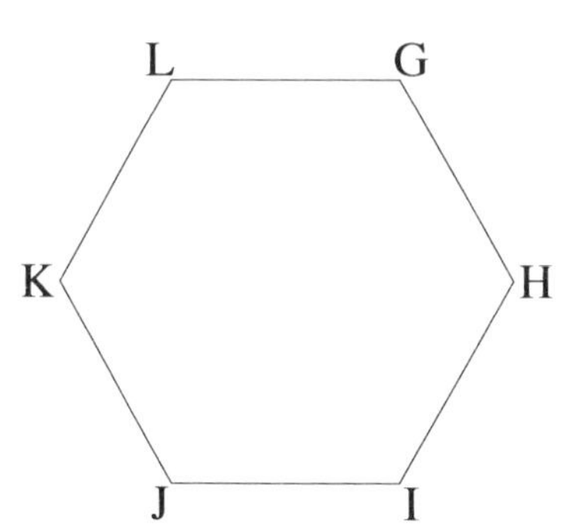

☐ A. $\overrightarrow{O}$

☐ B. $\overrightarrow{GJ}$

☐ C. $2\overrightarrow{GJ}$

☐ D. $4\overrightarrow{GH}$

19) If $\tan\theta = -\frac{3}{4}$ and $\cos\theta < 0$, then what is the value of $\sin\theta =$?

☐ A. $\frac{3}{5}$

☐ B. $-\frac{4}{5}$

☐ C. $-\frac{3}{5}$

☐ D. $\frac{4}{5}$

20) If $GHIJKL$ is a regular hexagon, find the value of $\overrightarrow{GH} + \overrightarrow{GI} + \overrightarrow{GK} + \overrightarrow{GL}$.

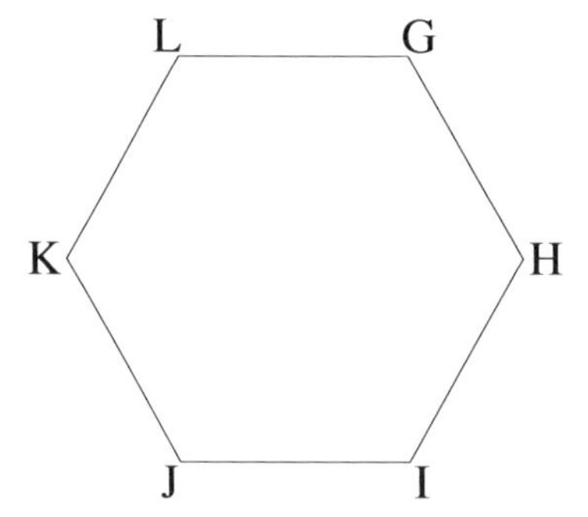

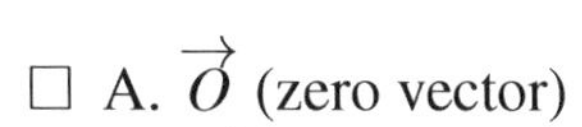

☐ A. $\overrightarrow{O}$ (zero vector)

☐ B. $\overrightarrow{GJ}$

☐ C. $2\overrightarrow{GJ}$

☐ D. $4\overrightarrow{GH}$

21) Two poles are standing opposite each other on either side of a straight road. One pole is 120 meters tall. From the top of this pole, the angles of depression to the top and bottom of the other pole are 45° and 60°, respectively. What is the height of the other pole? (Choose the nearest option)

☐ A. 62

☐ B. 51

☐ C. 125

☐ D. 100

22) What is the equation of a circle with center $(2,-3)$ and perimeter 6π?

☐ A. $(x-2)^2+(y+3)^2=6\pi$
☐ B. $(x-2)^2+(y+3)^2=9$
☐ C. $(x+2)^2+(y-3)^2=3$
☐ D. $(x+2)^2+(y-3)^2=3\pi$

23) What is the value of $\cos 45^{\circ}$?

☐ A. $\frac{\sqrt{2}}{2}$
☐ B. $\frac{1}{2}$
☐ C. $-\frac{1}{2}$
☐ D. $\frac{\sqrt{3}}{2}$

24) If $(hog)(x)=x+1$, how might $h(x)$ and $g(x)$ be defined?

☐ A. $h(x)=x-1$ and $g(x)=x+3$
☐ B. $h(x)=x+1$ and $g(x)=x-3$
☐ C. $h(x)=x-1$ and $g(x)=-x+3$
☐ D. $h(x)=x-2$ and $g(x)=x+3$

25) How many x-intercepts does the graph of $y=\frac{x+2}{2-x^2}$ have?

☐ A. 0
☐ B. 1
☐ C. 2
☐ D. 3

26) What is the value of a_2 in the geometric sequence $\left\{\frac{1}{2},a_2,8,\cdots\right\}$?

☐ A. -2
☐ B. -1
☐ C. 4
☐ D. -3

27) Which graph represents the function $y=-\log(x+2)$?

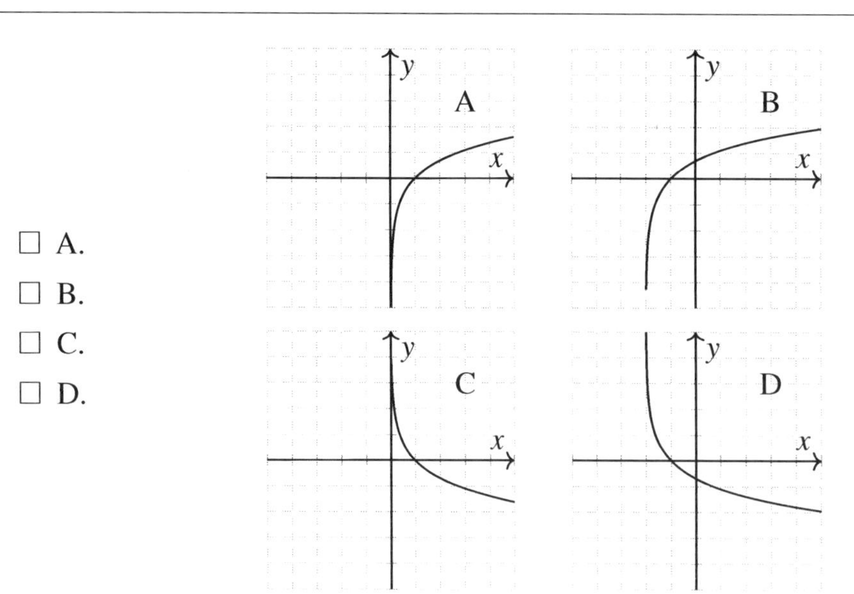

- ☐ A.
- ☐ B.
- ☐ C.
- ☐ D.

28) What is the minimum value of $3\cos^2(\beta)+7\sin^2(\beta)$?

- ☐ A. 3
- ☐ B. 7
- ☐ C. 4
- ☐ D. 10

Section I. Part B: 12 questions (40 minutes). A graphing calculator is required

29) If $C \times D = \begin{bmatrix} -1 & 6 \end{bmatrix}$, where C and D are matrices, identify C and D.

- ☐ A. $C = \begin{bmatrix} 3 & 1 \end{bmatrix}$ and $D = \begin{bmatrix} -1 & 2 \end{bmatrix}$
- ☐ B. $C = \begin{bmatrix} 3 & 1 \end{bmatrix}$ and $D = \begin{bmatrix} 0 & 2 \\ -1 & 0 \end{bmatrix}$
- ☐ C. $C = \begin{bmatrix} 3 & 1 \end{bmatrix}$ and $D = \begin{bmatrix} 1 & 0 \\ 1 & 2 \end{bmatrix}$
- ☐ D. $C = \begin{bmatrix} 3 \\ 0 \end{bmatrix}$ and $D = \begin{bmatrix} 1 \\ 2 \end{bmatrix}$

30) A regular hexagon can be divided into 6 equilateral triangles. The diagram below shows one of the equilateral triangles. Which one is the height, h, of the equilateral triangle below?

☐ A. $\sqrt{3}x$

☐ B. $\sqrt{2}x$

☐ C. $\frac{\sqrt{3}}{2}x$

☐ D. $\frac{\sqrt{2}}{2}x$

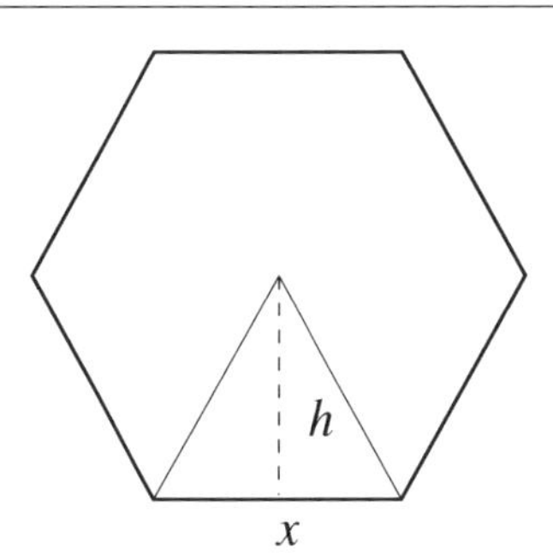

31) In an arithmetic sequence where the 4-th term is 16 and the 8-th term is 32, what is the sum of the first 12 terms of this series?

☐ A. 288

☐ B. 324

☐ C. 312

☐ D. 396

32) Which statement best describes the two functions $h(x) = x^2 + 4x + 4$ and $j(x) = -2x^2 + 4$?

☐ A. They have no common points.

☐ B. They have the same x-intercepts.

☐ C. The maximum of $h(x)$ is the same as the minimum of $j(x)$.

☐ D. They have the same y-intercept.

33) If $10^{\log 7} = y$, what is the value of y?

☐ A. 10

☐ B. 7

☐ C. 3

☐ D. 10^7

34) If $x = 3\left(\log_2 \frac{1}{64}\right)$, what is the value of x?

☐ A. -20

☐ B. $\frac{1}{4}$

☐ C. -18

☐ D. $-\frac{1}{4}$

35) What is the value of x in the equation $\log_4(x+3) - \log_4(x-3) = 2$.

☐ A. 0

☐ B. $\frac{17}{5}$

☐ C. 17

☐ D. $-\frac{17}{5}$

36) What is the center and radius of a circle with the equation $(x+2)^2+(y-5)^2=7$?

☐ A. $(-2,5)$ and 7

☐ B. $(-2,-5)$ and $\sqrt{7}$

☐ C. $(2,-5)$ and 7

☐ D. $(-2,5)$ and $\sqrt{7}$

37) Which statement best describes the relationship between $h(x)=3x^2+2x-4$ and $j(x)=-2x^2+3x+5$?

☐ A. The maximum of $h(x)$ is less than the minimum of $j(x)$.

☐ B. The minimum of $h(x)$ is less than the maximum of $j(x)$.

☐ C. The maximum of $h(x)$ is greater than the minimum of $j(x)$.

☐ D. The minimum of $h(x)$ is greater than the maximum of $j(x)$.

38) Consider the function $g(x)=\frac{1}{(x-5)^2+6(x-5)+9}$. For what value of x is the function $g(x)$ undefined?

☐ A. -1

☐ B. 1

☐ C. 2

☐ D. -5

39) A man on a cliff observes a speedboat heading towards the shore. It takes 10 minutes for the angle of elevation to change from $30°$ to $60°$. How soon will the speedboat reach the shore? (Choose the nearest option)

☐ A. 4 minutes

☐ B. 5 minutes

☐ C. 8 minutes

☐ D. 10 minutes

40) The determinant of which matrix is equal to 5?

☐ A. A=$\begin{bmatrix} 3 & 2 \\ 1 & 4 \end{bmatrix}$

☐ B. B=$\begin{bmatrix} 1 & 4 & 2 \\ 0 & 2 & 3 \\ 5 & 0 & 1 \end{bmatrix}$

☐ C. C=$\begin{bmatrix} 2 & -1 \\ 3 & 1 \end{bmatrix}$

☐ D. D=$\begin{bmatrix} 2 & 5 & 0 \\ 0 & 3 & 1 \\ 2 & -2 & 0 \end{bmatrix}$

Section II. Part A: 2 questions (30 minutes). A graphing calculator is required.

41) Jane started a savings account with \$600. The following graph shows the amount of money at the end of each month. Calculate the rate of change of money per month from the second month to the sixth month.

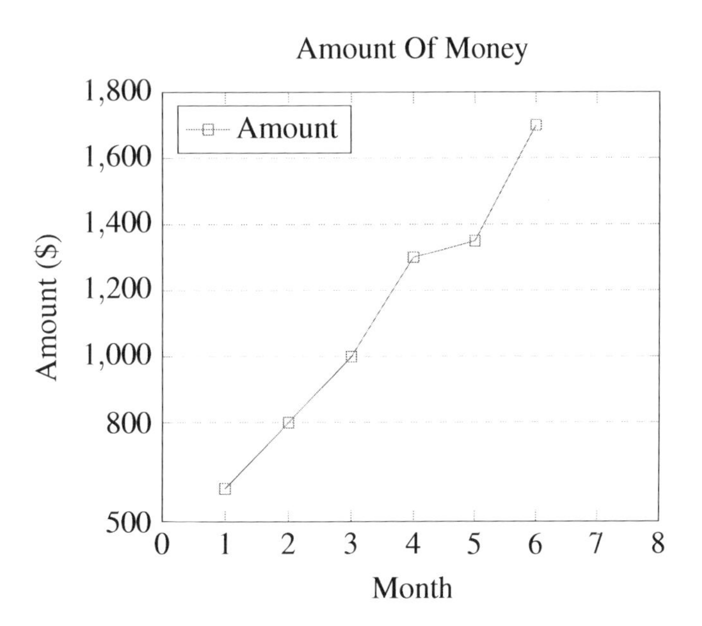

42) A wildlife research group monitors an eagle population by counting the number of nesting sites where the eagles lay their eggs. The table below shows the number of nesting sites for several years since 2015. The data can be modeled by an exponential function.

Number of Years Since 2015, x	Number of Nesting Sites, $n(x)$
0	50,000
1	40,000
2	32,000
3	25,600
4	20,480
5	16,384
6	13,107

Determine the best model for this data. Assume an exponential decay model of the form $n(x) = ab^x$.

Section II. Part B: 2 questions (30 minutes). No calculator is permitted.

43) Consider the quadratic functions given below:

$$f(x)=\frac{5}{3}x^2-7,\quad g(x)=-3x^2+2,\quad h(x)=\frac{1}{2}x^2+3.$$

Evaluate the following statements about these functions and determine whether they are true or false:

A. The graphs of two of these functions have a maximum point.

B. The graphs of all these functions have the same axis of symmetry.

C. The graph of two of these functions intersect each other on the x-axis.

D. The graphs of all these functions have different y-intercepts.

E. The functions do not intersect with each other.

44) Consider the function $f(x)=x^2(1-x)$. Analyze the following statements about $f(x)$:

A. $f(x)$ is an even function.

B. When $x\to\infty$, then $f(x)\to\infty$.

C. The domain of $f(x)$ is all real numbers except 1.

D. $f(x)$ has two real roots.

19.2 Answer Keys

1) B. $C = \begin{bmatrix} 3 & 1 \end{bmatrix}$ and $D = \begin{bmatrix} 0 & 2 \\ -1 & 0 \end{bmatrix}$

2) C. $\frac{\sqrt{3}}{2}x$

3) C. 312

4) A. $\frac{c}{a}$

5) C. Domain: $\{x \in \mathbb{R} : x \neq k\pi \quad \forall k \in \mathbb{Z}\}$, Range: $[-3,3]$

6) D. They have the same y-intercept.

7) B. 7

8) D. 0

9) D. $\sin\left(x+\frac{\pi}{4}\right)-1$

10) D. 0

11) D. $\mathbb{R}$ and $[-5,+\infty)$

12) C.

13) A. $a=3$, $b=1$, $c=-2$

14) B. 72°

15) C. -18

16) B. $\frac{17}{5}$

17) A. $\frac{1-y^2}{y}$

18) D. $(-2,5)$ and $\sqrt{7}$

19) B. The minimum of $h(x)$ is less than the maximum of $j(x)$.

20) C. 2

21) B. 5 minutes

22) C. $\begin{bmatrix} 2 & -1 \\ 3 & 1 \end{bmatrix}$

23) B. $\cos\theta < \cos^2\theta$

24) D. $y=-2x^2+3$

25) B. 13.9 cm

26) A. $g(x)=-x^2+6x-8$

27) C. $\frac{2}{3}$

28) D.

29) A. $(-1,-4)$

30) C. $2\overrightarrow{GJ}$

31) A. $\frac{3}{5}$

32) C. $2\overrightarrow{GJ}$

33) B. 51

34) B. $(x-2)^2+(y+3)^2=9$

35) A. $\frac{\sqrt{2}}{2}$

36) D. $h(x)=x-2$ and $g(x)=x+3$

37) B. 1

38) A. -2

39) D. $y=-\log(x+2)$

40) A. 3

41) See answer details

42) See answer details.

43) Responses to the statements:

A. False

B. True

C. False

D. True

E. False

44) Evaluations of the statements:

A. False.

B. False.

C. False.

D. True.

19.3 Answers with Explanation

1) To find the correct matrices C and D, we first need to check if matrix multiplication is possible. The number of columns in C must equal the number of rows in D. Let us examine each option:

Options A and D: Matrix multiplication is not possible because the number of columns in C does not equal the number of rows in D.

In other options, matrix multiplication is possible.

Option B:

$$C \times D = \begin{bmatrix} 3 & 1 \end{bmatrix} \begin{bmatrix} 0 & 2 \\ -1 & 0 \end{bmatrix} = \begin{bmatrix} (3)(0)+(1)(-1) & (3)(2)+(1)(0) \end{bmatrix} = \begin{bmatrix} -1 & 6 \end{bmatrix}.$$

This matches the given product $\begin{bmatrix} -1 & 6 \end{bmatrix}$.

Option C:

$$C \times D = \begin{bmatrix} 3 & 1 \end{bmatrix} \begin{bmatrix} 1 & 0 \\ 1 & 2 \end{bmatrix} = \begin{bmatrix} (3)(1)+(1)(1) & (3)(0)+(1)(2) \end{bmatrix} = \begin{bmatrix} 4 & 2 \end{bmatrix}.$$

This result also does not match the given product. Therefore, the correct option is B.

2) Considering a regular hexagon where each side length is x, it can be decomposed into six equilateral triangles with side length x. The height h of an equilateral triangle is determined using the properties of a 30-60-90 right triangle, which results from drawing an altitude from a vertex to the opposite side. The altitude bisects the side opposite the vertex, creating two right triangles with side lengths x, $\frac{x}{2}$, and h. Applying Pythagoras' theorem:

$$h^2 + \left(\frac{x}{2}\right)^2 = x^2.$$

Simplifying this, we get:

$$h^2 = x^2 - \left(\frac{x}{2}\right)^2 = x^2 - \frac{x^2}{4} = \frac{3x^2}{4} \Rightarrow h = \sqrt{\frac{3x^2}{4}} = \frac{\sqrt{3}}{2}x.$$

Therefore, the height h of the equilateral triangle is $\frac{\sqrt{3}}{2}x$.

3) To find any term in an arithmetic sequence, use the formula $a_n = a_1 + d(n-1)$. We have $a_4 = 16$ and $a_8 = 32$. Therefore:

$$a_4 = a_1 + d(4-1) \Rightarrow a_1 + 3d = 16,$$

and

$$a_8 = a_1 + d(8-1) \Rightarrow a_1 + 7d = 32.$$

Solve the following equation:

$$\begin{cases} a_1 + 3d = 16, \\ a_1 + 7d = 32. \end{cases}$$

Subtracting the first equation from the second, we get:

$$(a_1 + 7d) - (a_1 + 3d) = 32 - 16 \Rightarrow a_1 + 7d - a_1 - 3d = 16 \Rightarrow 4d = 16 \Rightarrow d = 4.$$

By substituting $d = 4$ in the first equation, we get:

$$a_1 + 3d = 16 \Rightarrow a_1 + (3 \times 4) = 16 \Rightarrow a_1 = 4.$$

Then, use this formula:

$$S_n = \frac{n}{2}(2a_1 + d(n-1)).$$

Therefore:

$$S_{12} = \frac{12}{2}(2(4) + 4(12-1)) = 6(8+44) = 312.$$

4)

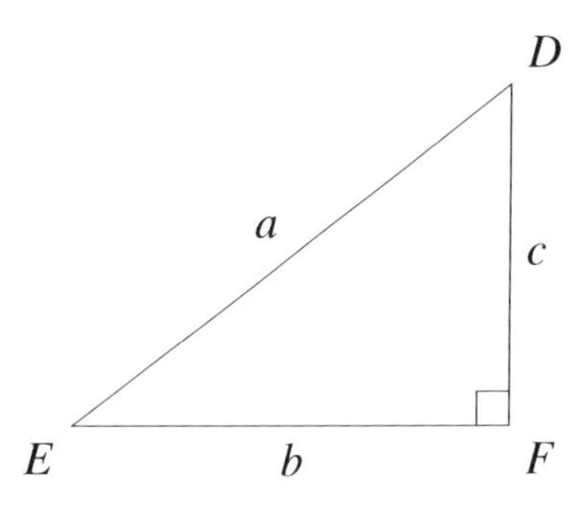

In the right triangle DEF with $\angle F = 90°$, we can express $\cos(E)$ and $\cot(D)$ using the sides of the triangle:

$$\cos(E) = \frac{\text{adjacent}}{\text{hypotenuse}} = \frac{b}{a}.$$

Similarly:

$$\cot(D) = \frac{\text{adjacent}}{\text{opposite}} = \frac{c}{b}.$$

Hence, $\cos(E)\cot(D) = \left(\frac{b}{a}\right)\left(\frac{c}{b}\right) = \frac{c}{a}$.

5) The function $g(x) = \frac{3}{\cos(x - \frac{\pi}{2})}$ involves the cosine function. The cosine function has a period of 2π and is undefined at $\frac{\pi}{2} + k\pi$ for any integer k. Thus, the domain of $g(x)$ is $\mathbb{R}$ minus the points where $\cos(x - \frac{\pi}{2}) = 0$, which are $x = k\pi$ for every integer k. Therefore,

$$D_g = \{x \in \mathbb{R} : x \neq k\pi \quad \forall k \in \mathbb{Z}\}.$$

The range of the cosine function is $[-1, 1]$. Since we have $\frac{3}{\cos(x-\frac{\pi}{2})}$, the range will be all real numbers between -3 and 3, Hence, the range is $[-3, 3]$.

6) - For $h(x)$: $h(x) = x^2 + 4x + 4$ is a parabola opening upwards with its vertex at $(-2, 0)$. The y-intercept is found when $x = 0$, giving $h(0) = 4$. The x-intercepts are the solutions to $x^2 + 4x + 4 = 0$, which are $x = -2$ (a repeated root).
- For $j(x)$: $j(x) = -2x^2 + 4$ is a downward-opening parabola with its vertex at $(0, 4)$. The y-intercept is also $j(0) = 4$. The x-intercepts are the solutions to $-2x^2 + 4 = 0$, which are $x = -\sqrt{2}$ and $x = \sqrt{2}$.

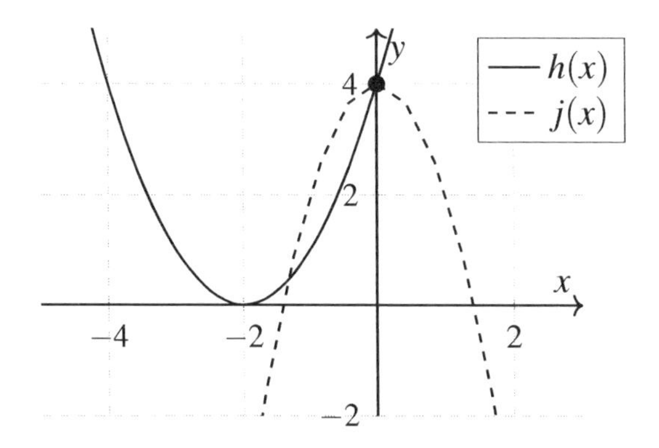

Both functions have the same y-intercept at $y = 4$, but different x-intercepts and different orientations of their parabolas. $h(x)$ has a minimum at $(-2, 0)$ and $j(x)$ has a maximum at $(0, 4)$. Therefore, the correct answer is D.

7) If $a^y = x$, then $\log_a x = y$. Therefore:

$$10^{\log 7} = y \Rightarrow \log y = \log 7 \Rightarrow y = 7.$$

8) We apply the addition and subtraction formulas for sine and cosine:

$$\cos(45^\circ + \theta) = \cos 45^\circ \cos\theta - \sin 45^\circ \sin\theta = \frac{1}{\sqrt{2}}\cos\theta - \frac{1}{\sqrt{2}}\sin\theta,$$

$$\sin(45^\circ - \theta) = \sin 45^\circ \cos\theta - \cos 45^\circ \sin\theta = \frac{1}{\sqrt{2}}\cos\theta - \frac{1}{\sqrt{2}}\sin\theta.$$

Subtracting these two expressions, we get:

$$\cos(45^\circ+\theta)-\sin(45^\circ-\theta)=\left(\frac{1}{\sqrt{2}}\right)\cos\theta-\left(\frac{1}{\sqrt{2}}\right)\sin\theta-\left(\frac{1}{\sqrt{2}}\right)\cos\theta+\left(\frac{1}{\sqrt{2}}\right)\sin\theta=0.$$

Therefore, the answer is D.

9) The graph of this sine function begins at its midpoint and moves upwards, indicative of the sine function. The amplitude is 1, so the coefficient in front of the sine function is 1. The period of the graph is standard (2π), requiring no adjustment. The graph is shifted $\frac{\pi}{4}$ units to the left, suggesting a phase shift, so the function is $y=\sin\left(x+\frac{\pi}{4}\right)$. Furthermore, the graph is shifted down by 1 unit, which adds -1 outside the function. Thus, the equation of the graph is $y=\sin\left(x+\frac{\pi}{4}\right)-1$. Hence, the correct answer is D, $y=\sin\left(x+\frac{\pi}{4}\right)-1$.

10) Let us evaluate each option:
A: $\cos^2\left(\frac{\pi}{6}\right)=\left(\frac{\sqrt{3}}{2}\right)^2=\frac{3}{4}$ and $1-\left(\frac{\pi}{6}\right)^4\approx 0.92$.
B: $\cos^2\left(\frac{\pi}{2}\right)=0^2=0$ and $1-\left(\frac{\pi}{2}\right)^4\approx -5.088$.
C: $\cos^2(1)=(\cos(1))^2\approx 0.292$ and $1-1^4=0$.
D: $\cos^2(0)=1^2=1$ and $1-0^4=1$.
The only value that satisfies the equation $\cos^2(x)=1-x^4$ is $x=0$. Therefore, the correct answer is D.

11) The domain of $g(x)$ is all real numbers. For the range, note that x^2 is never negative, so x^2-5 is never less than -5. Therefore, the range of $g(x)$ can be written as $[-5,+\infty)$.

12) Considering that $(h\circ k)(x)=h(k(x))$, then:

$$h(k(x))=h(3)=(-2\times 3)+3=-3.$$

Therefore, the equation of the composition is $y=-3$, which is a constant function. Among the provided graphs, Graph C, which represents the line $y=-3$, is the correct answer.

13) First, observe the amplitude of the secant function in the graph. The amplitude is 3, as the difference between the peaks and valleys of the curve is 6 units. Hence, $a=3$.
Next, determine the period of the function. The period is the horizontal distance covered by one complete cycle of the graph. The graph completes one cycle in an interval of 2π, indicating that $b=1$ (period $=\frac{2\pi}{b}$).
Finally, c represents the vertical shift. The graph is shifted down by 2 units, as indicated by the position of the central line of the secant graph relative to the x-axis. Thus, $c=-2$.
Therefore, the correct answer is A. $a=3$, $b=1$, $c=-2$.

14) The problem is solved using trigonometric relationships to find the angle of elevation. We need to calculate the angle that the line of sight from the child's eyes to the apple makes with the horizontal.

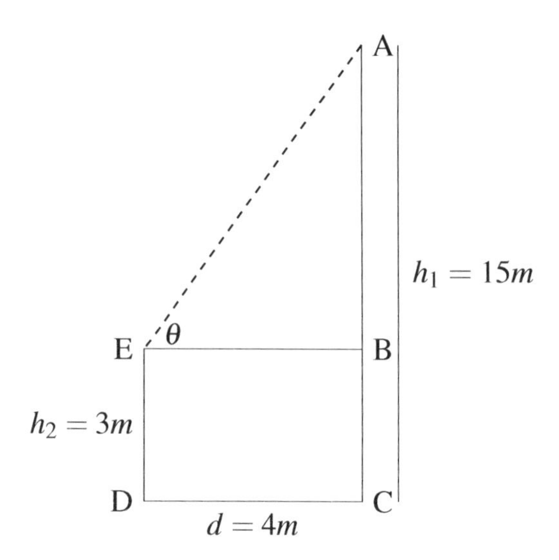

Let us denote the height of the branch as $h_1 = 15\ ft$, the height of the child as $h_2 = 3\ ft$, and the distance of the child from the tree as $d = 4\ ft$. The tangent of the angle θ is given by:

$$\tan\theta = \frac{h_1 - h_2}{d} = \frac{15-3}{4} = \frac{12}{4} = 3.$$

To find the angle, we take the arctan (inverse tangent) of 3. Using a calculator, $\arctan(3)$ is approximately 72°.

15) We can use the property of logarithms $\log_a \frac{x}{y} = \log_a x - \log_a y$. So, $x = 3\left(\log_2 \frac{1}{64}\right)$ becomes:

$$x = 3(\log_2 1 - \log_2 64).$$

Since $\log_2 1 = 0$, we have:

$$x = 3(-\log_2 64).$$

Now, using the property $\log_a x^b = b \cdot \log_a x$, we can simplify $\log_2 64$ as follows:

$$\log_2 64 = \log_2 2^6 = 6\log_2 2 = 6.$$

Therefore $x = 3 \times (-6) = -18$.

16) We can use the property of logarithms $\log_a \frac{x}{y} = \log_a x - \log_a y$. So, we have:

$$\log_4(x+3) - \log_4(x-3) = 2 \Rightarrow \log_4 \frac{x+3}{x-3} = 2.$$

Now, we can rewrite 2 as $\log_4 4^2$:

$$\log_4 \frac{x+3}{x-3} = \log_4 4^2.$$

Using the property $\log_a b = \log_a c \Rightarrow b = c$, we have:

$$\frac{x+3}{x-3} = 4^2 = 16(x-3) \Rightarrow x+3 = 16x-48 \Rightarrow 15x = 51 \Rightarrow x = \frac{51}{15} = \frac{17}{5}.$$

17) Recall that $\sec(\theta) = \frac{1}{\cos(\theta)}$ and $\sin(90-\theta) = \cos(\theta)$. Therefore,

$$\sec(30) - \sin(60) = \frac{1}{\cos(30)} - \sin(90-30) = \frac{1}{\cos(30)} - \cos(30) = \frac{1-\cos^2(30)}{\cos(30)}.$$

Substituting the given $\cos(30) = y$ into the expression $\frac{1-\cos^2(30)}{\cos(30)}$, we get $\frac{1-y^2}{y}$. Therefore, the correct option is A.

18) The standard form of a circle's equation is $(x-h)^2 + (y-k)^2 = r^2$, where the center is at (h,k) and the radius is r. For the circle with equation $(x+2)^2 + (y-5)^2 = 7$, we have:

$$(x-(-2))^2 + (y-5)^2 = 7,$$

the center is at $(-2,5)$ and the radius is $\sqrt{7}$ (since $r^2 = 7$, so $r = \sqrt{7}$).

19) First, analyze the function $h(x) = 3x^2 + 2x - 4$, which opens upwards. Completing the square, we rewrite it as:

$$h(x) = 3\left(x + \frac{1}{3}\right)^2 - \frac{13}{3}.$$

Hence, the minimum value of $h(x)$ is $-\frac{13}{3}$.
Next, consider $j(x) = -2x^2 + 3x + 5$, which opens downwards. Completing the square, it can be rewritten as:

$$j(x) = -2\left(x - \frac{3}{4}\right)^2 + \frac{49}{8}.$$

Thus, the maximum value of $j(x)$ is $\frac{49}{8}$. Since $-\frac{13}{3} < \frac{49}{8}$, the minimum of $h(x)$ is less than the maximum of $j(x)$.

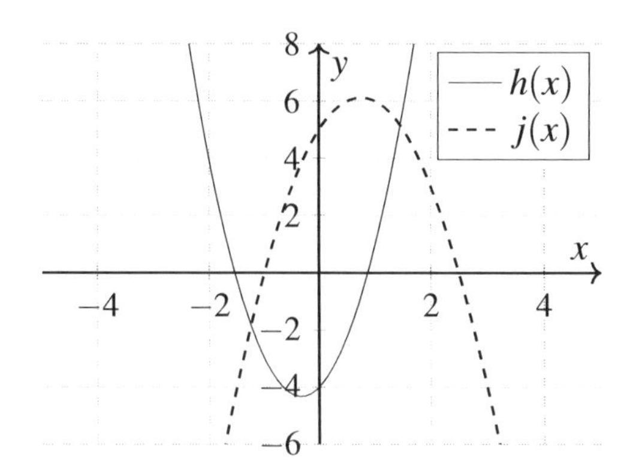

20) The function $g(x)$ is undefined when its denominator is zero, i.e., $(x-5)^2+6(x-5)+9=0$. Simplifying, we get:

$$x^2-10x+25+6x-30+9=0 \Rightarrow x^2-4x+4=0 \Rightarrow (x-2)^2=0.$$

Thus, the function is undefined when $x=2$.

21)

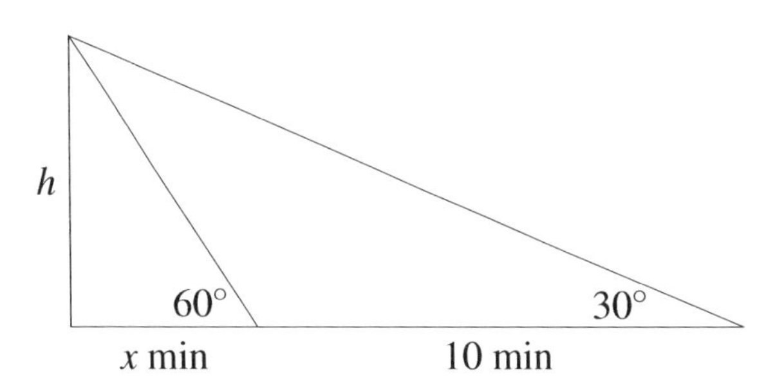

Given the angles and time for the angle of elevation to change, we use the tangent of the angles of elevation. Let h be the height of the cliff. We have:

$$\tan 30^\circ = \frac{h}{x+10} \Rightarrow h = (x+10)\tan 30^\circ.$$

On the other hand, we have:

$$\tan 60^\circ = \frac{h}{x} \Rightarrow h = x\tan 60^\circ.$$

Therefore,

$$(x+10)\tan 30^\circ = x\tan 60^\circ.$$

Solving the equation for x, we find:

$$x = \frac{10\tan 30^\circ}{\tan 60^\circ - \tan 30^\circ} \approx 5.$$

So, the boat will reach the shore in approximately 5 minutes.

22) Calculate the determinant for each matrix:

For A=$\begin{bmatrix} 3 & 2 \\ 1 & 4 \end{bmatrix}$: $\det(A) = (3)(4) - (2)(1) = 12 - 2 = 10$.

For B=$\begin{bmatrix} 1 & 4 & 2 \\ 0 & 2 & 3 \\ 5 & 0 & 1 \end{bmatrix}$: $\det(B) = 1(2\times1 - 3\times0) - 4(0\times1 - 3\times5) + 2(0\times0 - 2\times5) = 42$.

For C=$\begin{bmatrix} 2 & -1 \\ 3 & 1 \end{bmatrix}$: $\det(C) = (2)(1) - (-1)(3) = 2 + 3 = 5$.

For D=$\begin{bmatrix} 2 & 5 & 0 \\ 0 & 3 & 1 \\ 2 & -2 & 0 \end{bmatrix}$: $\det(D) = 2(3\times0 - ((-2)\times1)) - 5(0\times0 - 2\times1) + 0(0\times -2 - 3\times2) = 14$.

So, the correct option is C.

23) For $90° < \theta < 180°$, we have $-1 < \cos\theta < 0$ which implies $0 < \cos^2\theta < 1$. Thus, it is clear that $\cos\theta$ is always less than $\cos^2\theta$. Therefore, the correct answer is B.

24) According to the graph, the function is a parabola that opens downward, and its vertex appears to be at $(0,3)$. We need a function that opens downwards and has a vertex at $(0,3)$. Examining each option:
A: $y = 3x^2 + 1$ opens upwards, not matching the graph.
B: $y = -x^2 - 1$ opens downward but has a vertex at $(0,-1)$, not matching the graph.
C: $y = x^2 - 3$ opens upwards, not matching the graph.
D: $y = -2x^2 + 3$ opens downward and has a vertex at $(0,3)$, which matches the graph.
Therefore, the correct function is $y = -2x^2 + 3$.

25) In a right triangle DEF, where $\angle E = 90°$, $DE = 12\ cm$, and $\angle D = 30°$, we can calculate the length of side DF using the cosine function. The cosine of an angle in a right triangle is the ratio of the length of the side adjacent to the angle to the length of the hypotenuse. Here, side DE is adjacent to $\angle D$ and DF is the hypotenuse. Therefore, rearranging the formula, we can express DF as follows:

$$\cos(D) = \frac{DE}{DF} \Rightarrow DF = \frac{DE}{\cos(D)} = \frac{12}{\cos(30)^\circ} \approx 13.9\ cm.$$

So, the length of side DF is approximately $13.9\ cm$.

26) The graph represents a parabola opening downward. To identify the function, we can analyze the vertex and the direction of the parabola.
A: $g(x) = -x^2 + 6x - 8$ opens downward. Completing the square, we rewrite it as $-(x-3)^2 + 1$. So, the vertex of this parabola is $(3,1)$ and it matches the graph.

B: $g(x) = x^2 + 6x - 8$ opens upward, not matching the graph.
C: $g(x) = -x^2 - 6x - 8$ opens downward. Completing the square, we rewrite it as $-(x+3)^2 + 1$. So, the vertex of this parabola is $(-3, 1)$, not matching the graph.
D: $g(x) = x^2 - 6x - 8$ opens upward, not matching the graph.
Therefore, the correct function represented by the graph is $g(x) = -x^2 + 6x - 8$.

27) From the given information, $\tan\theta = \frac{5}{12}$ represents the ratio of the opposite side to the adjacent side in a right triangle. Let's assume the opposite side is 5 and the adjacent side is 12. Then, using the Pythagorean theorem, the hypotenuse h of the triangle is:

$$h = \sqrt{5^2 + 12^2} = \sqrt{25 + 144} = \sqrt{169} = 13.$$

Now, $\sin\theta$ is the ratio of the opposite side to the hypotenuse, so:

$$\sin\theta = \frac{5}{13}.$$

Next, we substitute $\sin\theta$ into the expression $\frac{1-\sin\theta}{1+\sin\theta}$:
Substituting the value of $\sin\theta$ into this, we get:

$$\frac{\sqrt{1 - \frac{5}{13}}}{\sqrt{1 + \frac{5}{13}}} = \frac{\sqrt{\frac{8}{13}}}{\sqrt{\frac{18}{13}}} = \sqrt{\frac{8}{18}} = \sqrt{\frac{4}{9}} = \frac{2}{3}.$$

So, the correct answer is C.

28) To draw the graph of $y < -3x - 3$, first graph the line $y = -3x - 3$. Since the inequality sign is "<", the line should be dashed. The slope is -3 and the y-intercept is -3. Then, choose a testing point, such as the origin $(0, 0)$, and substitute the values into the inequality:

$$y < -3x - 3 \Rightarrow 0 < -3(0) - 3 \Rightarrow 0 < -3.$$

This is incorrect, as 0 is not less than -3. Therefore, the area below the line represents the solution, which is correctly depicted in Option D.

29) The vertex of a parabola in the form $y = a(x-h)^2 + k$, is (h, k). In this case, the parabola is in the form $y = (x+1)^2 - 4$, which is equivalent to $y = (x - (-1))^2 + (-4)$. Comparing this to the general form, we can see that $h = -1$, and $k = -4$. Therefore, the vertex is $(-1, -4)$.

30) Since the quadrilateral GHJK is a rectangle, we have: $\overrightarrow{GH} + \overrightarrow{GK} = \overrightarrow{GJ}$. Also, GIJL is a rectangle. So,

$\overrightarrow{GI}+\overrightarrow{GL}=\overrightarrow{GJ}$. Now, by substituting we get:

$$\overrightarrow{GH}+\overrightarrow{GI}+\overrightarrow{GK}+\overrightarrow{GL}=\left(\overrightarrow{GH}+\overrightarrow{GK}\right)+\left(\overrightarrow{GI}+\overrightarrow{GL}\right)=\overrightarrow{GJ}+\overrightarrow{GJ}=2\overrightarrow{GJ}.$$

31) We are given that $\cos\theta<0$. In a standard Cartesian coordinate system, cosine is negative in the second and third quadrants. On the other hand, $\tan\theta$ is negative $(-\frac{3}{4})$, which means that θ is in the second or fourth quadrant. The last two expressions imply that θ is in the second quadrant and $\sin\theta$ is positive, i.e., $\sin\theta>0$. Now, we need to find the value of $\cos\theta$. Since we know that $\tan\theta=-\frac{3}{4}$, we can use the Pythagorean identity for tangent:

$$\tan^2\theta+1=\sec^2\theta.$$

Substitute the given value:

$$\left(-\frac{3}{4}\right)^2+1=\sec^2\theta. \Rightarrow \sec\theta=\pm\frac{5}{4}.$$

Recall that $\sec\theta=\frac{1}{\cos\theta}$, $\sin^2\theta+\cos^2\theta=1$, and $\cos\theta<0$, so:

$$\cos\theta=-\frac{4}{5}\Rightarrow\sin\theta=\pm\frac{3}{5}.$$

But in the second quadrant, we have $\sin\theta>0$, so we accept $\sin\theta=\frac{3}{5}$.

32) Since the quadrilateral GHJK is a rectangle, we have: $\overrightarrow{GH}+\overrightarrow{GK}=\overrightarrow{GJ}$. Also, GIJL is a rectangle. So, $\overrightarrow{GI}+\overrightarrow{GL}=\overrightarrow{GJ}$. Now, by substituting we get:

$$\overrightarrow{GH}+\overrightarrow{GI}+\overrightarrow{GK}+\overrightarrow{GL}=\left(\overrightarrow{GH}+\overrightarrow{GK}\right)+\left(\overrightarrow{GI}+\overrightarrow{GL}\right)=\overrightarrow{GJ}+\overrightarrow{GJ}=2\overrightarrow{GJ}.$$

33) Let the height of the other pole be h meters. Using the tangent of the angles of depression, we can find h.

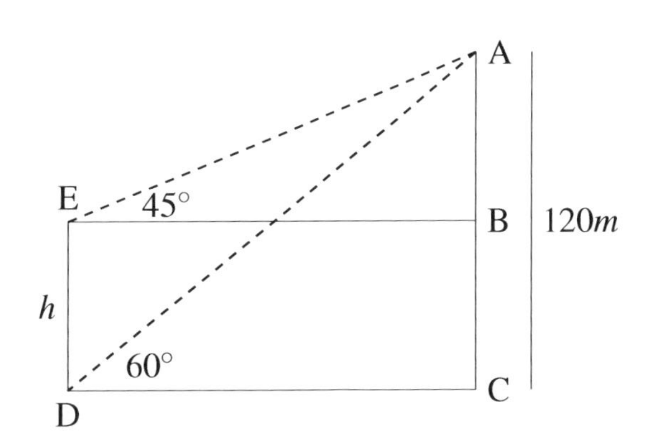

For the angle of 60°, the tangent ratio gives us:

$$\tan 60^\circ = \frac{AC}{DC} = \frac{120}{DC},$$

where DC is the distance between the two poles. From this,

$$DC = \frac{120}{\tan 60^\circ} = \frac{120}{\sqrt{3}} \approx 69.$$

Next, for the angle of 45°,

$$\tan 45^\circ = \frac{AB}{EB} = \frac{AB}{DC} = \frac{120-h}{DC} \Rightarrow 1 = \frac{120-h}{69}.$$

Solving for h, we get $h \approx 51$

34) The equation of a circle in standard form is $(x-h)^2+(y-k)^2=r^2$, where r is the radius of the circle and (h,k) is the center. Given that the perimeter of the circle is 6π, then $r=3$ (since perimeter $P=2\pi r$). Therefore, $(x-2)^2+(y+3)^2=3^2$ is the answer.

35) The value of $\cos 45^\circ$ is $\frac{\sqrt{2}}{2}$.

36) Evaluate each option:

A: $h(g(x)) = (x+3)-1 = x+2$, which does not match $x+1$.

B: $h(g(x)) = (x-3)+1 = x-2$, which does not match $x+1$.

C: $h(g(x)) = (-x+3)-1 = -x+2$, which does not match $x+1$.

D: $h(g(x)) = (x+3)-2 = x+1$. Thus, this matches $x+1$.

Therefore, the correct combination is D: $h(x) = -x+1$ and $g(x) = x+3$.

37)

To determine the number of x-intercepts of the graph of the function $y = \frac{x+2}{2-x^2}$, we need to find the values of x for which y is zero. An x-intercept occurs where the function intersects the x-axis, which corresponds to points where $y=0$. Set y to zero and solve for x:

$$0 = \frac{x+2}{2-x^2}.$$

For a fraction to equal zero, its numerator must be zero, and its denominator must not be zero. Therefore, we solve the equation $x+2=0$, while ensuring that $2-x^2 \neq 0$. Solving $x+2=0$, we find: $x=-2$. Since the denominator $2-x^2$ is a square of a real number, it will never be zero, and thus does not impose any additional restrictions on the values of x. Consequently, there is only one x-intercept for the given function, located at

$x = -2$.

38) Use the geometric sequence formula $a_n = a_1 r^{n-1}$. Substitute $a_1 = \frac{1}{2}$ and $a_3 = 8$. So,

$$a_3 = \frac{1}{2} r^{3-1} = 8.$$

Therefore,

$$r^2 = 8 \times 2 = 16 \Rightarrow r = 4 \quad \text{or} \quad r = -4.$$

Then,

$$a_2 = a_1 r^{2-1} \Rightarrow a_2 = \frac{1}{2} \times 4 = 2 \quad \text{or} \quad a_2 = \frac{1}{2} \times (-4) = -2.$$

So, the correct option is A.

39) To find the correct graph, consider how the transformations affect the basic logarithmic function $y = \log(x)$. The term $+2$ inside the logarithm indicates a horizontal shift 2 units to the left. The negative sign outside the logarithm indicates a reflection over the x-axis. Thus, the graph of $y = -\log(x+2)$ is the graph of $y = \log(x)$ shifted 2 units to the left and reflected over the x-axis. Therefore, the correct answer is D.

40) The expression, $y = 3\cos^2(\beta) + 7\sin^2(\beta)$, can be written as:

$$3\cos^2(\beta) + 7(1 - \cos^2(\beta)) = 7 - 4\cos^2(\beta).$$

The value of the function cosine is $-1 \leq \cos(\beta) \leq 1$, then $0 \leq \cos^2(\beta) \leq 1$. So, we have:

$$-4 \leq -4\cos^2\beta \leq 0.$$

Finally, we get:

$$3 \leq 7 - 4\cos^2\beta \leq 7.$$

Therefore, the minimum value is 3.

41) According to the graph, the amounts at the end of each month are as follows:

- Month 1: $600
- Month 2: $800
- Month 3: $1000
- Month 4: $1300
- Month 5: $1350
- Month 6: $1700

To find the rate of change from the second month to the sixth month, calculate the slope of the line connecting these points.

$$\text{Rate of Change} = \frac{\text{Change in Money}}{\text{Change in Time}} = \frac{1700-800}{6-2} = \frac{900}{4} = 225.$$

Therefore, the rate of change of money per month from the second month to the sixth is \$225 per month.

42) The general form of an exponential decay model is $n(x) = ab^x$, where:

- a is the initial amount,
- b is the base or growth factor.

From the table, $a = 50,000$ (the number of nesting sites in the first year). Using the data from the first two years, calculate b as follows:

$$40,000 = 50,000 \cdot b^1 \implies b = \frac{40,000}{50,000} = 0.8$$

Therefore, the model is:

$$n(x) = 50,000 \cdot 0.8^x$$

This model suggests that the number of nesting sites decreases by 20% each year.

43)

A. $f(x)$ and $h(x)$ have minimum points because their leading coefficients are positive, whereas $g(x)$ has a maximum point as its leading coefficient is negative.

B. All parabolas represented by these functions are symmetric about the y-axis, i.e., $x = 0$.

C. Intersection on the x-axis would imply the functions cross where $y = 0$. Solving for x in these equations does not yield any real solutions indicating an intersection on the x-axis.

D. Each function has a unique y-intercept based on the constant term in each equation: -7 for $f(x)$, 2 for $g(x)$, and 3 for $h(x)$.

E. While not intersecting on the x-axis, the quadratic nature of these functions ensures they will intersect elsewhere in the coordinate plane.

44)

A. An even function satisfies $f(-x) = f(x)$. Calculating $f(-x) = (-x)^2(1-(-x)) = x^2(1+x)$ which is not equal to $f(x)$. Hence, $f(x)$ is not even.

B. As $x \to \infty$, $f(x) = x^2(1-x) \to -\infty$ because the $-x^3$ term dominates.

C. The domain of a polynomial function is all real numbers, including 1. The statement is incorrect as $f(1) = 1^2(1-1) = 0$ is defined.

D. The function has three real roots, at $x = 0, 1$ (from $1-x=0$), and another root from $x^2 = 0$. Therefore,

$f(x) = 0$ is satisfied at $x = 0, 0, 1$.

Author's Final Note

I hope you enjoyed this book as much as I enjoyed writing it. I have tried to make it as easy to understand as possible. I have also tried to make it fun. I hope I have succeeded. If you have any suggestions for improvement, please let me know. I would love to hear from you.

The accuracy of examples and practice is very important to me. We have done our best. But I also expect that I have made some minor errors. Constant improvement is the name of the game. If you find any errors, please let me know. I will fix them in the next edition.

Your learning journey does not end here. I have written a series of books to help you learn math. Make sure you browse through them. I especially recommend workbooks and practice tests to help you prepare for your exams.

I also enjoy reading your reviews. If you have a moment, please leave a review on Amazon. It will help other students find this book.

If you have any questions or comments, please feel free to contact me at drNazari@effortlessmath.com.

And one last thing: Remember to use online resources for additional help. I recommend using the resources on https://effortlessmath.com. There are many great videos on YouTube.

Good luck with your studies!

Dr. Abolfazl Nazari